Solutions Towards Zero Carbon Buildings

Solutions Towards Zero Carbon Buildings

Editors

Bruno Peuportier
Zsuzsa Szalay

MDPI

Basel • Beijing • Wuhan • Barcelona • Belgrade • Novi Sad • Cluj • Manchester

Editors

Bruno Peuportier
Centre Efficacité Énergétique
des Systèmes, Mines
Paris-PSL
Paris
France

Zsuzsa Szalay
Budapest University of
Technology and Economics
(BME)
Budapest
Hungary

Editorial Office
MDPI AG
Grosspeteranlage 5
4052 Basel, Switzerland

This is a reprint of articles from the Special Issue published online in the open access journal *Energies* (ISSN 1996-1073) (available at: https://www.mdpi.com/journal/energies/special_issues/SZCB).

For citation purposes, cite each article independently as indicated on the article page online and as indicated below:

Lastname, A.A.; Lastname, B.B. Article Title. *Journal Name* **Year**, *Volume Number*, Page Range.

ISBN 978-3-7258-2483-0 (Hbk)
ISBN 978-3-7258-2484-7 (PDF)
doi.org/10.3390/books978-3-7258-2484-7

Contents

About the Editors

Bruno Peuportier

Bruno Peuportier, PhD, senior scientist at Mines Paris-PSL and professor at Ecole des Ponts ParisTech, is involved in the eco-design of buildings and neighbourhoods. He has been working on the environmental performance of buildings since 1980 and has launched and coordinated the development of building energy simulation (COMFIE) and life cycle assessment (EQUER) software tools. He has coordinated several European projects in this field, published 5 books, 38 articles in scientific journals, and 115 international conference papers. He was also the French representative in IEA-EBC Annex 72 "Assessing Life Cycle Related Environmental Impacts Caused by Buildings".

Zsuzsa Szalay

Zsuzsa Szalay, PhD, associate professor at Budapest University of Technology and Economics, Hungary, focuses on energy-conscious architecture and the life cycle assessment of buildings. She has participated in several international research projects including IEA-EBC Annex 72 "Assessing Life Cycle Related Environmental Impacts Caused by Buildings" and coordinated several national projects on building life cycle assessments. She worked on the transposition of the energy performance of buildings directive (EPBD) in Hungary, the development of its calculation methodology, and its requirement system for nearly zero-energy buildings. She has published 26 international journal papers and 43 international conference papers.

Preface

According to the Sixth Assessment Report of the Intergovernmental Panel on Climate Change (contribution of WG III on mitigation), historical cumulative net CO2 emissions between 1850 and 2019 amount to about four-fifths of the total carbon budget for a 50% probability of limiting global warming to 1.5 °C. Progressing towards net-zero GHG emissions is therefore necessary, both in new constructions and by retrofitting existing buildings.

This Special Issue deals with novel solutions for reaching this objective and, particularly, new zero-carbon building concepts and low-carbon construction products, e.g., bio-based materials; low-carbon equipment for heating, cooling, hot-water production, ventilation, and lighting; integration of renewable energy systems in buildings; energy-efficient building management and control systems; and applications of IoT and/or AI in buildings.

This issue is aimed at researchers, teachers, and professionals in the construction and energy sectors.

Authors from 8 countries (Africa, Asia, Europe, and Oceania) have contributed, with inputs from different disciplines, including architecture, civil engineering, infrastructure, mechanical engineering, material sciences, and real estate.

Bruno Peuportier and Zsuzsa Szalay
Editors

energies

MDPI

Article

A Comparison between Solution-Based Synthesis Methods of ZrO_2 Nanomaterials for Energy Storage Applications

Maria Leonor Matias [1], Emanuel Carlos [1,*], Rita Branquinho [1], Hadassa do Valle [1], João Marcelino [1], Maria Morais [1], Ana Pimentel [1], Joana Rodrigues [2], Teresa Monteiro [2], Elvira Fortunato [1], Rodrigo Martins [1,*] and Daniela Nunes [1,*]

[1] CENIMAT | i3N, Department of Materials Science, School of Science and Technology, NOVA University Lisbon and CEMOP/UNINOVA, 2829-516 Caparica, Portugal

[2] Physics Department & I3N, Aveiro University, Campus Universitário de Santiago, 3810-193 Aveiro, Portugal

* Correspondence: e.carlos@fct.unl.pt (E.C.); rm@uninova.pt (R.M.); daniela.gomes@fct.unl.pt (D.N.); Tel.: +351-21-294-8562 (R.M. & D.N.); Fax: +351-21-294-8558 (R.M. & D.N.)

Citation: Matias, M.L.; Carlos, E.; Branquinho, R.; do Valle, H.; Marcelino, J.; Morais, M.; Pimentel, A.; Rodrigues, J.; Monteiro, T.; Fortunato, E.; et al. A Comparison between Solution-Based Synthesis Methods of ZrO_2 Nanomaterials for Energy Storage Applications. *Energies* **2022**, *15*, 6452. https://doi.org/10.3390/en15176452

Academic Editors: Bruno Peuportier and Zsuzsa Szalay

Received: 26 July 2022
Accepted: 31 August 2022
Published: 3 September 2022

Abstract: The present study is focused on the synthesis of zirconium dioxide (ZrO_2) nanomaterials using the hydrothermal method assisted by microwave irradiation and solution combustion synthesis. Both synthesis techniques resulted in ZrO_2 powders with a mixture of tetragonal and monoclinic phases. For microwave synthesis, a further calcination treatment at 800 °C for 15 min was carried out to produce nanopowders with a dominant monoclinic ZrO_2 phase, as attested by X-ray diffraction (XRD) and Raman spectroscopy. The thermal behavior of the ZrO_2 nanopowder was investigated by in situ XRD measurements. From the scanning electron microscopy (SEM) and transmission electron microscopy (TEM) images, the presence of near spherical nanoparticles was clear, and TEM confirmed the ZrO_2 phases that comprised the calcinated nanopowders, which include a residual tetragonal phase. The optical properties of these ZrO_2 nanopowders were assessed through photoluminescence (PL) and PL excitation (PLE) at room temperature (RT), revealing the presence of a broad emission band peaked in the visible spectral region, which suffers a redshift in its peak position, as well as intensity enhancement, after the calcination treatment. The powder resultant from the solution combustion synthesis was composed of plate-like structures with a micrometer size; however, ZrO_2 nanoparticles with different shapes were also observed. Thin films were also produced by solution combustion synthesis and deposited on silicon substrates to produce energy storage devices, i.e., ZrO_2 capacitors. The capacitors that were prepared from a 0.2 M zirconium nitrate-based precursor solution in 2-methoxyethanol and annealed at 350 °C exhibited an average dielectric constant (κ) of 11 ± 0.5 and low leakage current density of $3.9 \pm 1.1 \times 10^{-7}$ A/cm^2 at 1 MV/cm. This study demonstrates the simple and cost-effective aspects of both synthesis routes to produce ZrO_2 nanomaterials that can be applied to energy storage devices, such as capacitors.

Keywords: ZrO_2; nanomaterials; microwave irradiation; solution combustion; energy storage devices

1. Introduction

Research in nanotechnology has grown over in recent years since it opens up possibilities for the design and fabrication of novel materials, as well as flexible and smart devices at the nanometer scale [1,2]. In this regard, nanostructures based on metal oxides have been extensively studied due to their unique physical and chemical properties and great potential in a myriad of applications, including electronics, energy conversion and storage, biomedicine, catalysis and sensing [3–8].

ZrO_2 is a metal oxide that exhibits three crystalline phases under atmospheric pressure, depending on the synthesis route: monoclinic, which is thermodynamically the most stable form [9], tetragonal and cubic [10–12]. The cubic form of ZrO_2 is stable at high temperatures (usually above 2370 °C), while the tetragonal is stable within the range of 1170–2370 °C, and the monoclinic phase stabilizes below 1170 °C [4,9].

ZrO_2 exhibits excellent properties, such as good optical transparency in the visible and near-infrared spectral range [13,14], high thermal and chemical stabilities, mechanical strength and fracture toughness, low absorption of light, high index of refraction [15], high corrosion resistance [16], high ionic stability at high temperature, it is biocompatible [17] and presents relatively low leakage current [18] and high breakdown field [19]. Despite this, ZrO_2 presents a wide band gap value (theoretically estimated as ~5.42 eV for monoclinic, ~6.40 eV for tetragonal and ~5.55 eV for cubic phases [14,20,21]). The introduction of defects in its structure, for instance, through doping with cations or the reduction of particle size below a critical value, is reported to allow the stabilization of high-temperature phases (i.e., cubic and/or tetragonal phases) at RT [22,23].

Applications for this material range from ceramic industry, gas sensing, solar fuel cells and biomedical/catalytic fields to its incorporation in large-scale integrated circuits, owing to a high dielectric constant and as a gate dielectric in metal-oxide-semiconductor devices (MOS) [22,24–26]. In this regard, ZrO_2 has emerged as a potential alternative to replace low-κ SiO_2 in electronic devices [27]. It is reported that the dielectric constant of ZrO_2 films ranges from $\kappa = 20$ in the monoclinic phase to $\kappa = 37$–47 for the cubic and tetragonal phases, whereas in the case of amorphous ZrO_2, the dielectric constant values are between 14 and 25 [14].

Recently ZrO_2-based materials have been explored in the field of energy. Optimization of the light absorption towards sunlight in zirconia has shown enormous potential for its use in solar receivers since this refractory material possesses enhanced mechanical/thermal properties and oxidation resistance, being a competitive material to substitute the typically used silicon carbide (SiC) [28,29]. Other applications include refractory/metal production [30]. In addition, it has demonstrated high energy storage performance, for instance, by employing ferroelectric ZrO_2 thin films as energy storage capacitors [31] or antiferroelectric ZrO_2, a material for dynamic random access memories (DRAM) [32]. Moreover, ZrO_2 capacitors display ultra-high power density and ultra-fast charge/discharge rates, yet their energy storage densities are relatively lower than the traditional electrical energy-storage devices, such as supercapacitors and batteries [33]. Nevertheless, these devices can play a key role in energy saving and thus reducing the carbon dioxide (CO_2) footprint. When associated with renewable energy sources, capacitors can have a real impact on climate change, influencing almost all sectors of human life, including the chemical/petrochemical, automotive, textiles, wood/paper and construction sectors, whereas, for the latter, with the advent of the zero carbon buildings, it can reduce the global greenhouse gas emissions emitting buildings. The future of humanity relies on addressing the availability of using carbon-neutral energy sources and finding solutions to lower the CO_2 footprint, industrial residues and human contaminants with direct impact in public health and on the environment.

Most of the applications for ZrO_2 depend on its dimensions. It is well known that when the size of metal oxide materials is reduced to the nanoscale level, different properties are found from their bulk counterparts [34]. Novel characteristics owed to quantum size effects may occur when the cluster size is smaller than the Bohr exciton radius. Moreover, the properties of the surface start to play a relevant role that may overrule the bulk ones. Hence, the synthesis method is of paramount importance to control not only the size of ZrO_2 nanomaterials but also the crystalline phase, morphology and lattice defects that highly influence the physico-chemical properties of the produced nanomaterials [35].

ZrO_2 nanomaterials have already been obtained with different morphologies, such as thin films [14], nanoparticles [36], nanobelts [37], nanowires [38] and nanotubes [39], mainly through chemical approaches [40]. The most common methods for the synthesis of ZrO_2 nanostructures include co-precipitation [41], sol-gel [42] and solution combustion [35], sonochemical-assisted methods [43,44], chemical vapor deposition [45], emulsion processing routes [46,47], as well as hydrothermal [10,48] and microwave syntheses [49,50]. Lately, a continuous increase in research work has been observed related to fast continuous methods for controlling the synthesis of ZrO_2 nanomaterials [51–55]. Continuous hydrothermal flow synthesis (CHFS) processes are preferred due to the possibility of tailoring and obtaining unique nanoparticle properties that cannot be achieved using more conventional batch hydrothermal or other synthesis methods. These processes have shown potential in manufacturing reproducible and homogeneous nanomaterials, with high throughput, in addition to being energy efficient and eco-friendly, since for this last case, water is typically used rather than toxic organic solvents. Annealing treatments can also be avoided by employing these processes with superheated water conditions (typically in the range of 200–400 °C) [55–57]. Even though these systems offer several benefits, their high cost is a downside [57].

Several studies have reported the production of ZrO_2 nanomaterials using sol-gel, particularly solution combustion synthesis, which is included in the sol-gel method [58–60]. Solution combustion synthesis is an attractive technique for the preparation of ZrO_2 nanopowders and thin films, owing to its simplicity, energy and time savings, cost-effectiveness, versatility, higher purity compared with conventional sol-gel methods, low synthesis temperatures, and compatibility with flexible substrates and large scale production [61–63].

The synthesis of ZrO_2 nanomaterials using the hydrothermal method assisted by microwave irradiation has also been growing exponentially over recent years [49,64–66], mainly due to its several advantages, such as volumetric heating (the entire volume of solution is evenly heated, instead of relying on heat diffusion processes across the reaction vessels), and the reaction times can be shorter as it is possible to synthesize nanostructures in just a few minutes. Furthermore, it also allows accurate control of particle morphology and size by adjusting the microwave parameters [50,67]. Upon optimization of the synthesis parameters, such as temperature, pH, time and zirconium oxide precursors (e.g., zirconyl chloride, zirconyl hydroxide, zirconyl nitrate hydrate and zirconium alkoxides), different ZrO_2 phases can be obtained [68].

Solution-based ZrO_2 nanomaterials are a highly appealing alternative to physical methods due to their process simplicity, high throughput, reduced equipment cost, since no vacuum-based systems are required [69,70], and the possibility to fabricate optoelectronic devices, even at low temperatures and by using green solvents [70]. As indicated, a broad range of solution-based synthesis methods have been developed to prepare ZrO_2 nanomaterials; however, solution combustion synthesis and hydrothermal synthesis/microwave irradiation are typically preferred, and for that reason, these techniques are presented in this work. The same zirconium precursor (zirconium (IV) oxynitrate hydrate) was used in both synthesis techniques.

Several studies have already reported the dielectric properties of the ZrO_x films produced by different solution-based processes. For instance, Seon et al. fabricated ZrO_2 films via a non-hydrolytic sol-gel route at low temperatures, using 2-methoxyethanol (2-ME) as a solvent, with further annealing at 300 °C. The solution precursors chosen were zirconium chloride and zirconium isopropoxide, which were prepared in equimolar amounts, and acted, respectively, as a metal halide and a metal alkoxide. A breakdown voltage greater than 4 MV cm^{-1}, a high dielectric constant (near 10) and a low leakage current density of 5×10^{-8} A cm^{-2} at a field of 1 MV cm^{-1} were obtained [71]. Another study by Gong et al. showed that by employing a low-temperature annealing treatment at 160 °C, high-quality amorphous ZrO_2 dielectric films could be produced via a low-cost solution process. In this study, a solution was prepared using zirconium (IV) acetylacetonate as zirconium precursor and N, N-dimethylformamide as a solvent. Hydrolysis and condensation reactions occurred during stirring of the solution for 32 h at 90 °C. The films exhibited a leakage current of 3.6×10^{-5} A cm^{-2} at −3 V, a capacitance of ~117.1 nF cm^{-2} and a dielectric constant of 7.8, both at 1 KHz [72]. Wang et al. also fabricated high-κ ZrO_2-dielectric films using a lightware (LW) irradiation-induced chloride-based low-temperature solution route. Results demonstrated a great capacitance of 270 nF cm^{-2}, a high dielectric constant of 14.1 (at 100 Hz) and a low leakage current of 7.6×10^{-8} A cm^{-2}, under 2 MV/cm. The superior performance was attributed to the effective formation of the metal-oxygen (M-O) framework that enabled the elimination of oxygen defects [73]. Another study by Jung et al.

demonstrated the fabrication of ultrathin ZrO_x films by deep ultraviolet irradiation, which revealed a leakage current density as low as 10^{-11} A cm^{-2} at 1 MV/cm, a capacitance of 260 nF cm^{-2} (at 1 MHz), high dielectric constant (22) and good breakdown voltage (around 6 MV/cm) [74]. In addition, Luo et al. prepared high-quality ZrO_2 films using an oxygen-doped precursor solution (ODS). The ODS-ZrO_2 films showed a low leakage current density of 10^{-7} A cm^{-2} (at 2 MV/cm), high breakdown electric field (7.0 MV/cm) and dielectric constant of 19.5 [75]. However, the challenge still relies on the fabrication of metal-oxide films with a high-quality surface (a smooth surface with a dense network) at a low temperature and through a simple approach to guarantee a low leakage current density and high breakdown field [75].

In this study, ZrO_2 nanomaterials were produced in the form of powders or thin films through solution-based processes, i.e., a hydrothermal method assisted by microwave irradiation and solution combustion synthesis. The microwave synthesized powder was further calcinated at 800 °C for 15 min under atmospheric conditions. The ZrO_2 nanomaterials were characterized by XRD, Raman spectroscopy, SEM coupled with energy dispersive X-ray spectroscopy (EDS) and focused ion beam (FIB) and TEM. The thermal behavior of the nanopowder produced under microwave irradiation was investigated through in situ XRD, and these powders had their optical properties assessed through PL and PLE at RT. The ZrO_x thin films produced by the solution combustion synthesis were further tested as capacitors.

2. Materials and Methods

2.1. Hydrothermal Synthesis of ZrO_2 Nanoparticles Assisted by Microwave Irradiation

The ZrO_2 nanoparticle synthesis route was adapted from ref. [4]. In a typical synthesis, 50 mL of an aqueous (aq.) solution of 0.2 M of zirconium (IV) oxynitrate hydrate (Sigma-Aldrich, St. Louis, MO, USA, 99.9%, CAS: 14985-18-3, $ZrO(NO_3)_2{\cdot}xH_2O$) is mixed with 50 mL of an aq. solution of 0.4 M of sodium hydroxide (Labchem, CAS: 1310-73-2, NaOH). The reagents were used without any further purification. The 100 mL solution was left to stir for 30 min. The molar ratio of zirconium precursor and sodium hydroxide was kept at 1:2. Microwave synthesis was then carried out with a CEM microwave digestion system, Matthews, NC, USA (MARS one), and the applied microwave parameters were 1000 W, 230 $\pm$ 10 °C and 25 min. Afterwards, the previous solution was equally distributed into Teflon vessels of 75 mL (each vessel containing 20 mL of solution). Subsequently, the centrifugation of the resultant nanopowder was performed for 3 min at 4750 rpm and washed three times alternately with deionized water and isopropyl alcohol (IPA). Finally, the nanopowder was dried in a desiccator at 60 °C for 5 h. The yield was around 0.77 g of nanopowder/batch.

After microwave synthesis, the dried ZrO_2 nanopowder was further calcinated in an alumina ceramic crucible at 800 °C for 15 min using a Nabertherm furnace under atmospheric conditions. The calcination treatment aimed to guarantee the formation of the mostly thermodynamically stable ZrO_2 phase, i.e., the monoclinic phase.

2.2. Solution Combustion Synthesis

A solution with a concentration of 0.2 M of zirconium (IV) oxynitrate hydrate (Sigma-Aldrich, 99.9%, CAS: 14985-18-3) was prepared in 2-methoxyethanol (2-ME, ACROS Organics, 99%, $C_3H_8O_2$), and it was left to stir at room temperature for 2 h. Urea (Sigma-Aldrich, 98%, $CO(NH_2)_2$) was used as fuel for the combustion reaction and continued stirring for a minimum of 1 h. The molar proportion between urea and the zirconium oxide precursor was 5:3 to ensure the redox stoichiometry of the reaction. The powder was produced considering 10 mL of the combustion solution with the same concentration and transferred to an alumina ceramic crucible for further annealing in an air furnace at 350 °C for 1 h.

As for the thin films, and prior to the ZrO_x deposition, a 2.5 $\times$ 2.5 cm *p*-type single crystal 100-oriented silicon substrate (resistivity $\cong$ 1–2 $\Omega{\cdot}$cm) was cleaned with acetone for 10 min in an ultrasonic cleaning bath at 60 °C. This cleaning process was repeated with

IPA, followed by deionized water and then the substrates were dried using nitrogen flow. The adhesion properties of the thin films to the substrate were enhanced by exposing the substrate to UV-ozone treatment (PSD-UV Novascan system-Ames, IA, USA) for 15 min at a 5 cm distance from the UV lamp. Prior to solution casting, the precursor solution was filtered through a 0.22 µm polytetrafluoroethylene (PTFE) syringe filter and one layer was deposited by spin coating on the silicon (Si) substrate at 2000 rpm for 35 s. To evaporate the solvent and improve the densification, the thin films were later annealed at 350 °C on a hot plate for 30 min.

The metal-insulator-semiconductor (MIS) capacitors were fabricated considering the structure Al/ZrO_x/*p*-Si. The schematic of the MIS devices' fabrication is depicted in Figure 1. To optimize the ohmic contact, the capacitors were completed by thermal evaporation of an 80 nm aluminum bottom electrode on the Si substrate and 80 nm Al top electrodes by applying a circular shadow mask, which produced capacitors with an area of 0.2 mm^2.

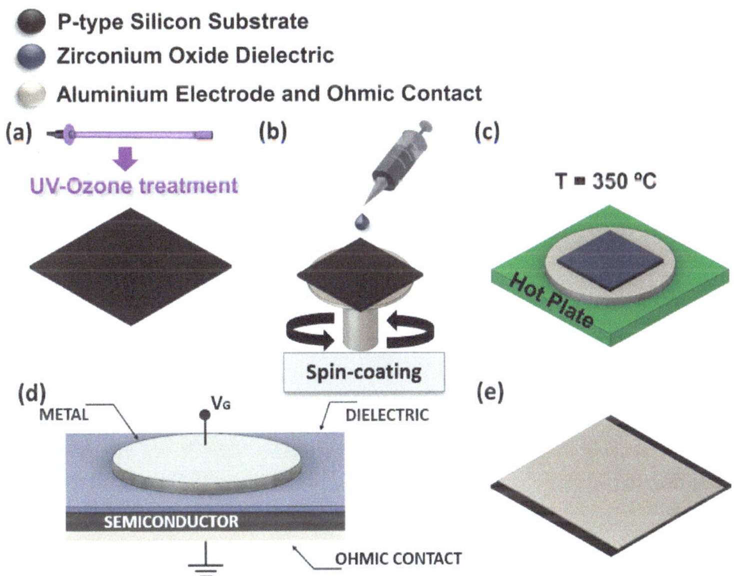

Figure 1. Steps to fabricate the MIS capacitor after the cleaning procedure: (**a**) surface treatment on silicon (Si) substrate by UV-ozone for 15 min; (**b**) deposition of the zirconium oxide precursor solution by spin coating technique; (**c**) annealing treatment of the substrate for film densification; (**d**) scheme of the device after the deposition of top electrodes (80 nm Al) by thermal evaporation; (**e**) scheme of the device after the deposition of the bottom electrode (80 nm Al) by thermal evaporation.

2.3. Characterization Techniques

XRD experiments were carried out using a PANalytical's X'Pert PRO MPD diffractometer (Almelo, The Netherlands) equipped with an X'Celerator 1D detector and using CuKα radiation (λ = 1.540598 Å). XRD data were recorded from 20° to 80° 2θ range with a step of 0.05°. The produced ZrO_2 nanopowders were measured in the Bragg–Brentano configuration. The crystalline phases in the produced nanopowders were identified by comparison with the International Centre for Diffraction Data (ICDD). The simulated monoclinic phase corresponds to ICDD file No. 00-037-1484, the simulated tetragonal phase to ICDD file No. 00-050-1089 and the simulated cubic phase to ICDD file No. 00-049-1642. In situ XRD experiments were performed in an MRI chamber from Bruker (Massachusetts, United States of America) at atmospheric conditions using a Pt–Rh foil as the heating element. The in situ diffractograms were recorded in the same 2θ range (20° to 80°) at temperatures of 30, 100, 200, 300, 400, 500, 600, 700 and 800 °C. The ZrO_2 nanopowder was kept at each temperature for 15 min, during which 15 consecutive scans were performed between 27° and 33° with a scanning step of 0.017° to investigate the evolution of the monoclinic and tetragonal phases. The temperature was increased at a rate of 50 °C/min.

The crystallinity of the produced ZrO_x thin films on silicon substrates annealed at 350 °C was also investigated using the same XRD equipment in grazing incidence mode (GIXRD, X'Pert PRO PANalytical) with a step of 0.2°/min and an angle of incidence of the X-ray beam fixed at 0.5° in the range of 10–60° (2θ).

SEM images were acquired using a Hitachi Regulus 8220 Scanning Electron Microscope (Mito, Japan) equipped with energy dispersive X-ray spectroscopy (EDS) equipment. For the FIB experiments, a Carl Zeiss AURIGA CrossBeam (FIB-SEM) workstation (Carl Zeiss MicrosPANI/TiO_2 copy GmbH, Oberkochen, Germany) was used. The inner structure of the ZrO_2 capacitor was observed by FIB, where Ga^+ ions were accelerated to 30 kV at 50 pA, and the etching depth was maintained at around 500 nm. TEM observations were performed with a Hitachi HF5000 field-emission transmission electron microscope (Mito, Japan) operated at 200 kV. A drop of the sonicated dispersions was deposited onto 200-mesh lacey-carbon copper grids and allowed to dry before observation. The average particle size and standard deviation were calculated from the dimensions of 50 nanoparticles with ImageJ software based on TEM images.

Raman spectroscopy measurements were conducted with an inVia Qontor confocal Raman microscope from Renishaw (Kingswood, UK). A 50 mW green diode operated at 532 nm was used as the excitation source, with an exposure time of 10 s and settings of 3 and 7 accumulations for ZrO_2 nanopowders before and after annealing treatment, respectively. The laser beam was focused with a long working distance (8.2 mm) 50 × Olympus objective and a 100 × Olympus objective with 0.35 mm working distance, respectively, for the samples before and after annealing treatment. The Raman spectra were recorded in the range of 110–800 cm^{-1} (as an extended scan). Several scans on different points of the ZrO_2 nanopowder's surface were recorded, and the present results are based on their average. Possible fluctuations of the Raman equipment were avoided with a previous calibration with a silicon wafer (521 cm^{-1} peak). All Raman measurements were performed at ambient conditions.

RT PL and PLE measurements were carried out in the microwave synthesized ZrO_2 nanopowders, before and after calcination, using a Fluorolog-3 Horiba Scientific set-up with a double additive grating Gemini 180 monochromator (1200 gr/mm and 2 × 180 mm) in the excitation and a triple grating iHR550 spectrometer in the emission (1200 gr/mm and 550 mm). A 450 W Xe lamp was used as the excitation source, and different excitation wavelengths were explored. The PLE data was obtained by monitoring the maximum of the PL emission.

The electrical and dielectric properties such as capacitance-voltage ($C - V$), capacitance-frequency ($C - f$) and current-voltage ($I - V$) measurements of the MIS capacitors ZrO_x-based were investigated by using a semiconductor parameter analyzer (Keysight B1500A) with a probe station (Cascade EPS150 Triax).

3. Results and Discussion

3.1. Structural and Optical Characterization of the ZrO_2 Nanopowders Produced under Microwave Irradiation

3.1.1. Raman Spectroscopy Measurements

ZrO_2 nanopowder was synthesized through a hydrothermal method assisted by microwave irradiation. Raman spectroscopy measurements were performed to investigate the crystalline phase of the synthesized ZrO_2 nanoparticles [76]. A further calcination process was also carried out at 800 °C for 15 min. Figure 2a,b show the Raman spectra in the 110–800 cm^{-1} range for the synthesized ZrO_2 nanopowders before and after the calcination treatment, respectively. Regarding the as-synthesized ZrO_2 powder, as seen in Figure 2a, broader Raman bands were observed when a comparison was made with the calcinated material (Figure 2b) [77]. Nevertheless, some of the peak positions of the vibrational modes occurred at the same frequencies, meaning that contributions of the same ZrO_2 polymorph were present in both materials. Tetragonal zirconia (D_{4h} point group) was expected to exhibit six Raman active vibrational modes at the Γ point of the first Brillouin zone with $A_{1g} + 2B_{1g} + 3E_g$ symmetries [78]. The frequencies of these modes were reported as 146 cm^{-1} (E_g), 270 cm^{-1} (A_{1g}), 318 cm^{-1} (B_{1g}), 458 cm^{-1} (E_g), 602 cm^{-1} (B_{1g}) and 648 cm^{-1} (E_g) [79]. For the case of the monoclinic ZrO_2 (C_{2h} point group), group theory analysis predicted 18 Raman active modes, $9A_g + 9B_g$ [80], with frequencies reported at 179 cm^{-1} (A_g), 190 cm^{-1} (A_g), 224 cm^{-1} (B_g), 305 cm^{-1} (A_g), 334 cm^{-1} (B_g), 348 cm^{-1} (A_g), 381 cm^{-1} (B_g), 476 cm^{-1} (A_g), 505 cm^{-1} (B_g), 536 cm^{-1} (B_g), 556 cm^{-1} (A_g), 616 cm^{-1} (B_g), 637 cm^{-1} (A_g) and a non-indexed peak at 757 cm^{-1} [77,81]. By comparing the spectra of the two materials in Figure 2, we were able to identify that the vibrational modes matched those reported for both crystalline phases [77,81]. In particular, the bandwidth of the Raman lines in the as-synthesized material was likely to be due to the overlap of the vibrational signatures from the tetragonal and monoclinic ZrO_2, as clearly identified by the presence of the 146 cm^{-1} and 265 cm^{-1} resonances and the bandshape asymmetry observed at 458 cm^{-1} and 648 cm^{-1}, which were assigned to the tetragonal phase. In contrast with Figure 2a, no bands related to the tetragonal ZrO_2 are visible in Figure 2b; however, its presence cannot be excluded. In addition, another band at 501 cm^{-1} (B_g) is visible, assigned to monoclinic ZrO_2 [77,81]. Cubic zirconia exhibits a characteristic broad band mode centered at 625 cm^{-1} due to disordered oxygen sub-lattice. This vibrational mode is not visible in Figure 2a,b. Moreover, the absence of the cubic phase was already expected since it is stable only at annealing temperatures near to ZrO_2 melting point or at RT by doping processes [82]. No other bands were detected [83].

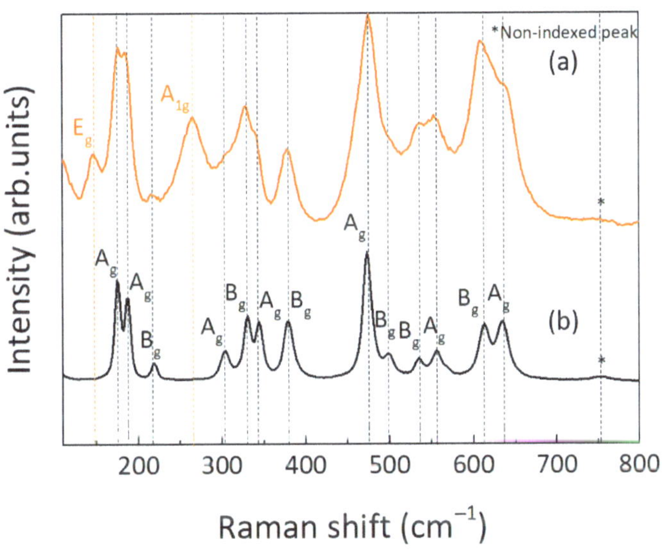

Figure 2. Raman spectra of the synthesized ZrO_2 nanopowders: (**a**) before and (**b**) after calcination at 800 °C for 15 min. The black and orange vertical dash lines represent the expected frequencies for the monoclinic and tetragonal ZrO_2, respectively, and according to [77,79,81].

3.1.2. X-ray Diffraction

XRD measurements were also performed to confirm the crystalline phase/phases obtained from Raman spectroscopy measurements. As seen in Figure 3a, the as-synthesized ZrO_2 nanopowder presents a mixture of ZrO_2 tetragonal and monoclinic phases, considering the ICDD card numbers 00-037-1484 and 00-050-1089 for the ZrO_2 monoclinic and tetragonal phases, respectively.

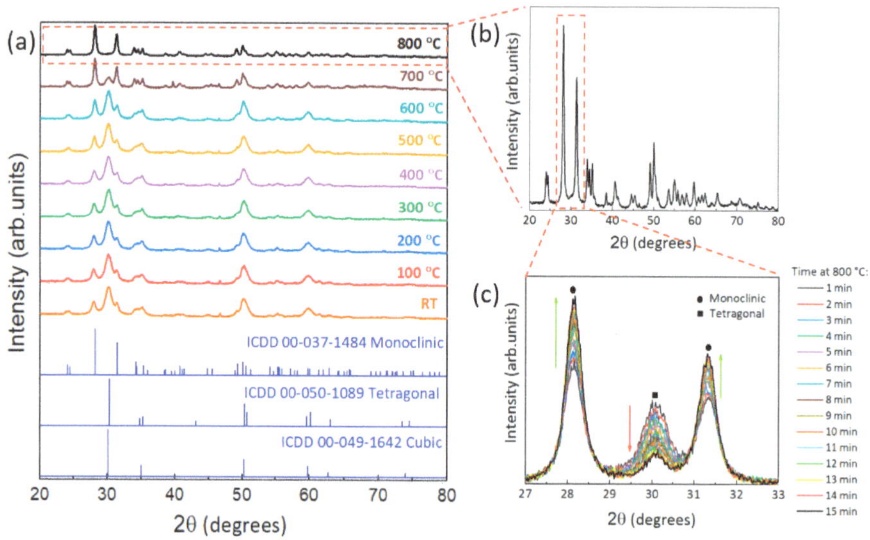

Figure 3. (**a**) In situ XRD diffractograms of the ZrO_2 nanopowder synthesized under microwave irradiation with different annealing temperatures ranging from RT to 800 °C with a plateau time of 15 min. For comparison, the simulated monoclinic, tetragonal and cubic zirconia structures are also presented. (**b**) XRD diffractogram of ZrO_2 nanopowder synthesized under microwave irradiation and calcinated at 800 °C for 15 min. (**c**) In situ XRD diffractograms of the ZrO_2 nanopowder at 800 °C for different annealing times. Green arrows represent an increase of monoclinic phase, while the red arrow indicates a decrease of tetragonal phase.

Calcination treatment after microwave synthesis is an important step in transforming an amorphous to crystalline phase or a single/mixture of crystalline phases into a specific crystalline phase, inducing optical and structural' modifications to the material [13,84]. In this regard, XRD diffractograms of the ZrO_2 nanopowder were recorded at different temperatures to infer the thermal behavior of the ZrO_2 nanomaterials and thus analyze the phase transition from tetragonal to monoclinic, Figure 3a. As can be observed, the intensity of the diffraction maximum at 30.176°, characteristic of the tetragonal phase, decreases with increasing temperature. At 800 °C, the peaks detected in the diffractogram of the ZrO_2 nanopowder (Figure 3b) were fully assigned to the monoclinic ZrO_2 phase (ICDD 00-037-1484). However, when performing XRD studies from 27° and 33° (Figure 3c) with a fixed temperature of 800 °C but different annealing times (each minute was recorded up to 15 min), it is evident that the conversion to monoclinic is accomplished at 800 °C with a minimal presence of the tetragonal phase. No other crystalline phases or impurities were detected with either of these two techniques.

The phase transformation of ZrO_2 tetragonal to monoclinic at 800 °C has been previously reported in the literature. Horti et al. [13] prepared ZrO_2 nanoparticles by chemical co-precipitation method and further annealed at different temperatures for 2 h, whose phase transition of tetragonal to monoclinic was seen to occur at the calcination temperature of 800 °C, with an evident mixture of both phases. In another study [85], ZrO_2 nanoparticles were produced by conventional and ultrasound-assisted precipitation in an alkaline medium followed by calcination at different temperatures ranging from 400 °C to 900 °C for 3 h. At 800 °C, a mixture of three ZrO_2 phases was reported, i.e., tetragonal-monoclinic-cubic phases. The present study has the advantage of having monoclinic ZrO_2 as the dominant phase after a fast calcination treatment of 15 min instead of hours.

3.1.3. Electron Microscopy

Figure 4 shows the SEM images of the ZrO_2 nanoparticles synthesized under microwave irradiation with a further calcination treatment at 800 °C for 15 min. It can be seen that the particles display a nearly spherical shape (Figure 4a), and from the EDS measurements, a homogeneous distribution of Zr (Figure 4b) and O (Figure 4c) is evident. No impurities were detected by EDS.

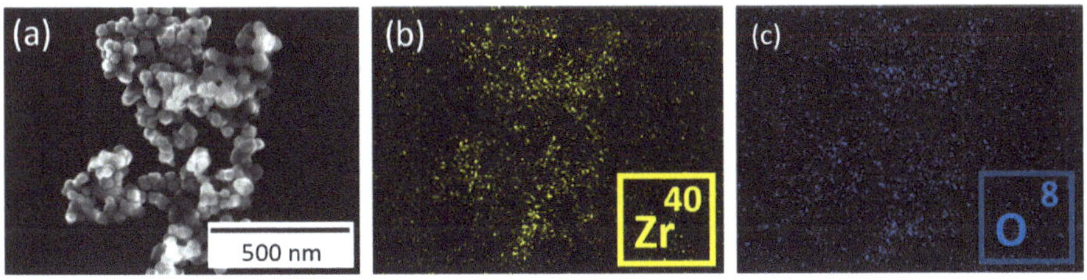

Figure 4. (**a**) SEM image of the ZrO_2 nanoparticles synthesized by microwave irradiation with a further calcination step at 800 °C for 15 min, together with the corresponding EDS maps of Zr (**b**) and O (**c**).

The atomic percentages (at.%) of Zr and O were also estimated from EDS spectra for the ZrO_2 nanopowders before and after the calcination treatment (Table 1). The atomic percentage values were close to those expected for pure zirconium oxide (the Zr/O ratio was 1:2 for the material before calcination and 1:2.7 after calcination).

Table 1. EDS point analysis of the ZrO_2 nanopowders before and after the calcination treatment at 800 °C for 15 min.

	ZrO_2 Nanopowder before the Calcination Treatment		**ZrO_2 Nanopowder after the Calcination Treatment**	
Elements	**Zr**	**O**	**Zr**	**O**
at.%	33.3	66.7	26.9	73.1

TEM measurements were also carried out for the ZrO_2 nanoparticles produced under microwave irradiation before and after calcination (Figure 5). The TEM images of the ZrO_2 nanoparticles before calcination demonstrate the presence of heterogeneous particles with an average particle size of 6.7 ± 1.9 nm. Abnormal large particles were also observed (Figure 5a,b). In terms of particle shape, microwave synthesis resulted in different nanostructures, including nanospheres, nanosquares and irregular-shaped nanoparticles. The heterogeneity detected is consistent with the XRD and Raman spectroscopy results, where it was shown that after microwave synthesis, there was a mixture of ZrO_2 phases, with the presence of both tetragonal and monoclinic ZrO_2 phases (Figures 2a and 3a). The lattice spacing of 0.3 nm measured in Figure 5c was consistent with the d-spacing of the (101) and (011) planes of the tetragonal ZrO_2. Furthermore, the fast Fourier transform (FFT) image carried out through the $[11\overline{1}]$ zone axis, attested the existence of pure tetragonal ZrO_2 nanocrystals.

After calcination, as expected, an expressive growth of the ZrO_2 nanoparticles and the presence of uniform nanoparticles in terms of size and shape were observed (Figure 5), exhibiting a nearly spherical shape, in agreement with SEM results (Figure 4a), and displaying an average particle size of 45.7 ± 9.9 nm. Figure 5e,f reveal that the measured lattice spacings of 0.32 and 0.37 nm are in good agreement with the d-spacing of the $(\overline{1}11)$ and (011) planes of the monoclinic ZrO_2 phase, respectively. The FFT image captured through the $[01\overline{1}]$ zone axis (Figure 5f) also confirmed the existence of pure monoclinic ZrO_2 nanocrystals. From TEM measurements, it was observed that the monoclinic ZrO_2

phase prevailed for the 800 °C calcinated material; however, some minor contaminations of tetragonal nanoparticles were also observed in accordance with the in situ XRD results (Figures 3c and S1).

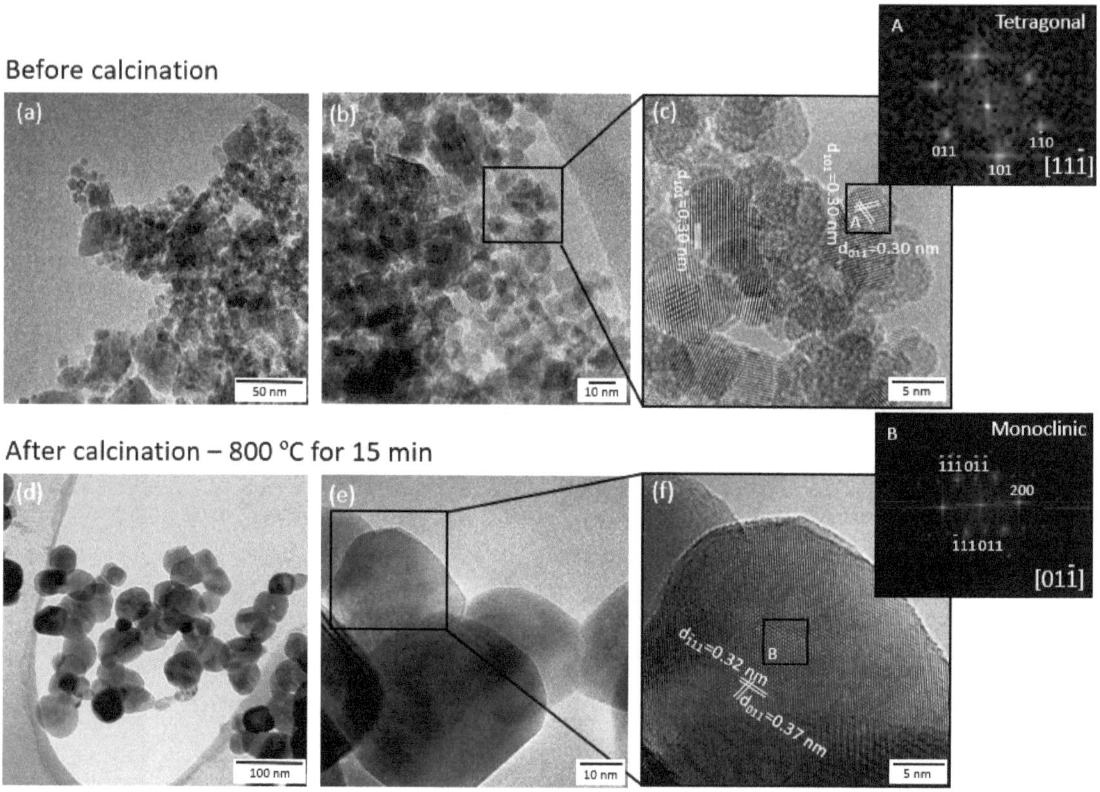

Figure 5. TEM images of the as-synthesized ZrO_2 nanoparticles before and after calcination at 800 °C for 15 min. (**a**) Bright-field TEM image of the ZrO_2 nanoparticles before calcination and (**b**) and (**c**) high-resolution TEM images of the ZrO_2 nanocrystals. (**d**) Bright-field TEM image of the ZrO_2 nanoparticles after calcination and (**e**,**f**) high-resolution TEM images of the ZrO_2 nanocrystals. The insets in (**c**,**f**) show the FFT images of the areas (black squares) indicated as (**A**,**B**), respectively.

3.1.4. Photoluminescence Measurements

Figure 6 depicts the results obtained for both materials before and after calcination, with clear differences in the PL spectra. In the case of the as-synthesized nanopowder without calcination (Figure 6a), an asymmetric band was observed in the visible spectral region, with a maximum at ~415 nm (~2.99 eV) when excited with different wavelengths in the UV range. After calcination in air (Figure 6b), the visible band experienced a broadening and an increase in its intensity. Under 240 nm photon excitation, its peak position was located at ~459 nm (~2.70 eV), whereas for longer wavelengths (260–290 nm), the PL maximum shifted to longer wavelengths, around ~475 nm (~2.61 eV). Indeed, the band observed at ~459 nm is likely to be composed of several recombination channels, including the ones which are preferentially excited with photons with wavelengths in the 260–290 nm range. The PLE monitored at the maximum of the observed PL bands revealed, in both cases, an excitation band at ~310 nm (~4.00 eV), a noticed shoulder at ~260 nm (~4.77 eV) and an excitation band maximum at ~240 nm (~5.17 eV), indicating that those are the preferential population paths for the observation of such emission.

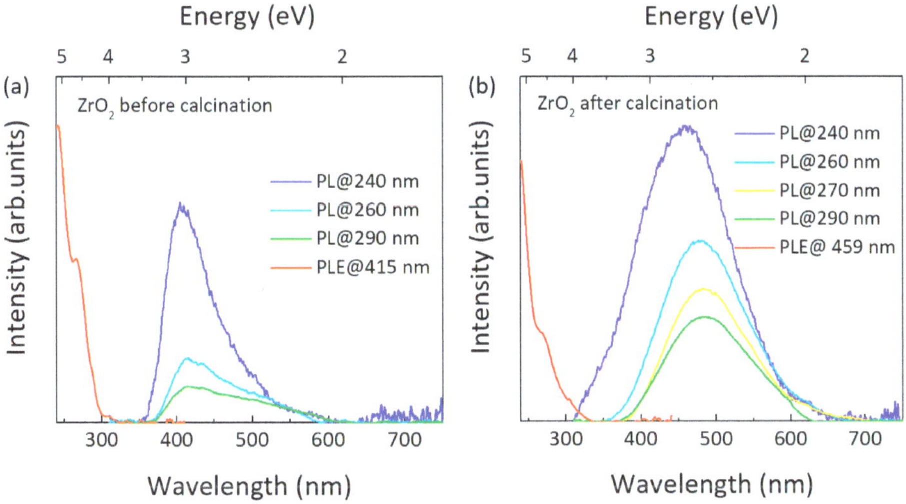

Figure 6. PLE/PL spectra acquired at RT on the ZrO_2 nanopowders (**a**) before and (**b**) after a calcination treatment at 800 °C for 15 min.

Broad visible bands were previously reported for the different undoped ZrO_2 polymorphs, namely the monoclinic one; however, the nature of the defects that originate from them is still not fully understood [86]. Among the suggested origins, oxygen vacancies and their complexes have been pointed out as the most probable defects, even though the presence of some impurities, such as Ti, have also been proposed [86–90]. For instance, Perevalov et al. [89] studied non-stoichiometric zirconia samples prepared at different oxygen pressures and inferred that the origin of the blue emission band peaked at 2.7 eV is likely associated with charged oxygen vacancy states since a decrease of the oxygen pressure during the grown leads to an increase in the intensity of the mentioned band. Moreover, slight shifts in the peak position (between 2.6 and 2.8 eV) were seen depending on the stoichiometry of the samples. In line with the present results, the reported PLE spectra also evidence a maximum at 5.1–5.2 eV, which they associated with the optical absorption of a defect created by oxygen vacancies. Nevertheless, according to the literature, the expected band gap energy for the ZrO_2 monoclinic phase predicted by theoretical models is in the range of ~5–6 eV [21,91–93], which is also in line with some experimental results [94–96]. Therefore, a likely explanation for the observed high energy excitation band that occurs in the vicinity of 240 nm (~5.17 eV) is that it corresponds to the band gap of the monoclinic ZrO_2 phase, in fair agreement with the values theoretically predicted for this crystalline phase. As mentioned, the produced nanoparticles are polyphasic in nature, exhibiting a contribution of the tetragonal ZrO_2 crystalline phase, which is expected to have a higher band gap energy than the monoclinic one. Therefore, the excitation shoulders/bands detected around ~260 nm and ~310 nm should be related to defects/impurities in electronic states inside the material's band gap.

Ashraf et al. [90] analyzed ZrO_2 monoclinic nano- and submicron crystals annealed at different temperatures and also observed broad emission bands from ~400 nm to 600 nm, whose spectral shape and intensity slightly varied depending on the annealing conditions and crystallite sizes. The authors assigned the PL emission centered at ~482 nm (~2.57 eV) to a transition involving F^+ centers, i.e., transitions from the valence band to local mid-gap states associated with singly occupied anion (oxygen) vacancies. At the shorter wavelength region, a shoulder at 422 nm (~2.94 eV) was also visible in the case of the nanoscale ZrO_2 before annealing and associated with a transition to singly ionized associated oxygen vacancy defects. After annealing, this shoulder disappeared, likely related to the dissociation of those centers into F^+ centers by the aid of atmospheric oxygen diffusion into the ZrO_2

particles [90]. A similar phenomenon may be occurring in the present materials, explaining the redshift observed in the peak position of the band after calcination. In the case of Ashraf et al. [90], further annealing in air at higher temperatures (1500 °C) promoted additional diffusion of oxygen into the crystals, leading to a decrease in the broad band PL intensity, strengthening the assumption of the involvement of oxygen vacancy defects in the origin of this band. Their PLE spectra showed an excitation band between ~250 nm and 350 nm, which suffered a blueshift from ~300 nm (~4.13 eV) to ~280 nm (~4.43 eV) after annealing. In 2021, Lokesha et al. [88] also studied monoclinic ZrO_2 samples, observing a broad band between 400 and 650 nm and peaked at 499 nm (~2.48 eV), which, in conjugation with other techniques, such as EPR, they attributed to the F^{2+} center related with aggregates of the singly occupied oxygen vacancies. These broad bands are frequently deconvoluted into different components (recombination channels), which can be associated with different charged states of the vacancy-related defect [88,90,97]. As a result, the existence of distinct charge states of the F-centers can account for the fact that three excitation maxima were observed in the present samples, subsequently leading to a shift in the band peak position depending on the excitation wavelength, particularly in the case of the calcinated material. Hence, the presence of oxygen vacancy-related defects seems to be a fair hypothesis for the nature of the defects involved in the PL emission identified in the present materials.

3.2. *Structural Characterization of the ZrO_2 Powder and ZrO_x Thin Films Produced by Solution Combustion Synthesis*

3.2.1. ZrO_2 Powder

X-ray Diffraction

Solution combustion synthesis is known to produce materials in powder form but also as thin films [58,60,98]. The ZrO_2 powder was obtained by conventional solution combustion synthesis, with zirconium oxynitrate and urea as precursors and 2-methoxyethanol (2-ME) as a solvent. The prepared solution was further annealed in air at 350 °C for 1 h to obtain the ZrO_2 powder. The XRD diffractogram of the ZrO_2 powder is shown in Figure 7. The ZrO_2 tetragonal phase is clearly identified with the peak at 30.176°, associated with the (111) diffraction plane [99] (ICDD 00-050-1089), along with other diffraction maxima that were assigned to the monoclinic phase (ICDD 00-037-1484). No impurities were detected, which suggests the presence of a highly pure ZrO_2 powder.

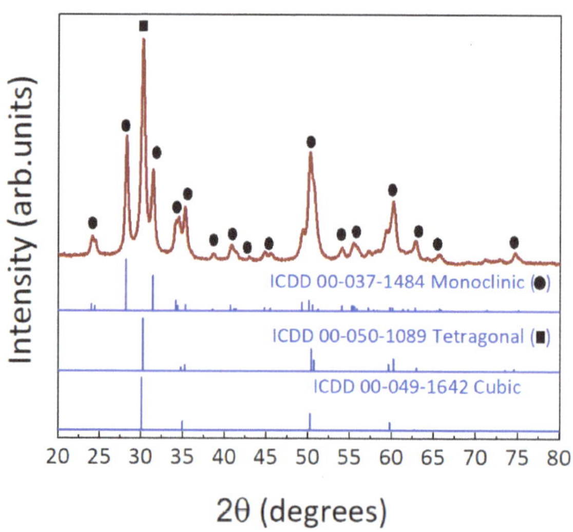

Figure 7. XRD diffractogram of ZrO_2 powder produced by solution combustion synthesis and annealed in air at 350 °C for 1 h (represented in red). For comparison, the simulated monoclinic, tetragonal and cubic zirconia structures are also presented.

3.2.2. Electron Microscopy

Figure 8 shows the SEM images of the ZrO_2 powder resultant from the solution combustion synthesis. The formation of micro-sized plate-like structures is clear, as seen in Figure 8a. Nevertheless, Figure 8b indicates the presence of nano-sized grains composing the micro-sized structure, and Figure S2 suggests the stacking of several nanolayers. Individual nanoparticles were also observed. Moreover, the void-like nature of the particle observed in Figure S2 can be associated with the escape of gaseous combustion products that were formed during combustion synthesis [58]. EDS measurements were also carried out, and, as observed for the ZrO_2 nanoparticles produced under microwave irradiation (Figure 4), the presence of a homogeneous distribution of Zr (Figure 8c) and O (Figure 8d) was clear. No impurities were detected by this technique.

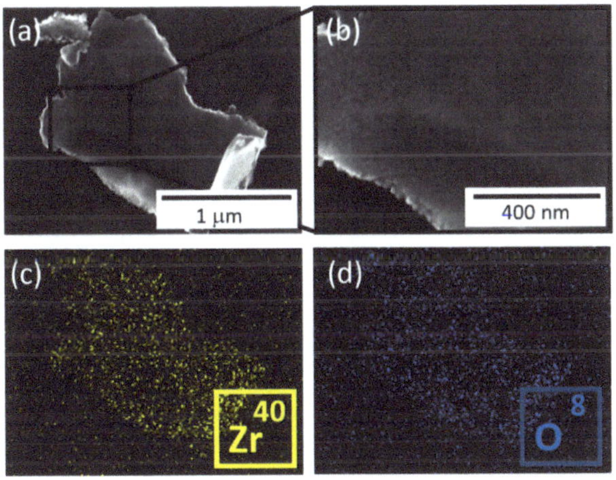

Figure 8. (**a**,**b**) SEM images of the ZrO_2 particle produced by the combustion synthesis method and after annealing in air at 350 °C for 1 h, together with the corresponding EDS maps of Zr (**c**) and O (**d**).

The atomic percentages of Zr and O were also estimated by EDS point analysis (Table 2). As observed for the nanopowders synthesized under microwave irradiation, the values were within the expected range for pure zirconium oxide (the Zr/O ratio was 1:2.1).

Table 2. EDS point analysis of the ZrO_2 nanopowder produced by the solution combustion synthesis.

Elements	Zr	O
at.%	31.9	68.1

The ZrO_2 powder produced by solution combustion synthesis was also observed by TEM, confirming the presence of individual ZrO_2 nanoparticles. From Figure 9a, nano-sized particles with different shapes can be observed, including nanospheres, nano squares and irregular-shaped nanoparticles (Figure 9a). Larger particles without a specific shape were also observed; nevertheless, the average particle size was 10 ± 7 nm. The lattice spacing of 0.3 nm was measured on an individual ZrO_2 nanoparticle (Figure 9b), which was consistent with the d-spacing of the (101) plane of the tetragonal ZrO_2. Nonetheless, as observed by XRD results, the powder was a mixture of tetragonal and monoclinic ZrO_2 phases.

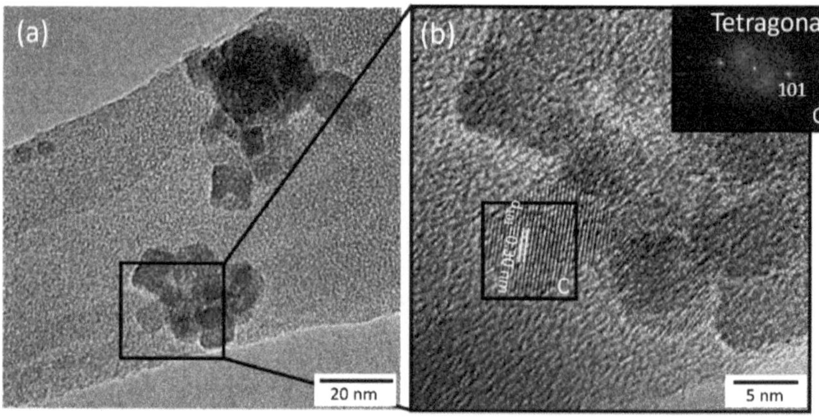

Figure 9. TEM images of the ZrO_2 powder produced with solution combustion synthesis and annealed in air at 350 °C for 1 h. (**a**) Bright-field TEM image of the ZrO_2 nanoparticles and (**b**) high-resolution TEM images of the ZrO_2 nanocrystals. The inset in (**b**) shows the FFT images of the area (black square) indicated as C.

3.2.3. ZrO_x Thin Films

X-ray Diffraction

The ZrO_x thin films were produced with the same solution used for producing the ZrO_2 powder by solution combustion. Spin coating was employed to deposit ZrO_x thin films on silicon substrates to be tested as capacitors. Crystalline dielectric films may provide, for instance, pathways for leakage current due to the grain boundaries, which results in a small on/off ratio, leading to poor capacitor performance. On the contrary, amorphous thin films have a smooth surface, high stability, low interface state density, as well as large-area uniformity [100,101].

Grazing angle mode was performed to analyze the crystalline structure of ZrO_x thin films annealed at 350 °C in air for 30 min on a hot plate. An annealing treatment at 350 °C was carried out since it was reported that annealing treatments in solution-processed high-κ ZrO_x thin films between 300–600 °C are expected to increase O^{-2} content in ZrO_x films, leading to a reduction of oxygen defects and/or hydroxyl groups and consequently of conduction paths [69,102].

As observed in Figure 10, some diffraction maxima are visible at $2\theta = 51.4°$ and $2\theta = 53.4°$, accompanied by a bump at around 55° (2θ), which correspond to forbidden diffraction planes of a (100)-oriented silicon wafer that appear when the grazing angle mode is used at certain phi angle [103]. Apart from that, no other diffraction maxima were identified, indicating the amorphous nature of the produced ZrO_x thin films, which is ideal for thin film capacitor applications.

3.3. Electrical Characterization of the ZrO_x Capacitors

ZrO_x thin films were implemented in metal-insulator-semiconductor (MIS) devices to evaluate their dielectric performance. Figure 11a show the typical C–V curve of the ZrO_x thin film capacitors with a small hysteresis. The devices present good stability with a small change of the capacitance over the frequency (1 kHz–100 kHz), as depicted in Figure 11b. To calculate the dielectric contact, a cross-section of the device to determine the ZrO_x thickness was performed, as shown in the inset of Figure 11b. The ZrO_x thin films present a capacitance per unit area of 672 ± 28 nF cm^{-2} and a dielectric constant of 11.0 ± 0.5 (both calculated at 1 kHz). These values are lower when compared with vacuum deposited crystalline films due to the lower film density and presence of pores as a consequence of the solution process [71,72,104–106]. Nevertheless, the devices present a good breakdown field of 2.8 ± 0.1 MV/cm and a low leakage current density of (3.9 ± 1.1) × 10^{-7} A/cm^2 at 1 MV/cm.

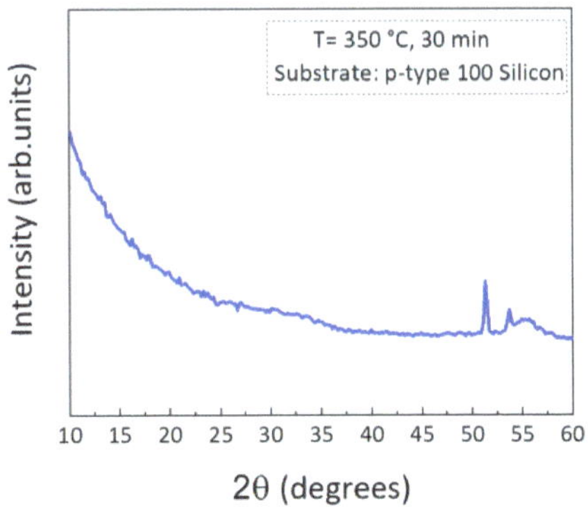

Figure 10. Grazing incidence XRD diffractogram of the spin-coated ZrO_x thin films annealed at 350 °C in air for 30 min and deposited on a silicon substrate.

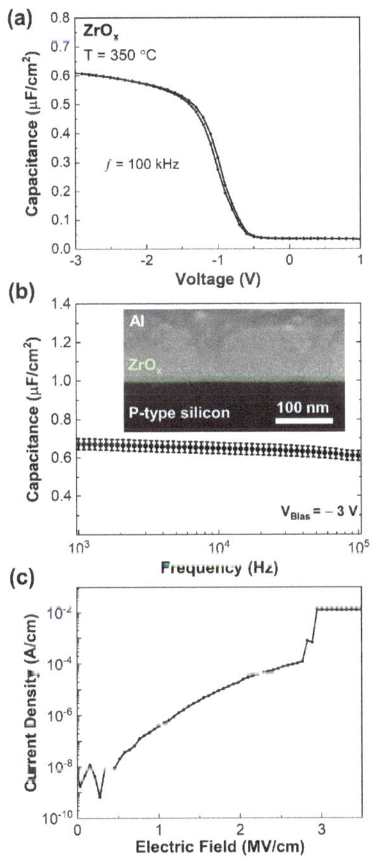

Figure 11. Electrical characteristics of solution-based ZrO_x MIS capacitors produced at 350 °C: (**a**) typical capacitance-voltage characteristics at 100 kHz; (**b**) average of capacitance versus frequency characteristics, and the inset shows the cross-section SEM of the devices; (**c**) typical leakage current density versus the breakdown field.

Figure 11c shows a typical current density versus breakdown field dependence. The results are in accordance with solution-based ZrO_x thin film capacitors [71]. Once low leakage current density was reached (<10^{-5} A/cm^2), it opens the window for future applications such as gate insulators for solution-based thin film transistors.

4. Conclusions

This work demonstrated that both the hydrothermal method assisted by microwave irradiation and solution combustion synthesis are reliable and simple production routes for ZrO_2 nanomaterials, including nanoparticles and thin films. Moreover, the low-cost character of the study is evidenced by the inexpensive apparatus required to produce and deposit ZrO_x thin films and thus fabricate energy storage devices. In both synthesis routes, XRD analysis revealed that the as-synthesized powders had a mixture of ZrO_2 monoclinic and tetragonal phases. A calcination treatment at 800 °C for 15 min was imposed on the microwave synthesized nanopowder to promote the conversion into single-phase materials. Nevertheless, in situ XRD and TEM analysis revealed a minor presence of the ZrO_2 tetragonal phase after the calcination treatment. Different nanostructures were observed after synthesis in both production routes. In the case of the microwave synthesized material, after calcination, near-spherical nanoparticles were observed, while plate-like structures composed of nano-sized grains were identified for the combustion synthesis. The effect of calcination on the optical properties of the microwave synthesized nanopowders was evaluated by PL and PLE, showing a clear broadening and an increased intensity of the ZrO_2 band observed in the visible spectral region after calcination. This band was tentatively associated with the presence of oxygen vacancy-related defects. Solution combustion synthesis also resulted in thin films, and XRD analysis suggested their amorphous nature. SEM and FIB measurements of the ZrO_x capacitors revealed that the thin films present a thickness of 14.2 ± 0.1 nm, which leads to a dielectric constant of 11.0 ± 0.5 (calculated at 1 kHz). The devices also presented a good breakdown field of 2.8 ± 0.1 MV/cm and a low leakage current density of $(3.9 \pm 1.1) \times 10^{-7}$ A/cm^2 at 1 MV/cm. From this study, the production of ZrO_2 nanomaterials with two simple, cost-effective and easily upscaling synthesis routes and their integration on capacitors can be highlighted.

Supplementary Materials: The following supporting information can be downloaded at: https://www.mdpi.com/article/10.3390/en15176452/s1, Figure S1: High-resolution TEM image of the ZrO_2 nanocrystals calcinated at 800 °C for 15 min. The inset shows the FFT images of the area (black square) indicated as A; Figure S2: SEM image of a ZrO_2 particle produced by solution combustion synthesis and annealed at 350 °C.

Author Contributions: M.L.M. was responsible for producing and characterizing microwave synthesized materials, as well as writing the manuscript. H.d.V. was responsible for the production and characterization of the ZrO_x films, and J.M. was responsible for the combustion powder synthesis. M.M. and A.P. were responsible for the XRD measurements. D.N., R.B. and E.C. were responsible for the overall scientific orientation and revising the manuscript. D.N. performed the TEM observations and analysis. T.M. and J.R. were responsible for the optical measurements, respective analyses and revising of the manuscript. E.F. and R.M. were responsible for supervising all the processes and obtaining funding. Part of the work resulted from the Master of Science thesis of H.d.V., titled: Effect of eco-friendly solvents in solution-based ZrO_x dielectrics, developed at the Nova School of Science and Technology. All authors have read and agreed to the published version of the manuscript.

Funding: This work was financed by national funds from FCT-Fundação para a Ciência e a Tecnologia, I.P., in the scope of the projects UI/BD/151292/2021 (Ph.D. research scholarship), LA/P/0037/2020, UIDP/50025/2020 and UIDB/50025/2020 of the Associate Laboratory Institute of Nanostructures, Nanomodelling and Nanofabrication-i3N, but also the 2021.03825.CEECIND. The work was also partially funded by the Nanomark collaborative project between INCM (Imprensa Nacional-Casa da Moeda) and CENIMAT/i3N. Acknowledgments also go to the EC project SYNERGY H2020-WIDESPREAD-2020-5, CSA, proposal n° 952169, EMERGE-2020-INFRAIA-2020-1, proposal n°

101008701, and to the European Community's H2020 program under grant agreement No. 787410 (ERC-2018-AdG DIGISMART).

Data Availability Statement: The authors confirm that the data supporting the findings of this study are available within the article and its Supplementary Materials.

Conflicts of Interest: The authors declare no conflict of interest.

References

1. Shah, N.A.; Gul, M.; Abbas, M.; Amin, M. Synthesis of Metal Oxide Semiconductor Nanostructures for Gas Sensors. In *Gas Sensors*; IntechOpen: London, UK, 2019; ISBN 978-1-78985-160-1.
2. Roco, M.C.; Mirkin, C.A.; Hersam, M.C. Nanotechnology research directions for societal needs in 2020: Summary of international study. *J. Nanoparticle Res.* **2011**, *13*, 897–919. [CrossRef]
3. Chavali, M.S.; Nikolova, M.P. Metal oxide nanoparticles and their applications in nanotechnology. *SN Appl. Sci.* **2019**, *1*, 1–30. [CrossRef]
4. Ahmad, T.; Shahazad, M.; Phul, R. Hydrothermal synthesis, characterization and dielectric properties of zirconia nanoparticles. *Mater. Sci. Eng. Int. J.* **2017**, *1*, 100–104. [CrossRef]
5. Nunes, D.; Pimentel, A.; Gonçalves, A.; Pereira, S.; Branquinho, R.; Barquinha, P.; Martins, R. Metal Oxide Nanostructures for Sensor Applications. *Semicond. Sci. Technol.* **2019**, *34*, 1–178. [CrossRef]
6. Nunes, D.; Pimentel, A.; Santos, L.; Barquinha, P.; Pereira, L.; Fortunato, E.; Martins, R.; Pimentel, A.; Barquinha, P.; Pereira, L.; et al. *Metal Oxide Nanostructures: Synthesis, Properties and Applications*, 1st ed.; Korotcenkov, G., Ed.; Elsevier: Amsterdam, The Netherlands, 2018; ISBN 9780128115121.
7. Nunes, D.; Fortunato, E.; Martins, R. Flexible nanostructured TiO_2-based gas and UV sensors: A review. *Discov. Mater.* **2022**, *2*, 1–23. [CrossRef]
8. Nunes, D.; Pimentel, A.; Branquinho, R.; Fortunato, E.; Martins, R. Metal oxide-based photocatalytic paper: A green alternative for environmental remediation. *Catalysts* **2021**, *11*, 504. [CrossRef]
9. Vaidya, S.; Ahmad, T.; Agarwal, S.; Ganguli, A.K. Nanocrystalline Oxalate/Carbonate Precursors of Ce and Zr and Their Decompositions to CeO_2 and ZrO_2 Nanoparticles. *J. Am. Ceram. Soc.* **2007**, *90*, 863–869. [CrossRef]
10. Behbahani, A.; Rowshanzamir, S.; Esmaeilifar, A. Hydrothermal Synthesis of Zirconia Nanoparticles from Commercial Zirconia. *Procedia Eng.* **2012**, *42*, 908–917. [CrossRef]
11. Van Tran, T.; Nguyen, D.T.C.; Kumar, P.S.; Din, A.T.M.; Jalil, A.A.; Vo, D.V.N. Green synthesis of ZrO_2 nanoparticles and nanocomposites for biomedical and environmental applications: A review. *Environ. Chem. Lett.* **2022**, *20*, 1309–1331. [CrossRef]
12. Hassan, N.S.; Jalil, A.A. A review on self-modification of zirconium dioxide nanocatalysts with enhanced visible-light-driven photodegradation of organic pollutants. *J. Hazard. Mater.* **2022**, *423 (Pt A)*, 126996. [CrossRef]
13. Liu, B.; Wang, C.; Chen, Y.; -Structural, A.; Properties of, L.Y.; Zirconia Nanopowders Korsunska, C.-C.N.; Baran, M.; Polishchuk, Y.; Horti, N.C.; Kamatagi, M.D.; et al. Structural and optical properties of zirconium oxide (ZrO_2) nanoparticles: Effect of calcination temperature. *Nano Express* **2020**, *1*, 1–10. [CrossRef]
14. Boratto, M.H.; Lima, J.V.M.; Scalvi, L.V.A.; Graeff, C.F.O. Low-temperature ZrO_2 thin films obtained by polymeric route for electronic applications. *J. Mater. Sci. Mater. Electron.* **2020**, *31*, 16065–16072. [CrossRef]
15. Namavar, F.; Wang, G.; Cheung, C.L.; Sabirianov, R.F.; Zeng, X.C.; Mei, W.N.; Bai, J.; Brewer, J.R.; Haider, H.; Garvin, K.L. Thermal stability of nanostructurally stabilized zirconium oxide. *Nanotechnology* **2007**, *18*, 1–6. [CrossRef]
16. Abd El-Lateef, H.M.; Khalaf, M.M. Corrosion resistance of ZrO_2–TiO_2 nanocomposite multilayer thin films coated on carbon steel in hydrochloric acid solution. *Mater. Charact.* **2015**, *108*, 29–41. [CrossRef]
17. Septawendar, R.; Sutardi, S.; Karsono, U.; Sofiyaningsih, N. A Low-Cost, Facile Method on Production of Nano Zirconia and Silica from Local Zircon in a Large Scale Using a Sodium Carbonate Sintering Technology. *J. Aust. Ceram. Soc.* **2016**, *52*, 92–102.
18. Xifeng, L.; Enlong, X.; Jianhua, Z. Low temperature solution processed zirconium oxide gate insulators for thin-film transistors. *IEEE Trans. Electron Devices* **2013**, *60*, 3413–3416. [CrossRef]
19. Chang, J.P.; Lin, Y.S. Dielectric property and conduction mechanism of ultrathin zirconium oxide films. *Appl. Phys. Lett.* **2001**, *79*, 3666. [CrossRef]
20. Zandiehnadem, F.; Murray, R.A.; Ching, W.Y. Electronic structures of three phases of zirconium oxide. *Physica B+C* **1988**, *150*, 19–24. [CrossRef]
21. Králik, B.; Chang, E.K.; Louie, S.G. Structural properties and quasiparticle band structure of zirconia. *Phys. Rev. B Condens. Matter* **1998**, *57*, 7027–7036. [CrossRef]
22. Poungchan, G.; Ksapabutr, B.; Panapoy, M. One-step synthesis of flower-like carbon-doped ZrO_2 for visible-light-responsive photocatalyst. *Mater. Des.* **2016**, *89*, 137–145. [CrossRef]
23. Soares, M.R.N. Development of Zirconia Based Phosphors for Application in Lighting and as Luminescent Bioprobes. Ph.D. Thesis, University of Aveiro, Aveiro, Portugal, 2016. Available online: https://ria.ua.pt/handle/10773/15884 (accessed on 21 July 2022).

24. Nova, C.V.; Reis, K.A.; Pinheiro, A.L.; Dalmaschio, C.J.; Chiquito, A.J.; Teodoro, M.D.; Rodrigues, A.D.; Longo, E.; Pontes, F.M. Synthesis, characterization, photocatalytic, and antimicrobial activity of ZrO_2 nanoparticles and Ag@ZrO_2 nanocomposite prepared by the advanced oxidative process/hydrothermal route. *J. Sol-Gel Sci. Technol.* **2021**, *98*, 113–126. [CrossRef]
25. Stolzenburg, P.; Freytag, A.; Bigall, N.C.; Garnweitner, G. Fractal growth of ZrO_2 nanoparticles induced by synthesis conditions. *CrystEngComm* **2016**, *18*, 8396–8405. [CrossRef]
26. Rafiq Hussain Siddiqui, M.; Ibrahim Al-Wassil, A.; Mohmmed Al-Otaibi, A.; Mohamad Mahfouz, R. Effects of Precursor on the Morphology and Size of ZrO_2 Nanoparticles, Synthesized by Sol-gel Method in Non-aqueous Medium. *Mater. Res.* **2012**, *15*, 986–989. [CrossRef]
27. Pang, Q.; Chen, H.; Wang, X.; Wang, T.; Wang, D.; Feng, S.; Lu, H.; Li, Q. Field Effect Transistors Based on In Situ Fabricated Graphene Scaffold–ZrO_2 Nanofilms. *Adv. Electron. Mater.* **2018**, *4*, 1–6. [CrossRef]
28. Sani, E.; Sciti, D.; Capiani, C.; Silvestroni, L. Colored zirconia with high absorbance and solar selectivity. *Scr. Mater.* **2020**, *186*, 147–151. [CrossRef]
29. Sani, E.; Mercatelli, L.; Sans, J.L.; Sciti, D. Optical properties of black and white ZrO_2 for solar receiver applications. *Sol. Energy Mater. Sol. Cells* **2015**, *140*, 477–482. [CrossRef]
30. Silvestroni, L.; Capiani, C.; Sciti, D.; Sani, E. Coloring zirconium oxide for novel energy saving industrial applications. *Renew. Energy* **2022**, *190*, 223–231. [CrossRef]
31. Silva, J.P.B.; Silva, J.M.B.; Sekhar, K.C.; Palneedi, H.; Istrate, M.C.; Negrea, R.F.; Ghica, C.; Chahboun, A.; Pereira, M.; Gomes, M.J.M. Energy storage performance of ferroelectric ZrO_2 film capacitors: Effect of HfO_2:Al_2O_3 dielectric insert layer. *J. Mater. Chem. A* **2020**, *8*, 14171–14177. [CrossRef]
32. Pešić, M.; Hoffmann, M.; Richter, C.; Mikolajick, T.; Schroeder, U. Nonvolatile Random Access Memory and Energy Storage Based on Antiferroelectric Like Hysteresis in ZrO_2. *Adv. Funct. Mater.* **2016**, *26*, 7486–7494. [CrossRef]
33. Wang, Y.; Wang, Y.; Zeng, H.; Wei, X. Ultra-high energy storage density of transparent capacitors based on linear dielectric ZrO_2 thin films with the thickness scaled up to hundreds nanometers. *Appl. Phys. Lett.* **2022**, *120*, 023904. [CrossRef]
34. Nunes, D.; Vilarigues, M.; Correia, J.B.; Carvalho, P.A. Nickel–carbon nanocomposites: Synthesis, structural changes and strengthening mechanisms. *Acta Mater.* **2012**, *60*, 737–747. [CrossRef]
35. Vidya, Y.S.; Anantharaju, K.S.; Nagabhushana, H.; Sharma, S.C.; Nagaswarupa, H.P.; Prashantha, S.C.; Shivakumara, C. Danithkumar Combustion synthesized tetragonal ZrO_2: Eu^{3+} nanophosphors: Structural and photoluminescence studies. *Spectrochim. Acta Part A Mol. Biomol. Spectrosc.* **2015**, *135*, 241–251. [CrossRef] [PubMed]
36. Bumajdad, A.; Nazeer, A.A.; Al Sagheer, F.; Nahar, S.; Zaki, M.I. Controlled Synthesis of ZrO_2 Nanoparticles with Tailored Size, Morphology and Crystal Phases via Organic/Inorganic Hybrid Films. *Sci. Rep.* **2018**, *8*, 3695. [CrossRef] [PubMed]
37. Jiang, C.; Wang, F.; Wu, N.; Liu, X. Up- and down-conversion cubic zirconia and hafnia nanobelts. *Adv. Mater.* **2008**, *20*, 4826–4829. [CrossRef]
38. Cao, H.; Qiu, X.; Luo, B.; Liang, Y.; Zhang, Y.; Tan, R.; Zhao, M.; Zhu, Q. Synthesis and Room-Temperature Ultraviolet Photoluminescence Properties of Zirconia Nanowires. *Adv. Funct. Mater.* **2004**, *14*, 1–4. [CrossRef]
39. Stępień, M.; Handzlik, P.; Fitzner, K. Synthesis of ZrO_2 nanotubes in inorganic and organic electrolytes by anodic oxidation of zirconium. *J. Solid State Electrochem.* **2014**, *18*, 3081–3090. [CrossRef]
40. Mangla, O.; Roy, S. Monoclinic Zirconium Oxide Nanostructures Having Tunable Band Gap Synthesized under Extremely Non-Equilibrium Plasma Conditions. *Proceedings* **2019**, *3*, 10. [CrossRef]
41. Yue, M.; Cui, M.; Zhang, N.; Long, Z.; Huang, X. Characterization of CeO_2-ZrO_2 mixed oxides prepared by two different co-precipitation methods. *J. Rare Earths* **2013**, *31*, 251–256. [CrossRef]
42. Lim, H.S.; Ahmad, A.; Hamzah, H. Synthesis of zirconium oxide nanoparticle by sol-gel technique. *AIP Conf. Proc.* **2013**, *1571*, 812. [CrossRef]
43. Zinatloo-Ajabshir, S.; Salavati-Niasari, M. Synthesis of pure nanocrystalline ZrO_2 via a simple sonochemical-assisted route. *J. Ind. Eng. Chem.* **2014**, *20*, 3313–3319. [CrossRef]
44. Liang, J.; Jiang, X.; Liu, G.; Deng, Z.; Zhuang, J.; Li, F.; Li, Y. Characterization and synthesis of pure ZrO_2 nanopowders via sonochemical method. *Mater. Res. Bull.* **2003**, *38*, 161–168. [CrossRef]
45. Beer, S.M.J.; Samelor, D.; Abdel Aal, A.; Etzkorn, J.; Rogalla, D.; Turgambaeva, A.E.; Esvan, J.; Kostka, A.; Vahlas, C.; Devi, A. Direct liquid injection chemical vapor deposition of ZrO_2 films from a heteroleptic Zr precursor: Interplay between film characteristics and corrosion protection of stainless steel. *J. Mater. Res. Technol.* **2021**, *13*, 1599–1614. [CrossRef]
46. Chang, Y.; Li, X. Bin Preparation of ZrO_2 spherical nanometer powders by emulsion processing route. *Trans. Nonferrous Met. Soc. China* **2006**, *16*, s332–s336. [CrossRef]
47. Lee, M.H.; Tai, C.Y.; Lu, C.H. Synthesis of spherical zirconia by reverse emulsion precipitation. *Korean J. Chem. Eng.* **1999**, *16*, 818–822. [CrossRef]
48. Zhao, N.; Pan, D.; Nie, W.; Ji, X. Two-phase synthesis of shape-controlled colloidal zirconia nanocrystals and their characterization. *J. Am. Chem. Soc.* **2006**, *128*, 10118–10124. [CrossRef]
49. Guo, M.; Wang, G.; Zhao, Y.; Li, H.; Tang, K.; Zhao, Y.; Burgess, K. Preparation of Nano-ZrO_2 powder via a microwave-assisted hydrothermal method. *Ceram. Int.* **2021**, *47*, 12425–12432. [CrossRef]

50. Gonell, F.; Portehault, D.; Julián-López, B.; Vallé, K.; Sanchez, C.; Corma, A. One step microwave-assisted synthesis of nanocrystalline WO_x–ZrO_2 acid catalysts. *Catal. Sci. Technol.* **2016**, *6*, 8257–8267. [CrossRef]
51. Bondioli, F.; Bonamartini Corradi, A.; Ferrari, A.M.; Leonelli, C. Synthesis of Zirconia Nanoparticles in a Continuous-Flow Microwave Reactor. *J. Am. Ceram. Soc.* **2008**, *91*, 3746–3748. [CrossRef]
52. Kim, J.R.; Myeong, W.J.; Ihm, S.K. Characteristics of CeO_2–ZrO_2 mixed oxide prepared by continuous hydrothermal synthesis in supercritical water as support of Rh catalyst for catalytic reduction of NO by CO. *J. Catal.* **2009**, *263*, 123–133. [CrossRef]
53. Li, Q.; Liu, L.; Wang, Z.; Wang, X. Continuous Hydrothermal Flow Synthesis and Characterization of ZrO_2 Nanoparticles Doped with CeO_2 in Supercritical Water. *Nanomaterials* **2022**, *12*, 668. [CrossRef]
54. Hobbs, H.; Briddon, S.; Lester, E. The synthesis and fluorescent properties of nanoparticulate ZrO_2 doped with Eu using continuous hydrothermal synthesis. *Green Chem.* **2009**, *11*, 484–491. [CrossRef]
55. Dudnik, E.V. Modern methods for hydrothermal synthesis of ZrO_2-based nanocrystalline powders. *Powder Metall. Met. Ceram.* **2009**, *48*, 238–248. [CrossRef]
56. Liu, L.; Wang, S.; Zhang, B.; Jiang, G.; Yang, J. Supercritical hydrothermal synthesis of nano-ZrO_2: Influence of technological parameters and mechanism. *J. Alloy. Compd.* **2022**, *898*, 16287. [CrossRef]
57. Darr, J.A.; Zhang, J.; Makwana, N.M.; Weng, X. Continuous Hydrothermal Synthesis of Inorganic Nanoparticles: Applications and Future Directions. *Chem. Rev.* **2017**, *117*, 11125–11238. [CrossRef]
58. Ponkumar, S.; Janaki, K.; Prakashbabu, D.; B Ramalingam, H.; Munirathnam, K.; Hari Krishna, R. Solution Combustion Synthesis of ZrO_2:Tb^{3+} Nanophosphors Viable for WLEDs. *Mater. Today Proc.* **2018**, *5*, 10717–10721. [CrossRef]
59. Samantaray, S.; Mishra, B.G.; Pradhan, D.K.; Hota, G. Solution combustion synthesis and physicochemical characterization of ZrO_2-MoO_3 nanocomposite oxides prepared using different fuels. *Ceram. Int.* **2011**, *37*, 3101–3108. [CrossRef]
60. Purohit, R.D.; Saha, S.; Tyagi, A.K. Combustion synthesis of nanocrystalline ZrO_2 powder: XRD, Raman spectroscopy and TEM studies. *Mater. Sci. Eng. B Solid-State Mater. Adv. Technol.* **2006**, *130*, 57–60. [CrossRef]
61. Carlos, E.; Martins, R.; Fortunato, E.; Branquinho, R. Solution Combustion Synthesis: Towards a Sustainable Approach for Metal Oxides. *Chem.—A Eur. J.* **2020**, *26*, 9099–9125. [CrossRef]
62. Branquinho, R.; Santa, A.; Carlos, E.; Salgueiro, D.; Barquinha, P.; Martins, R.; Fortunato, E. Solution Combustion Synthesis: Applications in Oxide Electronics. In *Developments in Combustion Technology*; IntechOpen: London, UK, 2016; ISBN 978-953-51-2669-0.
63. Moreira, M.; Carlos, E.; Dias, C.; Deuermeier, J.; Pereira, M.; Barquinha, P.; Branquinho, R.; Martins, R.; Fortunato, E. Tailoring IGZO Composition for Enhanced Fully Solution-Based Thin Film Transistors. *Nanomaterials* **2019**, *9*, 1273. [CrossRef]
64. Merten, D.; Broekaert, J.A.C.; Brandt, R.; Jakubowski, N. Analysis of ZrO_2 powders by microwave assisted digestion at high pressure and ICP atomic spectrometry. *J. Anal. At. Spectrom.* **1999**, *14*, 1093–1098. [CrossRef]
65. Iqbal, Z.; Sadiq, S.; Sadiq, M.; Khan, I.; Saeed, K. Effect of Microwave Irradiation on the Catalytic Activity of Tetragonal Zirconia: Selective Hydrogenation of Aldehyde. *Arab. J. Sci. Eng.* **2021**, *47*, 5841–5848. [CrossRef]
66. Srikalyani, C.H.; Sultana, S. Microwave Synthesis of ZrO_2 Nanoparticles and its Reinforcement in Geo-Polymer gel. *Int. J. ChemTech Res.* **2017**, *10*, 1–6.
67. Goharshadi, E.K.; Hadadian, M. Effect of calcination temperature on structural, vibrational, optical, and rheological properties of zirconia nanoparticles. *Ceram. Int.* **2012**, *38*, 1771–1777. [CrossRef]
68. Padovini, D.S.S.; Pontes, D.S.L.; Dalmaschio, C.J.; Pontes, F.M.; Longo, E. Facile synthesis and characterization of ZrO_2 nanoparticles prepared by the AOP/hydrothermal route. *RSC Adv.* **2014**, *4*, 38484–38490. [CrossRef]
69. Liu, G.X.; Liu, A.; Meng, Y.; Shan, F.K.; Shin, B.C.; Lee, W.J.; Cho, C.R. Annealing dependence of solution-processed ultra-thin ZrO_x films for gate dielectric applications. *J. Nanosci. Nanotechnol.* **2015**, *15*, 2185–2191. [CrossRef]
70. Valle, H.R.L. Effect of Eco-Friendly Solvents in Solution-Based ZrO_x Dielectrics. Master's Thesis, FCT-UNL, Caparica, Portugal, 2019. Available online: https://run.unl.pt/handle/10362/91651 (accessed on 21 July 2022).
71. Seon, J.B.; Cho, N.K.; Yoo, G.; Kim, Y.S.; Char, K. Solution-processed amorphous ZrO_2 gate dielectric films synthesized by a non-hydrolytic sol–gel route. *RSC Adv.* **2018**, *8*, 39115–39119. [CrossRef]
72. Gong, Y.; Zhao, K.; He, H.; Cai, W.; Tang, N.; Ning, H.; Wu, S.; Gao, J.; Zhou, G.; Lu, X.; et al. Solution processable high quality ZrO_2 dielectric films for low operation voltage and flexible organic thin film transistor applications. *J. Phys. D Appl. Phys.* **2018**, *51*, 115105. [CrossRef]
73. Wang, S.; Xia, G. A facile low-cost preparation of high-*k* ZrO_2 dielectric films for superior thin-film transistors. *Ceram. Int.* **2019**, *45*, 23666–23672. [CrossRef]
74. Jung, S.H.; Han, H.S.; Kim, Y.B.; Kim, D.S.; Deshpande, N.G.; Oh, S.J.; Choi, J.H.; Cho, H.K. Toward ultraviolet solution processed ZrO_x/IZO transistors with top-gate and dual-gate operation: Selection of solvents, precursors, stabilizers, and additive elements. *J. Alloy. Compd.* **2020**, *847*, 156431. [CrossRef]
75. Luo, C.; Huang, T.; Li, C.; Zhang, Y.; Zou, Z.; Li, Y.; Tao, R.; Gao, J.; Zhou, G.; Lu, X.; et al. Enhancement of electrical properties of solution-processed oxide thin film transistors using ZrO_2 gate dielectrics deposited by an oxygen-doped solution. *J. Phys. D Appl. Phys.* **2021**, *54*, 125101. [CrossRef]
76. Aminipoya, H.; Ghomi, A.B.; Niazi, A. Comparative synthesis of ZrO_2 nanoparticles by green and co-precipitation methods: The effect of template on structure. *Int. J. Nano Dimens* **2021**, *12*, 59–66.

77. Ji, P.; Wang, Z.; Shang, X.; Zhang, Y.; Liu, Y.; Mao, Z.; Shi, X. Direct observation of enhanced Raman scattering on nano-sized ZrO_2 substrate: Charge-transfer contribution. *Front. Chem.* **2019**, *7*, 245. [CrossRef] [PubMed]
78. Soares, M.R.N.; Nico, C.; Peres, M.; Ferreira, N.; Fernandes, A.J.S.; Monteiro, T.; Costa, F.M. Structural and optical properties of europium doped zirconia single crystals fibers grown by laser floating zone. *J. Appl. Phys.* **2011**, *109*, 013516. [CrossRef]
79. Merle, T.; Guinebretiere, R.; Mirgorodsky, A.; Quintard, P. Polarized Raman spectra of tetragonal pure ZrO_2 measured on epitaxial films. *Phys. Rev. B—Condens. Matter Mater. Phys.* **2002**, *65*, 1443021–1443026. [CrossRef]
80. Fernandez Lopez, E.; Sanchez Escribano, V.; Panizza, M.; Carnasciali, M.M.; Busca, G. Vibrational and electronic spectroscopic properties of zirconia powders. *J. Mater. Chem.* **2001**, *11*, 1891–1897. [CrossRef]
81. Ding, S.; Zhao, J.; Yu, Q. Effect of zirconia polymorph on vapor-phase ketonization of propionic acid. *Catalysts* **2019**, *9*, 768. [CrossRef]
82. Gopal, R.; Jain, J.; Goyal, A.; Gupta, D.K.; Nagar, M. Formation of nano-sized cubic zirconia by aqueous sol–gel route. *J. Aust. Ceram. Soc.* **2018**, *54*, 691–700. [CrossRef]
83. Basahel, S.N.; Ali, T.T.; Mokhtar, M.; Narasimharao, K. Influence of crystal structure of nanosized ZrO_2 on photocatalytic degradation of methyl orange. *Nanoscale Res. Lett.* **2015**, *10*, 73. [CrossRef]
84. Lopes, D. Low-Cost Upconversion Nanomaterials for Anti-Counterfeiting Solutions. Master's Thesis, FCT-UNL, Caparica, Portugal, 2020. Available online: https://run.unl.pt/handle/10362/113701 (accessed on 22 July 2022).
85. Prasad, K.; Pinjari, D.V.; Pandit, A.B.; Mhaske, S.T. Synthesis of zirconium dioxide by ultrasound assisted precipitation: Effect of calcination temperature. *Ultrason. Sonochemistry* **2011**, *18*, 1128–1137. [CrossRef]
86. Carvalho, J.M.; Rodrigues, L.C.V.; Felinto, M.C.F.C.; Nunes, L.A.O.; Hölsä, J.; Brito, H.F. Structure–property relationship of luminescent zirconia nanomaterials obtained by sol–gel method. *J. Mater. Sci.* **2015**, *50*, 873–881. [CrossRef]
87. Kiisk, V.; Puust, L.; Utt, K.; Maaroos, A.; Mändar, H.; Viviani, E.; Piccinelli, F.; Saar, R.; Joost, U.; Sildos, I. Photo-, thermo- and optically stimulated luminescence of monoclinic zirconia. *J. Lumin.* **2016**, *174*, 49–55. [CrossRef]
88. Lokesha, H.S.; Nagabhushana, K.R.; Singh, F.; Tatumi, S.H.; Prinsloo, A.R.E.; Sheppard, C.J. Unraveling the Charge State of Oxygen Vacancies in Monoclinic ZrO_2 and Spectroscopic Properties of ZrO_2:Sm $^{3+}$ Phosphor. *J. Phys. Chem. C* **2021**, *125*, 27106–27117. [CrossRef]
89. Perevalov, T.V.; Gulyaev, D.V.; Aliev, V.S.; Zhuravlev, K.S.; Gritsenko, V.A.; Yelisseyev, A.P. The origin of 2.7 eV blue luminescence band in zirconium oxide. *J. Appl. Phys.* **2014**, *116*, 244109. [CrossRef]
90. Ashraf, S.; Irfan, M.; Kim, D.; Jang, J.-H.; Han, W.-T.; Jho, Y.-D. Optical influence of annealing in nano- and submicron-scale ZrO_2 powders. *Ceram. Int.* **2014**, *40*, 8513–8518. [CrossRef]
91. Jiang, H.; Gomez-Abal, R.I.; Rinke, P.; Scheffler, M. Electronic band structure of zirconia and hafnia polymorphs from the GW perspective. *Phys. Rev. B* **2010**, *81*, 085119. [CrossRef]
92. Gallino, F.; Di Valentin, C.; Pacchioni, G. Band gap engineering of bulk ZrO_2 by Ti doping. *Phys. Chem. Chem. Phys.* **2011**, *13*, 17667–17675. [CrossRef]
93. Li, J.; Meng, S.; Niu, J.; Lu, H. Electronic structures and optical properties of monoclinic ZrO_2 studied by first-principles local density approximation + *U* approach. *J. Adv. Ceram.* **2017**, *6*, 43–49. [CrossRef]
94. Singh, H.; Sunaina; Yadav, K.K.; Bajpai, V.K.; Jha, M. Tuning the bandgap of m-ZrO_2 by incorporation of copper nanoparticles into visible region for the treatment of organic pollutants. *Mater. Res. Bull.* **2020**, *123*, 110698. [CrossRef]
95. Teeparthi, S.R.; Awin, E.W.; Kumar, R. Dominating role of crystal structure over defect chemistry in black and white zirconia on visible light photocatalytic activity. *Sci. Rep.* **2018**, *8*, 5541. [CrossRef]
96. Sakfali, J.; Ben Chaabene, S.; Akkari, R.; Dappozze, F.; Berhault, G.; Guillard, C.; Saïd Zina, M. High photocatalytic activity of aerogel tetragonal and monoclinic ZrO_2 samples. *J. Photochem. Photobiol. A Chem.* **2022**, *430*, 113970. [CrossRef]
97. Smits, K.; Grigorjeva, L.; Millers, D.; Sarakovskis, A.; Grabis, J.; Lojkowski, W. Intrinsic defect related luminescence in ZrO_2. *J. Lumin.* **2011**, *131*, 2058–2062. [CrossRef]
98. Sil, A.; Goldfine, E.A.; Huang, W.; Bedzyk, M.J.; Medvedeva, J.E.; Facchetti, A. Role of Fluoride Doping in Low-Temperature Combustion-Synthesized ZrO_x Dielectric Films. *ACS Appl. Mater. Interfaces* **2022**, *14*, 12340–12349. [CrossRef]
99. Pradhan, S.; Mishra, B.G. Catalytic application of SO_4^{2-}/Fe-ZrO_2 nanoparticles synthesized by a urea hydrolysis method for environmentally benign one pot synthesis of 1,8-dioxodecahydroacridines. *RSC Adv.* **2015**, *5*, 86179–86190. [CrossRef]
100. Cai, W.; Zhu, Z.; Wei, J.; Fang, Z.; Ning, H.; Zheng, Z.; Zhou, S.; Yao, R.; Peng, J.; Lu, X. A simple method for high-performance, solution-processed, amorphous ZrO_2 gate insulator TFT with a high concentration precursor. *Materials* **2017**, *10*, 972. [CrossRef] [PubMed]
101. Meng, Y.; Liu, G.; Liu, A.; Song, H.; Hou, Y.; Shin, B.; Shan, F. Low-temperature fabrication of high performance indium oxide thin film transistors. *RSC Adv.* **2015**, *5*, 37807–37813. [CrossRef]
102. Zhu, L.; He, G.; Lv, J.; Fortunato, E.; Martins, R. Fully solution-induced high performance indium oxide thin film transistors with ZrO_x high-k gate dielectrics. *RSC Adv.* **2018**, *8*, 16788–16799. [CrossRef]
103. Lee, H.; Zhang, X.; Kim, J.W.; Kim, E.J.; Park, J. Investigation of the Electrical Characteristics of Bilayer ZnO/In_2O_3 Thin-Film Transistors Fabricated by Solution Processing. *Materials* **2018**, *11*, 2103. [CrossRef]
104. Oluwabi, A.T.; Katerski, A.; Carlos, E.; Branquinho, R.; Mere, A.; Krunks, M.; Fortunato, E.; Pereira, L.; Oja Acik, I. Application of ultrasonic sprayed zirconium oxide dielectric in zinc tin oxide-based thin film transistor. *J. Mater. Chem. C* **2020**, *8*, 3730–3739. [CrossRef]

105. Zhu, C.; Liu, A.; Liu, G.; Jiang, G.; Meng, Y.; Fortunato, E.; Martins, R.; Shan, F. Low-temperature, nontoxic water-induced high-*k* zirconium oxide dielectrics for low-voltage, high-performance oxide thin-film transistors. *J. Mater. Chem. C* **2016**, *4*, 10715–10721. [CrossRef]
106. Boratto, M.H.; Congiu, M.; dos Santos, S.B.O.; Scalvi, L.V.A. Annealing temperature influence on sol-gel processed zirconium oxide thin films for electronic applications. *Ceram. Int.* **2018**, *44*, 10790–10796. [CrossRef]

energies

Article

Quantitative Evaluation of the Effects of Heat Island on Building Energy Simulation: A Case Study in Wuhan, China

Long Pei *, Patrick Schalbart and Bruno Peuportier

Mines Paris, PSL University, Centre for Energy Efficiency of Systems (CES), 75006 Paris, France; bruno.peuportier@mines-paristech.fr (B.P.)
* Correspondence: long.pei@mines-paristech.fr

Abstract: The climate data used for dynamic energy simulation of buildings located in urban regions are usually collected in meteorological stations situated in rural areas, which do not accurately represent the urban microclimate (e.g., urban heat island effect), and this might affect the simulation accuracy. This paper aims at quantitatively evaluating the effects of heat island on a high-rise building's energy performance based on the microclimate simulation tool ENVI-met and the building energy simulation tool COMFIE. However, the computation of microclimate models is time consuming; it is not possible to simulate every day of a year in a reasonable time. This paper proposes a method that generates hourly "site-specific climate data" to avoid long microclimate simulation times. A coupling method of ENVI-met and COMFIE was developed for more precise building energy simulation, accounting for the heat island effect. It was applied to a high-rise building in Wuhan, China. The results showed that the yearly average urban heat island effect intensity at the height of 3 m was estimated to be 0.55 °C and decreased with height. Compared to the simulation considering the outdoor temperature variation with the height and orientation, using the original climate data collected in rural areas led to an overestimation of the heating load by around 5.8% and an underestimation of the cooling load by around 8.7%. Compared to the weather file at the height of 3 m near the north facade neglecting the temperature variation along the height, the heating load was overestimated by 8.2% and the cooling load was underestimated by 10.8%. The methods proposed in this paper can be used for the more precise application of urban building energy simulation.

Keywords: heat island effect; high-rise building; building energy simulation; microclimate simulation

Citation: Pei, L.; Schalbart, P.; Peuportier, B. Quantitative Evaluation of the Effects of Heat Island on Building Energy Simulation: A Case Study in Wuhan, China. *Energies* **2023**, *16*, 3032. https://doi.org/10.3390/en16073032

Academic Editor: Elena Lucchi

Received: 22 February 2023
Revised: 22 March 2023
Accepted: 23 March 2023
Published: 26 March 2023

1. Introduction

Building energy consumption is one of the three main energy consumption domains including industry and transportation, accounting for approximately 36% of the global final energy in 2018 [1]. One potential way to reduce building energy consumption corresponds to decisions made during the design phase, which can be aided by dynamic building energy simulation (DBES) tools. The simulation results of a building's energy consumption are closely related to the accuracy of the weather file comprising 8760 h of various climatic parameters such as air temperature and solar radiation [2]. In present DBES tools, one mostly used weather file format is the Typical Meteorological Year (TMY) format [3], which consists of twelve Typical Meteorological Months (TMM) from the past decades [2]. These data usually come from meteorological stations located in the rural zones (e.g., an airport), which have a very different morphological form compared to the urban areas composed of urban settlements.

China is moving towards urbanisation and this will last for decades [4]. During this process, urban settlements are formed by replacing natural or agricultural land with urban environments. For example, the ground surface is covered with impermeable materials such as concrete and asphalt, and the vegetation area is sharply reduced [5]. This leads to specific microclimates in the urban areas. Urban microclimate can be defined as the local climate observed in urban areas, which can be significantly different from the climate of surrounding rural areas [6]. The urban microclimate involves the local climate characteristics between the near-ground atmosphere and the topsoil in a relatively small space, including temperature, solar radiation, wind, humidity, etc. [7]. The difference between an urban microclimate and a rural climate might bring potential simulation errors if the TMY weather file is directly used in an urban building's energy performance evaluation. To avoid this, it is essential to take the microclimate's effects into account in DBES.

The urban heat island (UHI) effect is a well-known phenomenon among the microclimate features. It describes the phenomenon that air temperature in urban areas is higher than the surrounding rural areas. It is reported in many urban areas regardless of the size and location [8–14]. It could be observed by the application of infrared thermography at different scales such as the city scale using satellites and the neighbourhood scale using aerial vehicles and rooftop observatories [15,16]. Drones are used more and more to detect the UHI effect or analyse urban heat fluxes [17–19]. The UHI effect can be quantitatively described by the urban heat island intensity (UHII) which is the urban air temperature minus the rural air temperature [14]. Occasionally, the urban air temperature can be lower than the rural air temperature, which is called the urban cool island (UCI) effect, and the temperature difference is the urban cool island intensity (UCII). Air temperature is one of the most important factors as it directly drives the operation of a cooling/heating system and influences the corresponding building cooling and heating energy consumption [20]. Many studies reported that the UHI effect increases the cooling energy consumption and decreases the heating energy consumption [20–23]. A weather input file considering the UHI effects can yield a more accurate evaluation of buildings' energy consumption.

Li et al. [20] summarised the procedure in evaluating UHI impacts on building energy consumption, which includes three steps: (1) preparing two temperature datasets with and without the UHI effect; (2) simulating/estimating building energy consumption respectively using these two temperature datasets; and (3) evaluating the impacts of UHI on building energy consumption by comparing the two results. The urban temperatures are mainly obtained by measurements or simulation. This paper focuses on the simulation of UHI, because as abovementioned the weather files in DBES are normally in TMY format, reflecting representative climates during a long period. On the contrary, measurements can only record the climate in a short time which might not be representative. This problem can be avoided using simulation.

Microclimate simulation tools (e.g., ENVI-met) can be used to model UHI effects. However, the calculation time is quite long for a standard PC (e.g., nearly 24 h to simulate one day in our case), such that it is not possible to obtain the hourly microclimate data for the whole year in a timely manner, especially if the energy performance optimisation of a building or a block is needed in the design phase. A method for the generation of hourly microclimate parameters could be beneficial. In addition to this, the new buildings are usually high-rise and dense in China. Contrary to low-rise buildings, little research has focused on the effect of UHI on a high-rise building in a densely built region. Normally, one measure of air temperature is used for the whole building in DBES, neglecting the temperature variation in terms of height and orientation. This might bring simulation errors for a high-rise building, especially where the UHI effect occurs.

This paper aims to investigate these two questions: what error is induced by using regional instead of microclimate data in DBES? How can DBES and microclimate simulation tools be chained? Firstly, a site-specific weather file generation method is proposed to account for the hourly UHI effect for a TMY weather file, avoiding long simulation time. Secondly, the coupling methodology of a microclimate simulation tool and a DBES tool is presented to perform a more accurate simulation of the building energy performance under the UHI effect. Finally, the methods are applied to a high-rise residential building in Wuhan (China) to quantitatively evaluate the effects of UHI on its energy performance.

2. Methodology

2.1. Site-Specific Weather File Generation Method

In this paper, the microclimate simulation tool used is ENVI-met v4.4.6, which is a three-dimensional (3D) numerical model initially developed by Bruse and Fleer [24] to analyse microclimates through the fundamental laws of fluids and thermodynamics. The interactions between buildings, soil, vegetation and atmosphere can be simulated with a typical spatial resolution between 0.5 m and 10 m and a timestep of 1–10 s. Every single plant and every urban structure can explicitly be simulated, making ENVI-met the perfect tool for urban planners, architects and urban climatologists who want to simulate the meteorological components of the urban environment [25]. This model has been validated by several studies [2,26–30] for different cities with different microclimate conditions. The DBES tool is COMFIE [31], which is a multizone model based on the finite volume method and modal reduction. It has been validated in several projects [32–36], including in the Wuhan context [37].

As mentioned in the introduction, it is not possible to obtain the hourly UHIIs for one whole year in ENVI-met with a standard PC in a reasonable calculation time. Therefore, a key point for the association of the DBES simulation tool and the microclimate tool is to propose a method which can generate a "site-specific weather file" containing the hourly UHIIs appropriately in reasonable simulation time.

The core idea of the proposed method to solve this problem is only simulating four representative days in ENVI-met to obtain their hourly UHIIs, and then deriving the hourly UHII of other days by linear interpolation. To better identify the UHII, one representative day is chosen for each season: extreme hot/cold days in summer/winter (in order to obtain potential largest/smallest UHII) and average temperature days in spring/autumn (in order to obtain reasonable interpolation between largest and smallest UHII). Once the hourly UHIIs for the whole year have been obtained, they could be added to the original weather file without considering the UHI effect to generate the site-specific weather file. It should be noted that, if not especially indicated, spring refers to March, April and May; summer refers to June, July and August; autumn refers to September, October and November; and winter refers to December, January and February.

The original weather file for DBES in COMFIE in this paper is an EnergyPlus Weather (EPW) format file downloaded from the website of EnergyPlus [38]. The reason why an EPW file was chosen is that it can be directly used in ENVI-met for microclimate simulation. The other reason is that this file also records the extreme air temperature weeks in summer and winter and the average air temperature weeks for all four seasons, which is convenient to select the representative days of the four seasons.

The proposed site-specific weather file generation method considering the UHI effect consists of four steps, as shown in Figure 1 (MATLAB was used in steps 1, 3 and 4):

Step 1: representative days selection. One representative day of the representative week (extreme hot/cold week for summer/winter and average week for spring/autumn, which are directly given in EPW file) of each season is chosen based on the day with the hourly temperature closest to the average hourly temperature of the week (see hereunder). Since this study focuses on the UHI effect, which is directly related to the ambient temperature, it is reasonable to select the representative days based on the air temperature. It should be noted that it is not the extreme hot/cold day in extreme hot/cold weeks for summer/winter, in order to be more representative. These four representative days can be abbreviated to ex-summer, ex-winter, av-spring and av-autumn;

Step 2: microclimate simulation. These four representative days are then simulated in ENVI-met to obtain the local microclimate parameters and the UHIIs;

Step 3: hourly UHII generation. The hourly UHIIs are obtained by linear interpolation for 24 h between two representative days;

Step 4: site-specific weather file generation. The air temperature of the site-specific weather file is generated by adding the hourly UHII to the original EPW file.

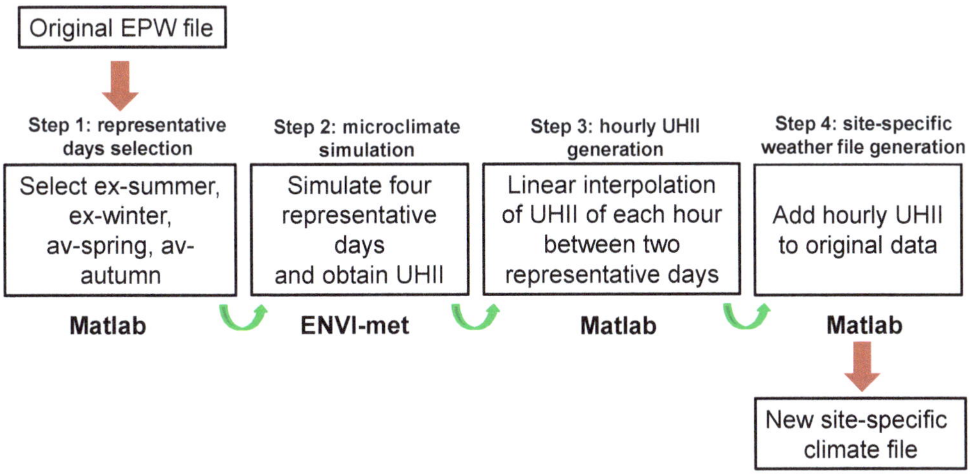

Figure 1. Algorithm of the proposed method to generate the site-specific weather file.

In step 1, in order to determine the representative day of the representative week of each season, the average temperature of the ith hour $T_{av,i}$ of the representative week is firstly to be identified:

$$T_{av,i} = \frac{\sum_{j=1}^{7} T_i^j}{7} \quad i = 1 \ldots 24 \tag{1}$$

where T_i^j is temperature of the ith hour of the jth day in the representative week.

The root mean square error (RMSE) and the mean absolute error (MAE) of each day for the 24 h are calculated by:

$$\text{RMSE}_j = \frac{\sqrt{\sum_{i=1}^{24} \left(T_i^j - T_{av,i}\right)^2}}{24} \tag{2}$$

$$\text{MAE}_j = \frac{\sum_{i=1}^{24} \left|T_i^j - T_{av,i}\right|}{24} \tag{3}$$

The day with the smallest RMSE is chosen as the representative day of each season; if RMSE is similar, the day with smallest MAE is selected.

In step 3, knowing the four representative days, the UHII of the ith hour of the kth day in the whole year can be calculated by linear interpolation. For example, the hourly UHII between the first (av-spring) and the second (ex-summer) representative days:

$$\text{UHII}_i^k = \frac{\text{UHII}_i^{k_{re2}} - \text{UHII}_i^{k_{re1}}}{k_{re2} - k_{re1}}(k - k_{re1}) + \text{UHII}_i^{k_{re1}} \tag{4}$$

where k_{re1} and k_{re2} are the day numbers of the 1st and 2nd representative day. The UHII between other representative days could be obtained by the same way.

It should be noted that the linear interpolation between two representative days to obtain the hourly UHII for the whole year is a highly simplified assumption. The hourly variation of UHII in cities is much more complex due to enormous factors such as complicated urban planning, various constructions, vegetation and climate conditions. The main aim of this study is to identify the UHI effect on building energy consumption for a long period (usually one year), rather than accurately estimate the UHI variation of one day. It mainly relies on the estimation of a general UHI feature, so there is a certain tolerance on the biases of hourly UHII [39].

2.2. Coupling Microclimate Simulation Tool with DBES Tool

The coupling methods of a microclimate tool and a DBES tool are classified into one-way coupling and double-way coupling [40]. The double-way coupling method considers the effect of the microclimate on the building, as well as the effect of the building on the microclimate (e.g., the anthropogenic emissions released outside due to air-conditioning or heating systems such as heat pumps), which is more realistic, but more complex. It requires the microclimate tools to deal with such heat sources, and needs user parameter inputs.

In this paper, the coupling of ENVI-met and COMFIE is a one-way coupling (ENVI-met v4.4.6 only allows one-way coupling), meaning the simulation results of ENVI-met are transferred to COMFIE but without feedback. The key point is modifying the weather input file from the original one (e.g., the meteorological data from rural stations) to the new one (which considers the urban microclimate). In COMFIE, a building is modelled by defining thermal zones according to orientations and heights. An identical temperature profile is used for the whole building, regardless of zone height and orientation. As mentioned in the introduction, this might bring simulation errors for a high-rise building in a dense block where the UHI effect occurs.

The coupling method proposed in this paper considers the air temperature variation with height and orientation, based on the microclimate simulation results. The main 3D model in ENVI-met (e.g., a district and its soil and atmosphere) is divided into $I \times J \times K$ grid cells with dimensions $\Delta x \times \Delta y \times \Delta z$, respectively. Buildings, vegetation and the digital elevation model are constricted to this grid cell. This means that a cell is either fully occupied by one of these obstacles or not at all. On the horizontal surface, Δx and Δy (usually from 0.5 m to 10 m) are constant for all the cells. In the vertical direction, the dimension Δz of the first n cells are identical, and afterwards the dimension of one cell is that of the previous cell multiplied by a scaling factor s ($s > 1$), as shown in Figure 2. This is because the cell close to the ground surface should have a smaller dimension in order to have a more accurate simulation of the interactions between the ground and the atmosphere.

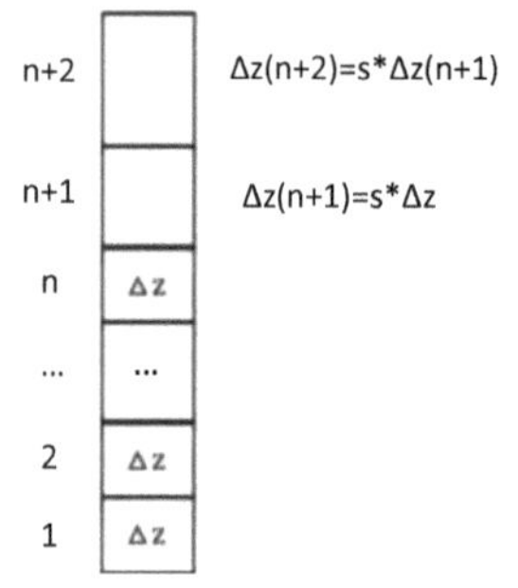

Figure 2. Cell definition in the vertical direction in ENVI-met.

The temperature profiles of four orientations (north, south, west and east) along N_h different heights are generated from ENVI-met. The value of N_h should respect the thermal zone definition regarding the height and in DBES tool and the cell definition in the vertical direction in ENVI-met. This will be described in detail in Section 3.3. The temperature profiles are inputted in the weather generation tool Meteocalc (which is integrated in COMFIE) and $4.N_h$ corresponding site-specific weather files are generated. Then, the simulation is performed $4.N_h$ times in COMFIE to obtain reference simulation results for each thermal zone. The whole coupling method is shown in Figure 3.

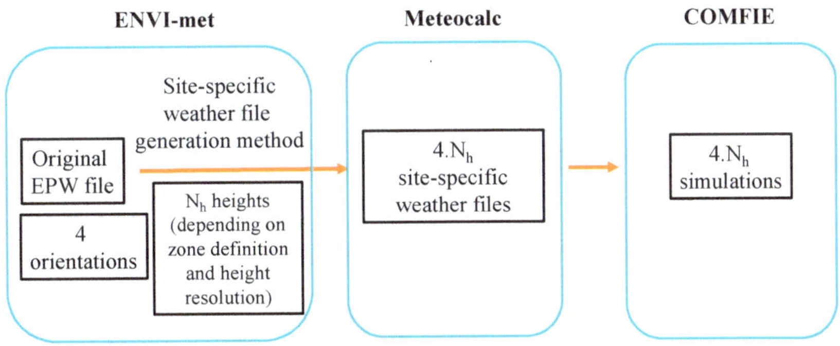

Figure 3. Method of coupling microclimate tool with the DBES tool.

3. Case Study

3.1. Basic Information

The studied case is located in Wuhan, the capital city of Hubei Province, China. Its topography is dominated by relatively flat land between 22 m and 27 m above sea level except the hilly areas sporadically distributed in suburban districts. Water bodies occupy a high percentage (>20%) of its territory. Wuhan's climate is humid subtropical with abundant rainfall and four distinctive seasons. Spring and autumn are generally mild, summer is hot and humid and winter is cold and dry.

The simulated high-rise building is unit 2 of building #2 (#2.2) in the Haishan Jingu (HSJG) block. HSJG consists of one office building (#1) and three residential buildings (#2, #3 and #4). Its latitude is 30°32′ north and longitude is 114°20′ east. This area is a business and commercial centre, mixed with many recently built high-rise offices, commercial and residential buildings and old buildings, as shown in Figure 4a. This area faces south by east with an angle of 20°. The overall plan is shown in Figure 4b. The studied building #2.2 consists of 34 floors with an average height 3 m for each floor. The net floor area of the building is around 12,500 m^2.

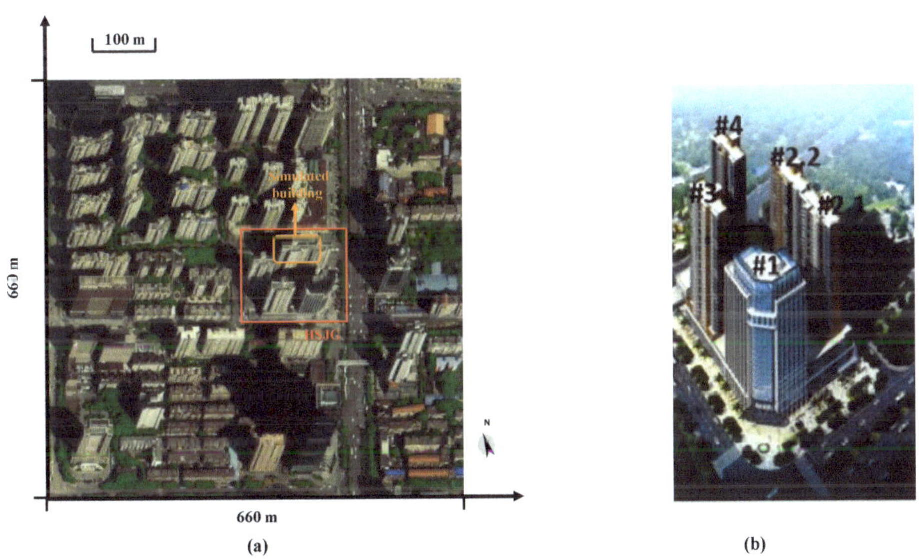

Figure 4. (**a**) Simulated area in ENVI-met and (**b**) HSJG block.

3.2. Microclimate Simulation Configuration

3.2.1. Representative Days Selection

The EPW file downloaded from the EnergyPlus website directly indicates the four representative weeks of the four seasons, as presented in Table 1. The extreme hot (in summer) and cold (in winter) weeks are the ones nearest to the maximal and minimal outdoor temperature in the corresponding periods. The average weeks in spring and autumn are the ones nearest to the average outdoor temperature in the corresponding periods. They are determined by a heuristic method in EnergyPlus. Using the site-specific weather file generation method, the representative day of each season was selected for the microclimate simulation. The day of each week with the smallest RMSE was chosen as the representative day for each season. The results are shown in Table 1.

Table 1. Representative days selection of each season.

Season	**Representative Week (Directly from EnergyPlus Website)**	**Representative Day**	**Abbreviation**	**RMSE (°C)**	**MAE (°C)**
Summer Extreme hot	5–11 August	9 August	Ex-summer	0.82	0.76
Winter Extreme cold	1–7 January	4 January	Ex-winter	1.11	0.90
Spring Average temperature	27 May–2 June	2 June (date is in June, but represents spring)	Av-spring	1.48	1.19
Autumn Average temperature	26 November–2 December	30 November	Av-autumn	1.25	1.02

It should be noted that instead of a normal air temperature (e.g., in EPW file), the air temperature in ENVI-met, $T_{air,pot}$, is the potential temperature. It can be converted from atmosphere pressure and normal temperature with the following Equation (5):

$$T_{air,pot} = T_{air}^{abs}\left(\frac{P_0}{P}\right)^{0.286} \quad (5)$$

where T_{air}^{abs} is the absolute normal air temperature (K), P_0 is the reference pressure (100,000 Pa in ENVI-met) and P is the atmosphere pressure (Pa).

3.2.2. Simulation Configuration

Locating the studied building approximately in the centre, the total area for microclimate simulation is a region of 660 m × 660 m, as shown in Figure 4a. Although it is very difficult to obtain detailed information regarding the surroundings, it is still possible to get a rough picture of the surroundings including the heights of the buildings, the green areas and the roads by using Google Maps and Google Earth. This information allows one to use ENVI-met to evaluate the site-specific climatic conditions. If detailed data of the surroundings can be obtained, a more accurate simulation can be performed. Accounting for the calculation speed and the model accuracy, 110 × 110 × 19 cells were used in this study, with a resolution of 6 m in the horizontal directions. In the vertical direction, the resolution of one cell is 6 m below 30 m (the first five cells) and afterwards the height of cell $n + 1$ is the height of cell n multiplied by 1.25 (n being the level of the cell from 6 to 19), as shown in Figure 2. In addition to this, eight empty cells were set at each border (which is called the "nesting area") to increase the stability and accuracy of the simulation. The microclimate simulation has better accuracy with a buffering time. A sensitivity analysis showed that the maximal and mean absolute temperature differences for representative days were smaller than 0.35 °C and 0.2 °C between with two and five buffering days, which concluded that a buffering time of two days is sufficient in this case study. Thus each simulation includes the two days ahead of the representative day (72 h in total) to obtain more precise simulation results.

The ground in the simulated area mainly contains four types of ground profiles: the sidewalk, the built area, the green space and the asphalt concrete road. The compositions of the ground profiles are from [30] and listed in Table 2.

Table 2. Composition and depth in the ground profiles in ENVI-met.

Sidewalk	Built Area	Green Space	Asphalt Concrete Road
Brick: 0~3 cm Concrete: 4~20 cm Sand: 21~30 cm Loam: 31~450 cm	Concrete: 0~20 cm Sand: 21~ 30 cm Loam: 31~450 cm	Sandy loam: 0~40 cm Loam: 41~450 cm	Asphalt concrete: 0~10 cm Sand: 11~40 cm Loam: 41~450 cm

The initial temperature and the relative humidity for the four layers of the soil are also needed: upper layer (0–20 cm), middle layer (20–50 cm), deep layer (50–200 cm) and bedrock layer (below 200 cm). Assuming the soil is semi-infinite and with unique physical properties, the soil temperatures were estimated by the method from Thiers [41]. The corresponding soil relative humidities of the four layers in ENVI-met were calculated from Wu et al. [42]. They are summarised in Table 3.

Table 3. Soil temperature and relative soil humidity of each soil layer in ENVI-met.

	Av-Spring	Ex-Summer	Av-Autumn	Ex-Winter
Upper layer (0–20 cm)	26 °C/77%	33.51 °C/85%	12.33 °C/84%	6.16 °C/80.5%
Middle layer (20–50 cm)	24.66 °C/80.5%	32.57 °C/81%	13.66 °C/83%	7.53 °C/83%
Deep layer (50–200 cm)	21.07 °C/82%	29.30 °C/79%	17.20 °C/81%	11.65 °C/84%
Bedrock layer (below 200 cm)	19 °C/82%	19 °C/79%	19 °C/81%	19 °C/84%

It should be noted that the density of the plants' leaves varies with the season in the simulation. Considering the aim of ENVI-met simulation is to obtain the outdoor air temperature, instead of the indoor thermal comfort or other parameters concerning the building envelope, the building's envelope configuration is not as important as other configurations, especially in a "one-way" coupling method. Therefore, the envelope characteristics of all the buildings were set to be identical, as shown in Table 4. The initial indoor temperature of the buildings in the beginning of the simulation (including the buffering time) was set in ENVI-met. Considering the DBES configurations presented in Section 3.3, the studied building was simulated with the original EPW file in a first step. Then, the initial indoor temperatures of the whole building for the four representative days were set as the volume average of all thermal zones: 24.8 °C, 26 °C, 18.2 °C and 18 °C for av-spring, ex-summer, av-autumn and ex-winter, respectively. They were kept constant during the simulation and identical for all other buildings in the district.

Table 4. Wall configuration of the buildings in ENVI-met.

External Wall	Roof
Ceramic tile 1 cm Insulation mortar 4 cm Aerated concrete block 20 cm	Ceramic tile 1 cm Extruded Polystyrene Board (XPS) 4 cm Concrete (C10) 20 cm

Receptors are the selected points inside the model area, where processes in the atmosphere and the soil are monitored in detail. They record the detailed simulation results for each height of the node on the vertical axis. As introduced in Section 2.2, in order to analyse the effects of the orientation on the building's energy performance, four receptors located at 3 m away from the simulated building's four facades were used: north (N), west (W), south (S) and east (E). The whole 3D model and the receptors are shown in Figure 5.

Figure 5. The 3D model of the simulated area and four receptors in ENVI-met.

3.3. DBES Configuration

According to the Chinese building energy standards [43], the heating and cooling thermostat setpoints are 18 °C and 26 °C for the whole year. The natural ventilation is 1 ach (air change per hour) because residents in Wuhan prefer opening windows for ventilation; an additional 0.2 ach accounts for the infiltration ventilation. The internal heat gain is 4.3 W/m^2 for the whole time. The occupancy is 0.02 person/m^2 from 8h00 to 17h00 from Monday to Friday, and the rest of the time it is 0.04 person/m^2.

In the studied building, each floor consists of four apartments named A, B, C, D (from west to east) and aisle. The 15th floor plan of building #2.2 is shown in Figure 6 as an example, and the plan of each floor is similar. The areas surrounded by dotted lines are the balconies functioning as integrated shadings in the simulation. Windows and doors are shown in blue and brown lines on the walls, respectively. The building was divided into 17 thermal zones based on height and orientation. Apartments A and D consist of the north zone and apartments B and C consist of the south zone for one or several floors. It should be noted that apartment C on the 15th floor was an exception, which contains three zones: living room, bedroom and other rooms, because experiments were conducted in a previous work [37] in this apartment and a more detailed zone definition was required. The studied case and its zones are shown in Figure 7 (blue and grey buildings are not simulated and are just considered as shading the studied building).

Combining the zone definition and the height of ENVI-met cells, six different heights were chosen for the DBES weather file generation. It should be noted that the height of one vertical cell refers to the height of its middle point in ENVI-met. Taking levels 2–6 as an example, heights range from 6 m to 18 m. The vertical cells 2 and 3 (which occupy 6–12 m and 12–18 m in the vertical direction respectively) in ENVI-met record the air temperatures at 9 m and 15 m (middle points of cells 2 and 3), which are located between level 2–6. Thus we use the average of the air temperatures at 9 m and 15 m in ENVI-met to generate the weather input file for levels 2–6 (which contain two zones). Considering four orientations (N, W, S, E) and six heights (from #1 to #6), 24 DBES weather files were generated as inputs for the energy simulation, as illustrated in Table 5. The abbreviation of one file can be expressed by orientation#height number (e.g., N#1).

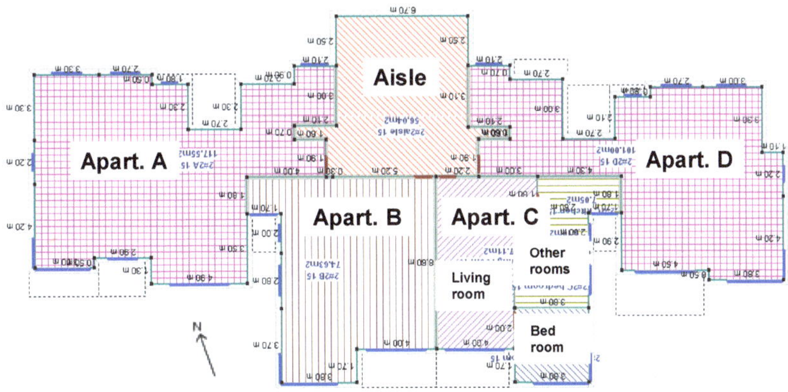

Figure 6. The 15th floor plan of the case study.

Figure 7. South view of HSJG and its zones in COMFIE.

Table 5. The 24 weather files for building energy simulation based on position (with the corresponding cells in ENVI-met).

Building Level	Level Height	Weather File Abbreviation of Four Orientations at Each Height	Height of the Weather File by Averaging the Heights of Vertical Cells in ENVI-met
1	3 m	N#1, W#1, S#1, E#1	3 m (**cell 1**)
2–6	6–18 m	N#2, W#2, S#2, E#2	(9 m (**cell 2**)+ 15 m (**cell 3**))/2 = 12 m
7–14	21–42 m	N#3, W#3, S#3, E#3	(21 m (**cell 4**) + 27 m (**cell 5**) + 33.75 m (**cell 6**) + 42.19 m (**cell 7**))/4 = 31 m
15–25	45–75 m	N#4, W#4, S#4, E#4	(52.73 m (**cell 8**) + 65.92 m (**cell 9**))/2 = 59.3 m
26–33	78–99 m	N#5, W#5, S#5, E#5	82.4 m (**cell 10**)
34	102 m	N#6, W#6, S#6, E#6	103 m (**cell 11**)

The 17 abovementioned thermal zones of the building are presented in Table 6. Based on the zone definition (Figure 6), the south zones are mainly affected by the air near the south facade, and the north zones are surrounded by the air in all the four orientations. To perform a precise energy performance simulation, the loads of each zone are calculated using the meteorological files according to the orientation and height, as shown in Table 6. The heating/cooling load of the north is the average of the corresponding simulation results from the four weather files of N, S, W and E. For the south zone, it is directly the simulation results from weather file S.

Table 6. Meteorological files to calculate the energy load of each zone.

Zone Number	Zone Name	Weather File
1	floor 1	(N#1 + S#1 + W#1 + E#1)/4
2	aisle	-
3	floor 2–6 north	(N#2 + S#2 + W#2 + E#2)/4
4	floor 2–6 south	S#2
5	floor 7–14 north	(N#3 + S#3 + W#3 + E#3)/4
6	floor 7–14 south	S#3
7	floor 15 C—living room (15)	S#4
8	floor 15 C—bedroom (15)	S#4
9	floor 15 C—other rooms (15)	S#4
10	floor 15 B	S#4
11	floor 15 north	(N#4 + S#4 + W#4 + E#4)/4
12	floor 16–25 north	(N#4 + S#4 + W#4 + E#4)/4
13	floor 16–25 south	S#4
14	floor 26–33 north	(N#5 + S#5 + W#5 + E#5)/4
15	floor 26–33 south	S#5
16	floor 34 north	(N#6 + S#6 + W#6 + E#6)/4
17	floor 34 south	S#6

4. Results and Discussion

4.1. Microclimate Simulation Results

Figures 8–11 show the potential air temperature maps of the simulated region for the four representative days at the height of 3 m at 13h00 (av-spring and ex-summer) and 14h00 (av-autumn and ex-winter) when the maximum UHIIs are observed. The whole simulation time for one representative day on a standard computer with an Intel i7-6700 CPU and 16 GB RAM is around 72 h (including the simulation of the buffering time of two days). The northeast region containing the asphalt road and the southwest region, respectively, show the highest and lowest UHI effect on all these four days. On ex-summer and av-autumn, the UHI effect is obvious along the north–south asphalt road; the north part has a stronger UHI effect. The temperature surrounding HSJG can vary over 1.5 °C on ex-summer, particularly higher on the side of the north–south asphalt road, but on other days the temperature difference is not very large (<0.5 °C). It should be noted that the representative day for spring is in June, resulting a higher air temperature compared to autumn.

4.2. UHI Results for the Four Representative Days

The heat island effect can be quantitatively described by UHII:

$$UHII = T_{air,ENVImet} - T_{air,EPW} \quad (6)$$

where $T_{air,ENVImet}$ is the air temperature (°C) converted from the potential air temperature in the ENVI-met simulation and $T_{air,EPW}$ is the air temperature (°C) in the EPW file.

The air temperature at the height of 3 m of the north receptor is shown in Figure 12. For the four representative days, all the simulated air temperatures are higher than the input air temperatures. The maximal UHII occurs around 13h00 for av-spring and ex-summer and 14h00 for av-autumn and ex-winter. The hourly average UHII ($UHII_{av}$) and maximal UHII ($UHII_{max}$) of these representative days are listed in Table 7. It can be inferred that the largest

UHII$_{max}$ occurs on the extreme hot summer day and the smallest UHII$_{max}$ occurs on the average autumn day. For the average UHII during one day, it is observed that ex-summer > av-spring > ex-winter > av-autumn.

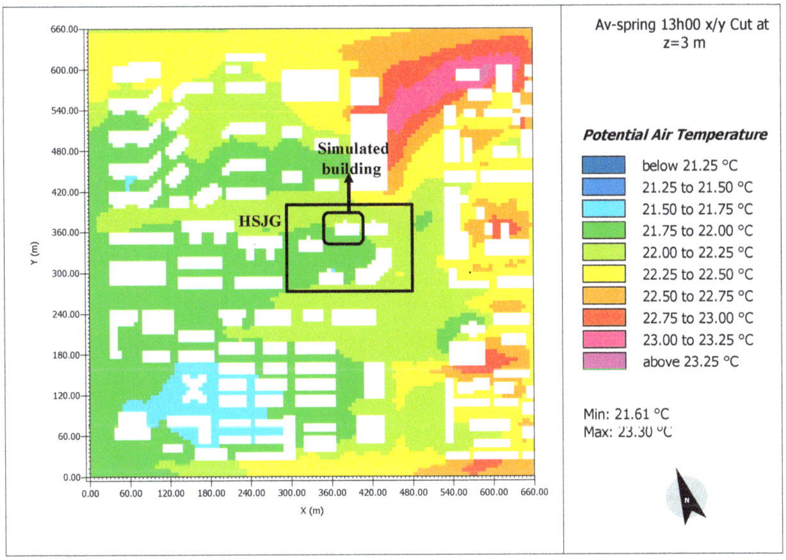

Figure 8. Potential temperature map of the simulated area for av-spring at the height of 3 m.

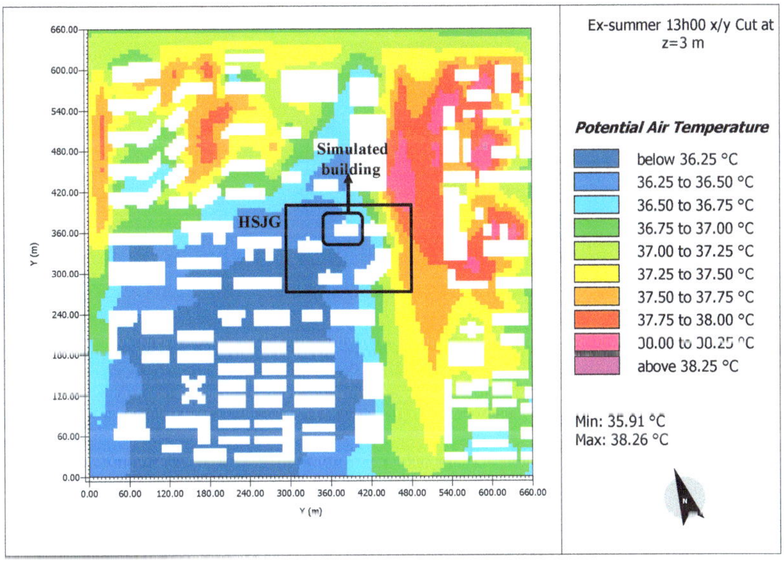

Figure 9. Potential temperature map of the simulated area for ex-summer at the height of 3 m.

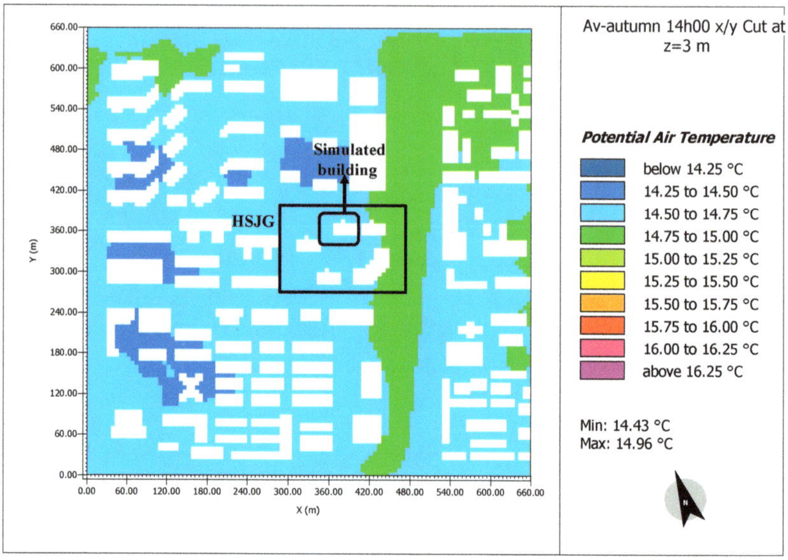

Figure 10. Potential temperature map of the simulated area for av-autumn at the height of 3 m.

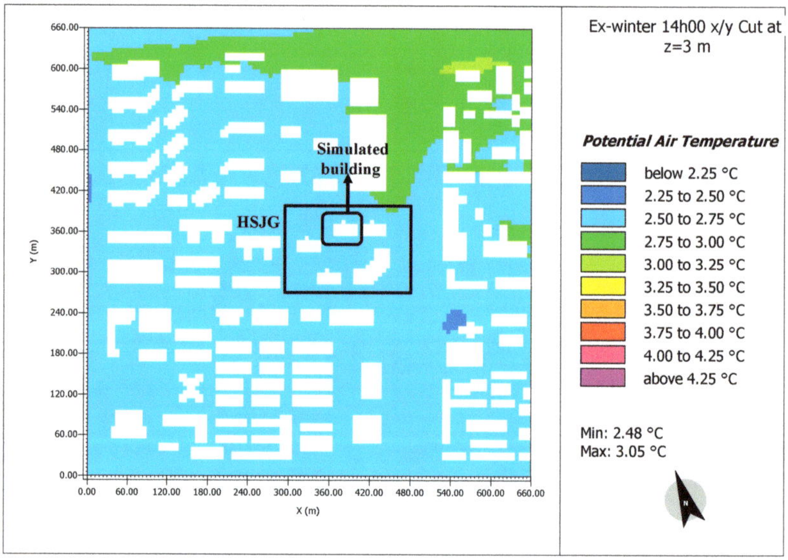

Figure 11. Potential temperature map of the simulated area for ex-winter at the height of 3 m.

Table 7. $UHII_{max}$ and $UHII_{av}$ at the height of 3 m of the north receptor.

Representative Day	$UHII_{max}$ (°C)	$UHII_{av}$ (°C)
Av-spring	1.35	0.58
Ex-summer	1.61	0.78
Av-autumn	0.58	0.34
Ex-winter	0.67	0.47

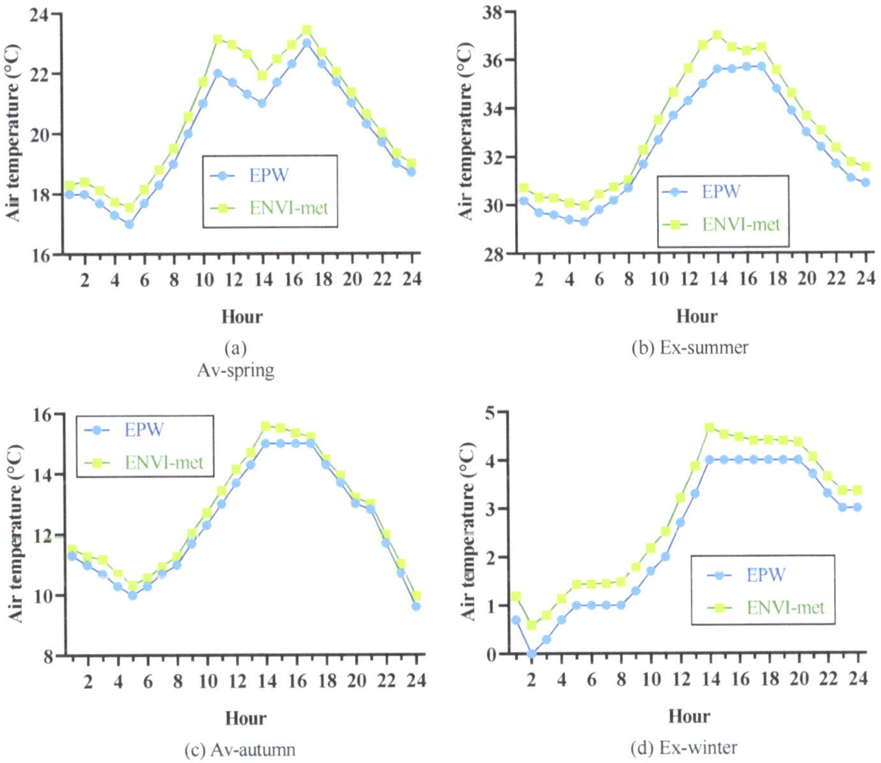

Figure 12. ENVI-met simulation results for the four representative days at the height of 3 m of the north receptor.

The air temperature along the height of the north receptor was investigated, as shown in Figure 13. In general, the air temperature decreases with the height. If the UHII is larger (e.g., 12h00 on av-spring), the air temperature decreases more sharply. The air temperature at the height #1 (3 m) is 1.1 °C higher than at height #6 (103 m). A more accurate building energy simulation should include the air temperature variation along the height. The air temperatures of the four orientations at 3 m are compared in Figure 14. The temperature difference is negligible for av-autumn and ex-winter for the four orientations; however the difference can be more obvious (maximal difference of 0.2 °C) from 11h00 to 14h00 on av-spring and ex-summer. This means the orientation might potentially have a larger influence on the cooling load than the heating load; it will be discussed in Section 4.4.

4.3. Generated Weather Files for COMFIE

Applying the method illustrated in Section 4.2, the meteorological files for building energy simulation were generated. The UHIIs at six different heights of four orientations were calculated and added to the original EPW file; meanwhile, the other meteorological parameters were kept the same. In total 24 meteorological files taking into account the local UHI effect were generated and were used to simulate the building's energy performance. The hourly temperature of N#1 is shown in Figure 15, compared to the original EPW. The yearly average UHII at the height of 3 m is estimated to be 0.55 °C (considering the average of four orientations).

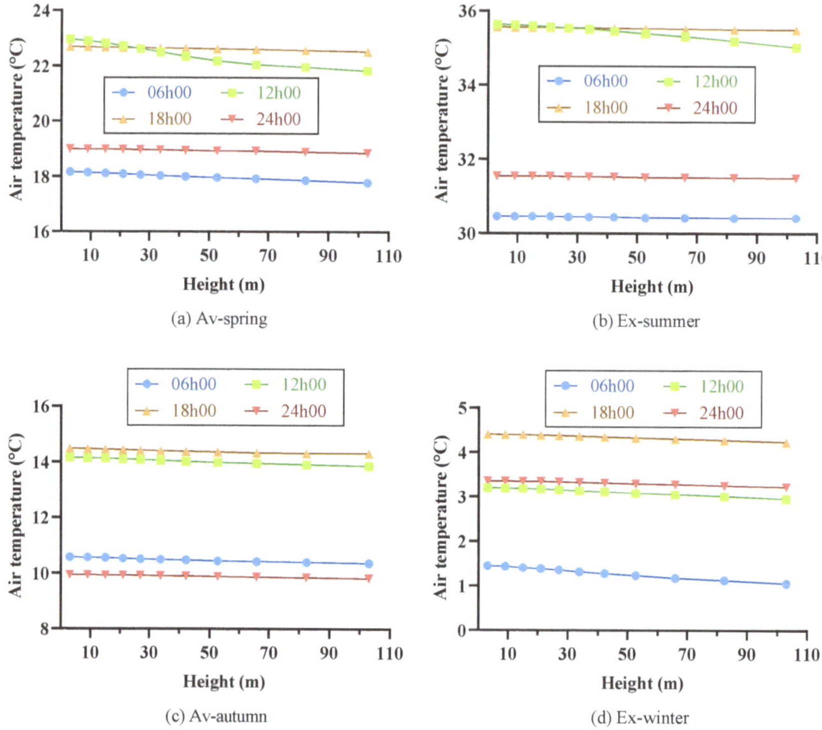

Figure 13. Air temperature evolution along height for four representative days of receptor N.

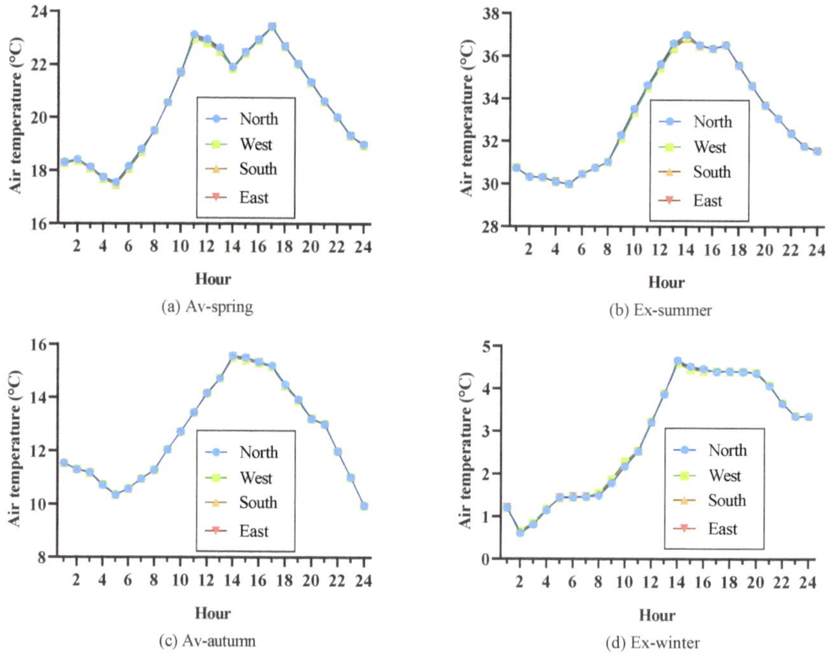

Figure 14. Air temperature of the four orientations at the height of 3 m for the four representative days.

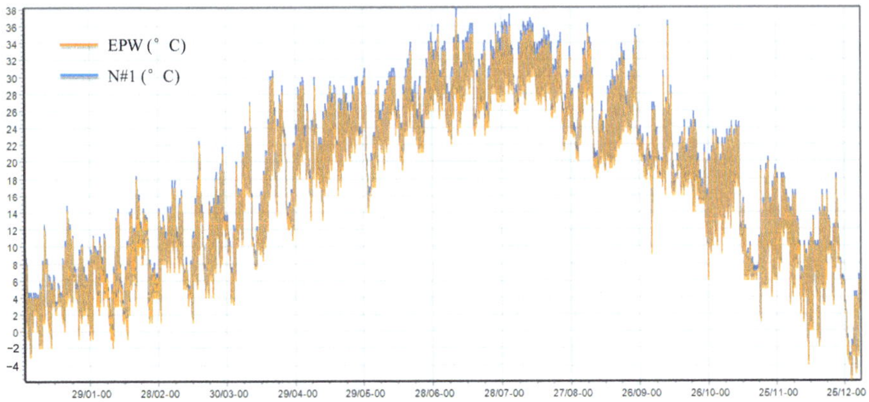

Figure 15. Air temperature of EPW and N#1.

The accumulations of hourly heat island or cool island intensity over TMY were calculated—namely, urban heat island degree-hours (UHIdh) [44], as shown in Figure 16. UHIdh measures how much and how long the air temperature at a site is higher than the reference suburb. At height #1, the values of UHIdh are ranked in the following order for all four orientations: summer > spring > autumn > winter. For the whole year, at height #1, the order of the number of UHIdh is N#1 (4929 °C hours) > E#1 (4802 °C hours) > W#1 (4736 °C hours) > S#1 (4680 °C hours). At height #6, the order changes to S#6 (2578 °C hours) > W#6 (2533 °C hours) > E#6 (2492 °C hours) > N#6 (2485 °C hours). UHIdh decreases with the height.

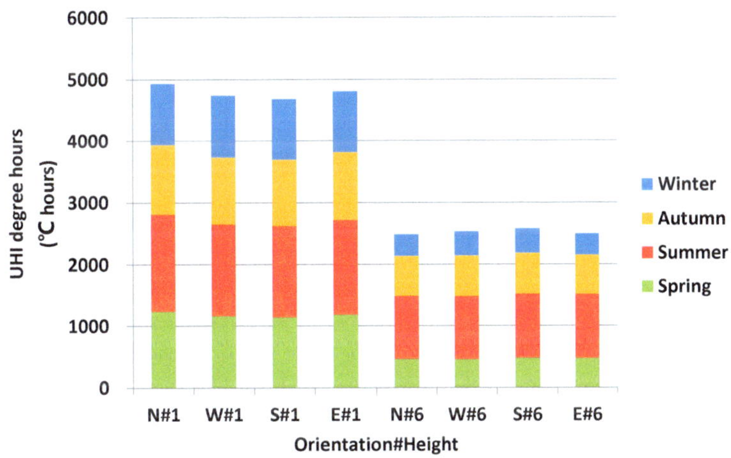

Figure 16. UHI degree-hours of four orientations at the height of 3 m and 103 m for four seasons.

The ratio of UHI degree-hours to heating/cooling degree-hours of the original EPW are presented in Figure 17, based on 18 °C for heating (i.e., Hdh18) and 26 °C for cooling (i.e., Cdh26). Hdh18 is the accumulation of hourly differences between 18 °C and the external temperature (which is smaller than 18 °C) during a period, and Cdh26 is the accumulation of hourly differences between the external temperature (which is larger than 26 °C) and 26 °C during a period. This ratio shows the potential of changing the heating/cooling load. For summer, N#1 has the largest ratio of around 29%. For winter, the four orientations have similar ratios, indicating that the heating load is not strongly influenced by the orientation. The ratio decreases with the height. At height #6, the ratio drops to below 2% in winter and to around 19% in summer. This indicates that a smaller UHI effect is foreseen with an increase of height.

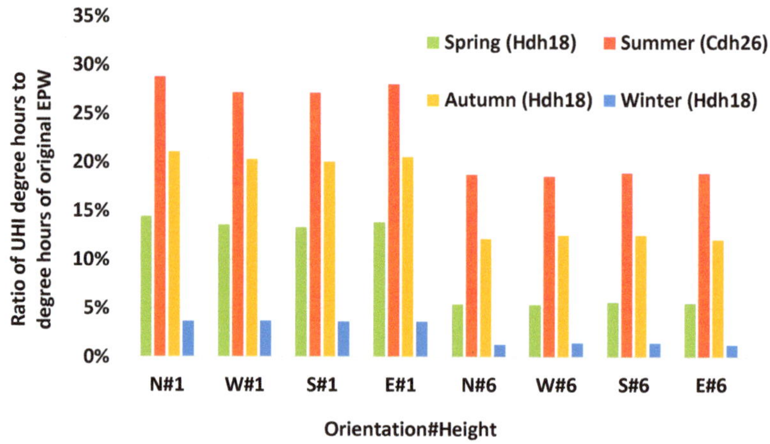

Figure 17. Ratio of UHI degree-hours to degree hours of original EPW.

4.4. Building Energy Simulation Results

4.4.1. DBES with Detailed Microclimate Data

The time needed for one dynamic building simulation is around 1 min. Applying the proposed methods, the energy simulation results with detailed microclimate data were obtained: the total heating and cooling loads are 575,000 kWh (46.2 kWh/m^2) and 610,000 kWh (49 kWh/m^2), respectively. Since this simulation considers the air temperature variation with height and orientation, it can be regarded as being closer to reality and it is, therefore, used as a reference (legend "Ref" in Figures 18 and 19) to compare the differences of other simulation results such as the simulation with the original EPW file.

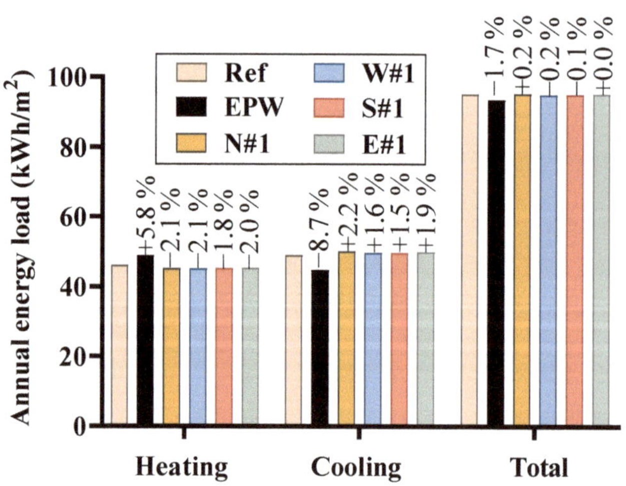

Figure 18. Energy loads for different orientations at the height of 3 m.

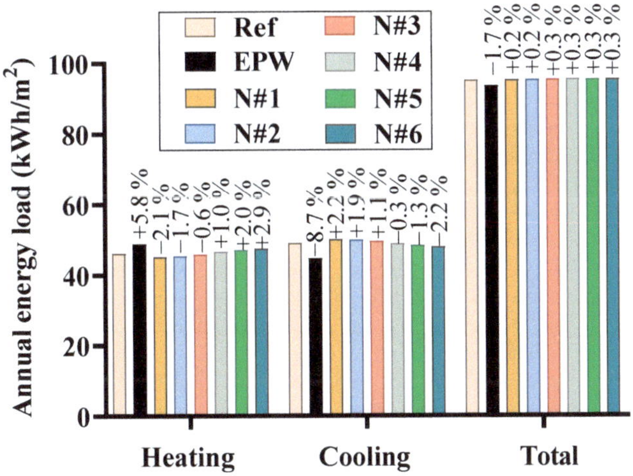

Figure 19. Annual loads of different heights for north orientation.

4.4.2. Impact of Orientation and Height

The simulation results of different meteorological files were compared including the original EPW and the reference simulation. Figure 18 shows the simulated annual loads of different orientation at a height of 3 m. Compared to the reference simulation, the simulation with the EPW file overestimates the heating load by 5.8% and underestimates the cooling load by 8.7%. However it can accurately predict the total energy load (1.7% smaller). This is mainly because the UHI effect reduces the heating load in winter and increases the cooling load in summer. In this study the difference between the heating and cooling loads is small. It should be noted that, in the simulation with the EPW file, the heating load is a bit larger than the cooling load, contrary to the reference simulation. At the height of 3 m, the orientation slightly affects the heating and cooling load, with a maximal variation of only 0.3% and 0.7%, respectively. The south orientation has a slightly more accurate simulation result than the other orientations. For the prediction of the total energy load, the four orientations all show good accuracy. In general, the effect of orientation is not important in this case study, especially if the uncertainty is considered. However, the effect might be larger in some other cases. For example, the geometry and layout of the block and vegetation might influence the air temperature distribution [45].

According to Figure 19, the air temperature drops along the height, yielding different energy loads. The building energy simulation results of different heights for north are illustrated in Figure 19. It can be inferred that compared to the reference simulation, the difference in heating load increases from −2.1% to +2.9% and the difference in cooling load decreases from +2.2% to −2.2% with height. The height has a minor influence on the total load, with a maximal difference of 0.3%. Compared to EPW, the differences of heating load and cooling load can reach, respectively, −7.5% and +12.1% for N#1. Many studies only consider the UHI effect observed near the ground (e.g., N#1), neglecting its variation along height. This brings a larger difference than the reference simulation considering the variation along the height. The smallest differences are from the largest height N#6, with differences of −2.8% and +7.2%. It can be indicated that the UHI's effects on the low-rise buildings is larger than the high-rise buildings. In this study, N#4 is the best representative meteorological file which can obtain the closest result compared to the reference simulation for both heating load (+1.0%) and cooling load (−0.3%).

It should be noted that, in our case study, the heating and cooling loads are almost balanced, resulting in a minor difference in the total energy load due to the microclimate. However, its effect could be stronger in other climate zones where the difference between the heating and cooling loads is large.

4.5. Discussion

Instead of simulating one representative day in each season in ENVI-met, more precise UHIIs could be obtained, such as by simulating the whole representative week in each season. However, this will largely increase the simulation time, which is not feasible in the design phase. We investigated this and found that the difference of the average UHII during the simulated 24 h between these two methods is acceptable (less than 10% on average). Therefore, simulating one representative day in each season is a compromise between the simulation time and accuracy. The method of generating the hourly UHII is based on the linear interpolation between four representative days in each season, which might not precisely evaluate the UHII for every hour. Although this is a good attempt to account for the local UHI effect on the building energy performance, a more accurate method generating the hourly UHII would be beneficial. Regardless, measurements of the local climate parameters could be helpful to validate the simulation results, especially if the local measurements could be collected in different heights and orientations for a high-rise building in a dense district. The energy performance of other uses of the building can also be investigated, such as the office buildings only operating in the daytime.

Although some studies—e.g., Chan [46]—reported a similar UHII measurement trend as Figure 12, in urban areas, a larger UHI effect is usually observed during the night period compared to the day time, which is contradictory to Figure 12. A longer buffering time of 3 days was used to investigate the possibility of improving the results. However the results still showed a limited heat island effect at night, which may be due to numerical approximation of ENVI-met. It would be useful to perform calculations again using a more powerful computer allowing for more precise simulation options to be chosen.

This paper focuses on the influence of air temperature. Other site-specific climate parameters such as relative humidity, solar radiation, longwave radiation and wind affecting the building energy loads were not considered in this study, which could be an interesting perspective in future research.

In this case study, the simulation using the EPW file predicts that the heating load is a bit larger than the cooling load, which is contrary to the reference simulation considering the UHI effect. A more precise evaluation of the load balance could be helpful for a more appropriate design of a reversible ground source heat pump. Therefore, considering the UHI effect in the building simulation could be beneficial for this.

This study only evaluates the effects of UHI on the energy consumption of the building. However there are other environmental impacts such as CO_2 emissions, human health and biodiversity. Moreover, the environmental impacts of electricity are different in winter and summer, because different electricity production technologies are used for different seasons [47]. Although the total energy consumption is almost identical, the variations in the heating and cooling loads might influence the environmental impacts. This is another interesting topic to be investigated.

5. Conclusions

China is under a state of rapid urbanisation. During this process, the urban environment is altered and shows different climate characteristics compared to rural areas. The simulation results of a building's energy consumption are highly dependent on the accuracy of the weather file comprising 8760 h of various climatic parameters. In present DBES tools, the most used weather data usually come from meteorological stations located in peripheral zones, which cannot reflect site-specific microclimate conditions such as the UHI effect. This might lead to simulation errors.

This study contributes to better accounting for the UHI effect in building energy simulation compared to using regional weather data. A site-specific weather file generation method was proposed to generate new local weather files accounting for hourly UHI effects by using the microclimate simulation tool ENVI-met. This method selects four representative days which are simulated in ENVI-met to obtain the hourly UHII. Afterwards the corresponding hourly UHII of other days in one year are evaluated by linear interpolation.

This method could avoid a long simulation time which is a big challenge for building energy performance optimisation. The method of coupling ENVI-met with the DBES tool COMFIE was introduced as well. This method combines the thermal zone definition in COMFIE and the height resolution in ENVI-met, providing more precise weather data for each zone in DBES. By applying this method, instead of the same weather file for all the building zones, air temperature variations along height and orientation are considered, yielding relatively more accurate simulation results, especially for high-rise buildings.

The methods were applied to a case study with a height of 102 m in Wuhan, China. A district area of 660 m × 660 m was modelled around the studied building. The microclimate simulation was performed for four representative days by ENVI-met. The hourly UHII was obtained by applying the proposed method and a set of site-specific weather files considering the UHI effect were generated for four orientations and six heights. The yearly average UHII at the height of 3 m was estimated to be 0.55 °C. The simulation using the original EPW file showed a heating load 5.8% larger, a cooling load 8.7% smaller and a total energy load 1.7% smaller compared to the reference simulation. When only the weather file at the height of 3 m near the north facade was used, neglecting the variation of UHII along the height, the heating load decreased by 7.5% and the cooling load increased by 12.1% compared to the EPW file. The UHI's effects on the low-rise buildings are larger than on the high-rise buildings. The methods proposed in this paper can be used for a more precise application of urban building energy simulation accounting for the UHI effect. The linear interpolation between two representative days to generate hourly UHII is a simplifying assumption, which should be improved in the case that a more precise UHII is expected. Including more climate features in these methods is another perspective.

Author Contributions: Conceptualization, L.P. and B.P.; methodology, L.P., P.S. and B.P.; software, L.P.; investigation, L.P.; validation, L.P., P.S. and B.P.; writing—original draft, L.P.; writing—review and editing, L.P., P.S. and B.P.; project administration, P.S. and B.P.; supervision, P.S. and B.P.; funding acquisition, B.P. All authors have read and agreed to the published version of the manuscript.

Funding: This research was funded by China Scholarship Council (CSC) grant number 201708420156 and the Chair ParisTech VINCI Eco-design of buildings and infrastructure.

Institutional Review Board Statement: Not applicable.

Data Availability Statement: The authors confirm that the data supporting the findings of this study are available within the article.

Acknowledgments: The authors would like to thank the financial support by China Scholarship Council (CSC) and the Chair ParisTech VINCI Eco-design of buildings and infrastructure.

Conflicts of Interest: The authors declare no conflict of interest.

Abbreviations

List of abbreviations

3D	Three-dimensional
ach	air change per hour
DBES	Dynamic building energy simulation
E	East
EPW	EnergyPlus Weather
HSJG	Haishan Jingu
MAE	Mean absolute error
N	North
RMSE	Root mean square error
S	South
TMM	Typical meteorological months
TMY	Typical meteorological year
UCI	Urban cool island
UCII	Urban cool island intensity

UHI	Urban heat island
UHII	Urban heat island intensity
W	West
List of symbols	
I	Number of cells on x-axis in ENVI-met model
i	Hour number in a day
J	Number of cells on y-axis in ENVI-met model
j	Day number in a representative week
K	Number of cells on z-axis in ENVI-met model
k	Day number in a year
n	Number of vertical cell
N_h	Number of different heights
P	Atmosphere pressure, Pa
P_0	Reference pressure, Pa
s	Scaling factor
T	Temperature, °C or K
UHIdh	Urban heat island degree-hours, °C hours
Superscripts and subscripts	
abs	Absolute
air	Air
av	Average
ENVImet	ENVI-met results
EPW	EPW file
max	Maximal
pot	Potential temperature
re1	1st representative day
re2	2nd representative day
List of Greek letters	
Δx	Resolution of cell on x-axis in ENVI-met model, m
Δy	The resolution of cell on y-axis in ENVI-met model, m
Δz	The resolution of cell on z-axis in ENVI-met model, m

References

1. Sun, Y.; Haghighat, F.; Fung, B.C.M. A Review of The-State-of-the-Art in Data-Driven Approaches for Building Energy Prediction. *Energy Build.* **2020**, *221*, 110022. [CrossRef]
2. Tsoka, S.; Tolika, K.; Theodosiou, T.; Tsikaloudaki, K.; Bikas, D. A Method to Account for the Urban Microclimate on the Creation of 'Typical Weather Year' Datasets for Building Energy Simulation, Using Stochastically Generated Data. *Energy Build.* **2018**, *165*, 270–283. [CrossRef]
3. Hall, I.J.; Prairie, R.R.; Anderson, H.E.; Boes, E.C. *Generation of a Typical Meteorological Year*; Sandia Labs.: Albuquerque, NM, USA, 1978.
4. Wei, Y.; Huang, C.; Li, J.; Xie, L. An Evaluation Model for Urban Carrying Capacity: A Case Study of China's Mega-Cities. *Habitat Int.* **2016**, *53*, 87–96. [CrossRef]
5. Maheshwari, B.; Pinto, U.; Akbar, S.; Fahey, P. Is Urbanisation Also the Culprit of Climate Change?—Evidence from Australian Cities. *Urban Clim.* **2020**, *31*, 100581. [CrossRef]
6. Toparlar, Y.; Blocken, B.; Maiheu, B.; van Heijst, G.J.F. Impact of Urban Microclimate on Summertime Building Cooling Demand: A Parametric Analysis for Antwerp, Belgium. *Appl. Energy* **2018**, *228*, 852–872. [CrossRef]
7. Li, C.; Zhou, J.; Cao, Y.; Zhong, J.; Liu, Y.; Kang, C.; Tan, Y. Interaction between Urban Microclimate and Electric Air-Conditioning Energy Consumption during High Temperature Season. *Appl. Energy* **2014**, *117*, 149–156. [CrossRef]
8. Oke, T.R. The Energetic Basis of the Urban Heat Island (Symons Memorial Lecture, 20 May 1980). *Q. J. R. Meteorol. Soc.* **1982**, *108*, 1–24.
9. Imhoff, M.L.; Zhang, P.; Wolfe, R.E.; Bounoua, L. Remote Sensing of the Urban Heat Island Effect across Biomes in the Continental USA. *Remote Sens. Environ.* **2010**, *114*, 504–513. [CrossRef]
10. Santamouris, M. Analyzing the Heat Island Magnitude and Characteristics in One Hundred Asian and Australian Cities and Regions. *Sci. Total Environ.* **2015**, *512–513*, 582–598. [CrossRef]
11. Zhou, W.; Wang, J.; Cadenasso, M.L. Effects of the Spatial Configuration of Trees on Urban Heat Mitigation: A Comparative Study. *Remote Sens. Environ.* **2017**, *195*, 1–12. [CrossRef]
12. Li, X.; Zhou, Y.; Asrar, G.R.; Imhoff, M.; Li, X. The Surface Urban Heat Island Response to Urban Expansion: A Panel Analysis for the Conterminous United States. *Sci. Total Environ.* **2017**, *605–606*, 426–435. [CrossRef] [PubMed]

13. Li, H.; Zhou, Y.; Li, X.; Meng, L.; Wang, X.; Wu, S.; Sodoudi, S. A New Method to Quantify Surface Urban Heat Island Intensity. *Sci. Total Environ.* **2018**, *624*, 262–272. [CrossRef] [PubMed]
14. Yang, X.; Peng, L.L.H.; Jiang, Z.; Chen, Y.; Yao, L.; He, Y.; Xu, T. Impact of Urban Heat Island on Energy Demand in Buildings: Local Climate Zones in Nanjing. *Appl. Energy* **2020**, *260*, 114279. [CrossRef]
15. Tejedor, B.; Lucchi, E.; Nardi, I. Application of Qualitative and Quantitative Infrared Thermography at Urban Level: Potential and Limitations. In *New Technologies in Building and Construction: Towards Sustainable Development*; Bienvenido-Huertas, D., Moyano-Campos, J., Eds.; Lecture Notes in Civil Engineering; Springer Nature: Singapore, 2022; pp. 3–19. ISBN 978-981-19189-4-0.
16. Martin, M.; Chong, A.; Biljecki, F.; Miller, C. Infrared Thermography in the Built Environment: A Multi-Scale Review. *Renew. Sustain. Energy Rev.* **2022**, *165*, 112540. [CrossRef]
17. Lagüela, S.; Díaz−Vilariño, L.; Roca, D.; Lorenzo, H. Aerial Thermography from Low-Cost UAV for the Generation of Thermographic Digital Terrain Models. *Opto-Electron. Rev.* **2015**, *23*, 78–84. [CrossRef]
18. Fabbri, K.; Costanzo, V. Drone-Assisted Infrared Thermography for Calibration of Outdoor Microclimate Simulation Models. *Sustain. Cities Soc.* **2020**, *52*, 101855. [CrossRef]
19. Cho, Y.-I.; Yoon, D.; Shin, J.; Lee, M.-J. Comparative analysis of the effects of heat island reduction techniques in urban heatwave areas using drones. *Korean J. Remote Sens.* **2021**, *37*, 1985–1999. [CrossRef]
20. Li, X.; Zhou, Y.; Yu, S.; Jia, G.; Li, H.; Li, W. Urban Heat Island Impacts on Building Energy Consumption: A Review of Approaches and Findings. *Energy* **2019**, *174*, 407–419. [CrossRef]
21. Sun, Y.; Augenbroe, G. Urban Heat Island Effect on Energy Application Studies of Office Buildings. *Energy Build.* **2014**, *77*, 171–179. [CrossRef]
22. Skelhorn, C.P.; Levermore, G.; Lindley, S.J. Impacts on Cooling Energy Consumption Due to the UHI and Vegetation Changes in Manchester, UK. *Energy Build.* **2016**, *122*, 150–159. [CrossRef]
23. Lowe, S.A. An Energy and Mortality Impact Assessment of the Urban Heat Island in the US. *Environ. Impact Assess. Rev.* **2016**, *56*, 139–144. [CrossRef]
24. Bruse, M.; Fleer, H. Simulating Surface–Plant–Air Interactions inside Urban Environments with a Three Dimensional Numerical Model. *Environ. Model. Softw.* **1998**, *13*, 373–384. [CrossRef]
25. Huttner, S. Further Development and Application Ofthe 3D Microclimate Simulation ENVI-Met. Ph.D. Thesis, Johannes Gutenberg University Mainz, Mainz, Germany, 2012.
26. López-Cabeza, V.P.; Galán-Marín, C.; Rivera-Gómez, C.; Roa-Fernández, J. Courtyard Microclimate ENVI-Met Outputs Deviation from the Experimental Data. *Build. Environ.* **2018**, *144*, 129–141. [CrossRef]
27. Ayyad, Y.; Sharples, S. Envi-MET Validation and Sensitivity Analysis Using Field Measurements in a Hot Arid Climate. In Proceedings of the IOP Conference Series: Earth and Environmental Science, Cardiff, Wales, 24–25 September 2019; Volume 329, p. 012040. [CrossRef]
28. Elwy, I.; Ibrahim, Y.; Fahmy, M.; Mahdy, M. Outdoor Microclimatic Validation for Hybrid Simulation Workflow in Hot Arid Climates against ENVI-Met and Field Measurements. *Energy Procedia* **2018**, *153*, 29–34. [CrossRef]
29. Sharmin, T.; Steemers, K. Understanding ENVI-Met (V4) Model Behaviour in Relation to Environmental Variables. In Proceedings of the PLEA 2017 Edinburgh: Design to Thrive, Edinburgh, UK, 2 July 2017; pp. 2156–2164.
30. Yang, X.; Zhao, L.; Bruse, M.; Meng, Q. Evaluation of a Microclimate Model for Predicting the Thermal Behavior of Different Ground Surfaces. *Build. Environ.* **2013**, *60*, 93–104. [CrossRef]
31. Peuportier, B.; Blanc-Sommereux, I. Simulation Tool with Its Expert Interface for the Thermal Design of Multizone Buildings. *Int. J. Sol. Energy* **1990**, *8*, 109–120. [CrossRef]
32. Peuportier, B. *COMFIE, Logiciel Pour L'architecture Bioclimatique, Quelques Applications Pour Les Vérandas*; Journée Technique GENEC (CEA): Marseille, France, 1993.
33. Peuportier, B. Bancs D'essais de Logiciels de Simulation Thermique. In Proceedings of the Journée Thématique IBPSA France—SFT 2005, Outil de Simulation Thermo-Aéraulique du Bâtiment, La Rochelle, France, 31 March 2005.
34. Brun, A.; Spitz, C.; Wurtz, E.; Mora, L. Behavioural Comparison of Some Predictive Tools Used in a Low-Energy Building. In Proceedings of the Eleventh International IBPSA Conference, Glasgow, Scotland, 27–30 July 2009.
35. Recht, T.; Munaretto, F.; Schalbart, P.; Peuportier, B. Analyse de la Fiabilité de COMFIE par Comparaison à Des Mesures. Application à un Bâtiment Passif. In Proceedings of the Conference IBPSA France-Arras-2014, Arras, France, 20 May 2014; p. 8.
36. Spitz, C. Analyse de la Fiabilite Des Outils de Simulation et Des Incertitudes de Métrologie Appliquée à L'efficacité Energétique Des Bâtiments. Ph.D. Thesis, Université de Grenoble, Grenoble, France, 2012.
37. Pei, L. The Study on Eco-Design of High-Rise Residential Buildings in Wuhan Based on Energy Simulation and Life Cycle Assessment. Master's Thesis, Huazhong University of Science & Technology, Wuhan, China, 2015.
38. EnergyPlus. Available online: https://energyplus.net/weather-location/asia_wmo_region_2/CHN/CHN_Hubei.Wuhan.574940_SWERA (accessed on 21 February 2023).
39. Yang, X.; Yao, L.; Peng, L.L.H.; Jiang, Z.; Jin, T.; Zhao, L. Evaluation of a Diagnostic Equation for the Daily Maximum Urban Heat Island Intensity and Its Application to Building Energy Simulations. *Energy Build.* **2019**, *193*, 160–173. [CrossRef]
40. Lauzet, N.; Rodler, A.; Musy, M.; Azam, M.-H.; Guernouti, S.; Mauree, D.; Colinart, T. How Building Energy Models Take the Local Climate into Account in an Urban Context—A Review. *Renew. Sustain. Energy Rev.* **2019**, *116*, 109390. [CrossRef]

41. Thiers, S. Bilans Energétiques et Environnementaux de Bâtiments à Energie Positive. Ph.D. Thesis, École Nationale Supérieure des Mines de Paris, Paris, France, 2008.
42. Wu, N.; Shi, P.; Zhu, G.; Pan, X. Changes of soil relative moisture content and influencing factor in the yangtze basin during 1992–2012. *Resour. Environ. Yangtze Basin* **2017**, *26*, 1001–1010. [CrossRef]
43. Ministry of Housing and Urban-Rural Development. *Design Standard for Energy Efficiency of Residential Buildings in Hot Summer and Cold Winter Zone*; China Architecture & Building Press: Beijing, China, 2010.
44. Yang, X.; Yao, L.; Jin, T.; Peng, L.L.H.; Jiang, Z.; Hu, Z.; Ye, Y. Assessing the Thermal Behavior of Different Local Climate Zones in the Nanjing Metropolis, China. *Build. Environ.* **2018**, *137*, 171–184. [CrossRef]
45. Huang, K.-T.; Li, Y.-J. Impact of Street Canyon Typology on Building's Peak Cooling Energy Demand: A Parametric Analysis Using Orthogonal Experiment. *Energy Build.* **2017**, *154*, 448–464. [CrossRef]
46. Chan, A.L.S. Developing a Modified Typical Meteorological Year Weather File for Hong Kong Taking into Account the Urban Heat Island Effect. *Build. Environ.* **2011**, *46*, 2434–2441. [CrossRef]
47. Peters, J.F.; Iribarren, D.; Juez Martel, P.; Burguillo, M. Hourly Marginal Electricity Mixes and Their Relevance for Assessing the Environmental Performance of Installations with Variable Load or Power. *Sci. Total Environ.* **2022**, *843*, 156963. [CrossRef] [PubMed]

energies

MDPI

Article

Prospective PCM–Desiccant Combination with Solar-Assisted Regeneration for the Indoor Comfort Control of an Office in a Warm and Humid Climate—A Numerical Study

Edson Manyumbu [1,2,*], Viktoria Martin [1] and Justin Ningwei Chiu [1]

1 Department of Energy Technology, KTH, Brinellvagen 68, SE-10044 Stockholm, Sweden; viktoria.martin@energy.kth.se (V.M.)
2 School of Engineering Sciences and Technology, Chinhoyi University of Technology, Chinhoyi 7724, Zimbabwe
* Correspondence: manyumbu@cut.ac.zw or manyumbu@ug.kth.se; Tel.: +263-773369914

Abstract: Favorable thermal conditions within buildings are a necessity. Mechanical air conditioning, although effective, contributes a significant percentage of the world's total energy use, which contributes to global warming. In addition, the refrigerants used in air conditioning also contribute to global warming. Passive means to provide thermal comfort have therefore been considered as alternative solutions. Phase-change materials (PCMs) have been considered as one passive cooling option. Although this option achieves a certain degree of effectiveness, especially in warm and dry climatic conditions, its effectiveness in warm humid climates is subdued due to its inability to handle humidity. In the present study, the suitability of a novel passive comfort provision strategy that combines a PCM and a desiccant is assessed. The passive system operates in a cycle of two phases: the moderating phase and the regenerating phase. For the proposed strategy, the regeneration process first involves the external desiccant bed, then night air drying using the regenerated external bed; the dried air subsequently regenerates the internal wall surface. The study involves the modeling of the proposed strategy and simulation of its performance. The simulation results indicate the significant potential for providing satisfactory comfort and health conditions through application of a combination of a desiccant and a PCM.

Keywords: comfort; passive; desiccant; phase-change materials; simulation

Citation: Manyumbu, E.; Martin, V.; Chiu, J.N. Prospective PCM–Desiccant Combination with Solar-Assisted Regeneration for the Indoor Comfort Control of an Office in a Warm and Humid Climate—A Numerical Study. *Energies* **2023**, *16*, 5391. https://doi.org/10.3390/en16145391

Academic Editors: Zsuzsa Szalay and Bruno Peuportier

Received: 27 May 2023
Revised: 28 June 2023
Accepted: 11 July 2023
Published: 14 July 2023

1. Introduction

Thermal comfort in offices is a high priority for human health and productivity [1–4], whereas the energy demand for the comfort provision is of concern [5–7], hence the need for low energy or passive ways for the comfort provision. Thermal comfort is largely a function of both the temperature and humidity in a codependent way [8]. At a given temperature, if the air is humid (high relative humidity), the perceived thermal sensation will be warmer than it would be with a relatively lower humidity. The control of both temperature and humidity is therefore a requirement for thermal comfort.

The energy consumption to ensure indoor thermal comfort conditions has been cited as one of the major contributors to the high energy utilization in buildings [5–7]. Energy production has been blamed for the adverse environmental impact that has caused enumerable problems, including climate change [5,9–13]. Furthermore, active conventional air-conditioning systems use refrigerants that have a global warming effect [10–14]. Therefore, research into comfort provision at zero or near-zero energy use through passive options has become imperative.

Indoor air exchanges thermal energy and moisture with the building envelope within the indoor space. The thermal energy and moisture exchange depend on the hygrothermal properties of the involved building envelope materials. To evaluate the performance of possible passive strategies, modeling and simulation of the exchanges between the air and these materials is necessary. Using PCMs for cooling in buildings has been widely researched [15–18] but only recently has there been some focus on materials that seek to combine the PCMs and desiccants for combined temperature–humidity regulation [19,20]. In Europe, T. Padfield [21] pioneered the study on the impact of moisture buffering in buildings. However, Rudd [22] pioneered the enhancement of the moisture buffering capacity of interior wall surfaces. Since both the humidity and temperature are important, especially in warm and humid (35 °C $\geq$ Ta $\geq$ 26 °C, RH $\geq$ 65%) climates, a strategy that seeks to address these two parameters cooperatively is the focus of this paper. The purpose of the study is to provide proof-of-concept feasibility with regard to combining a PCM and a desiccant to passively control the indoor temperature and humidity of the indoor climate. Regeneration (restoration of the PCM and desiccant to the state of being able to buffer the temperature and moisture, respectively) is key to the feasibility of the proposed passive strategy. Therefore, this study mainly focuses on the regeneration of the interior wall surfaces and the indoor climate control.

2. Materials and Methods

Modeling and simulation were applied to evaluate the effectiveness of a novel passive strategy for the indoor climate control of an office space in a warm and humid climate. The modeling and simulation were applied to a hypothetical office. Regeneration is key to the proposed passive strategy. For the PCM recharging (regeneration), night air ventilation is commonly applied; however, to include desiccant regeneration, a novel approach is called for. The proposed approach involves the use of low air humidity to regenerate hygroscopic internal desiccant-laden surfaces rather than the common approach of using elevated temperatures. The low humidity air is obtained from dry night air using an external desiccant bed. The external bed is solar dried during the day in a solar channel design. As shown in Figure 1, a solar channel, containing a silica gel bed as the absorber surface, is built onto an external wall (assumed to be not shaded). During the day, the desiccant bed is regenerated by solar radiation. At night, the dry desiccant bed dries night air on its way into the room space; this dried air is responsible for the regeneration of the internal desiccant. This solar-assisted regeneration (SAR) approach allows for the regeneration of both the PCM and the desiccant: the two materials required for the moderation of the temperature and humidity in the indoor space. Davison grade 11 silica gel was applied in the modeling and simulation as the desiccant (based on availability of performance data as well as the good performance within the desired humidity range). RT25HC (a paraffin-based PCM by Rubitherm Technologies GmbH, Berlin, Germany) was the PCM used (based on its superior latent heat capacity of 230 kJ/kg among PCMs with the desired phase change temperature range) [23]. Mathematical models were developed based on material properties from data in the literature and apposite data curve fitting using an excel spreadsheet. For the simulation, the office is assumed to accommodate two people performing desk work, two personal computers and a printer in operation, as well as lights for a period of 10 h. Sensible and latent loads were estimated through suitable models and data.

The initial step was to determine the moisture content of the solar-dried external silica gel. The next step was to establish the humidity and temperature of the air as it leaves the external silica gel 'bed'; this air is used for drying the internal silica gel surfaces. The subsequent step is to simulate the drying of the internal silica gel surfaces. The drying process of the internal silica gel and the regeneration of the PCM occur simultaneously. The regeneration of the PCM requires relatively low temperatures, whereas the silica gel regeneration needs relatively high temperatures. The regeneration of the silica gel is endothermic and thus provides a compensatory cooling effect towards achieving the desired coupled regenerations. Additionally, instead of relying on elevated temperatures for silica gel regeneration, the dryness of the air becomes the main driver for the regeneration. Simulation of the indoor air temperature considers the exothermic nature of the dehumidification process through moisture adsorption by the internal silica gel surfaces.

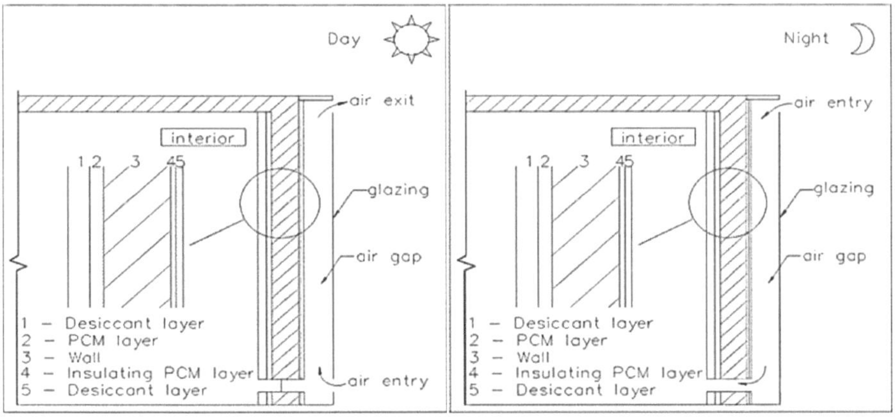

Figure 1. Proposed Regeneration Strategy for Interior and Exterior Desiccant.

The hypothetical office was a 3 m by 3 m in plan and the roof was considered to be a concrete slab type. It had a 1.8 m^2 door as well as a 1.6 m^2 window. We considered a plaster–PCM interior painted with desiccant-rich paint; the desiccant mass ratio in the paint was approximated to give 0.5 kg/m^2 of desiccant per surface painted. For the external desiccant, the limiting value for the mass of the desiccant was the maximum thickness that remained within the thin-layer definition. Hourly time steps were used since the hourly weather data was available and the hourly time steps provided an adequate simulation resolution.

2.1. External Silica Gel Drying Model

Mass and energy balance principles were applied in the modeling, together with the appropriate boundary and initial conditions. Instant equilibration (or instantaneous local equilibrium (ILE)) was assumed. The energy balance equation for the external silica gel bed system during bed regeneration using solar energy resulted in Equation (1). The energy balance considered the sorption heat. In Equation (1), T_f is the average channel air temperature, T_a is the ambient air temperature, $\dot{m}_a$ is the mass flow rate of air, η is the thermal efficiency, A_c is the surface area and Q is the sorption heat.

$$T_f = T_a + \frac{\eta I A_c + Q}{2\dot{m}_a}, \quad (1)$$

The rate of moisture uptake by the air passing over the silica gel is given in Equation (2).

$$\dot{m}_w = \dot{m}_a(X_{ao} - X_{ai}), \quad (2)$$

where $\dot{m}_w$ is the moisture transfer rate from the solid desiccant, $\dot{m}_a$ is the mass flow rate of air, X_{ao} is the specific humidity of air at the outlet and X_{ai} is the specific humidity of air at the inlet. The moisture transfer rate $\dot{m}_w$ based on the linear driving force (LDF) model is shown in Equation (3) [24]:

$$\dot{m}_w = \frac{m}{1+X} k(X - X_e), \quad (3)$$

where m is the mass of desiccant, X is the moisture content and X_e is the equilibrium moisture content. The equilibrium moisture content is correlated with the relative humidity according to Equation (4) [25]:

$$X_e = 0.25\varnothing^{0.6}, \quad (4)$$

The specific humidity of air leaving the desiccant bed can be expressed as:

$$X_{ao} = min\left[\left(X_{ai} + \frac{m}{\dot{m}_a(1+X)}k(X - X_e)\right) : X_{as}\right], \tag{5}$$

X_{as} is the saturated specific humidity of the air at outlet temperature. The saturation pressure as a function of temperature for the dry bulb temperature range 8 °C to 30 °C, based on steam tables, was (through curve fitting) found to be:

$$P_{sat} = 0.006e^{0.062T}, \tag{6}$$

According to thermodynamic relationship of moist air properties (specific humidity), X_a is a function of the saturation pressure P_{sat} and the relative humidity $\varnothing$ can be correlated as follows:

$$X_a = 0.003732\varnothing e^{0.062T} \rightarrow X_{as} = 0.003732e^{0.062T}, \tag{7}$$

since $\varnothing$=1, Equation (5) becomes:

$$X_{ao} = min\left[\left(X_{ai} + \frac{m}{\dot{m}}k(X - X_e)\right) : 0.003732e^{0.062T}\right], \tag{8}$$

The equations are then solved for the values of X_{ao} and the change in X after every time step (hourly basis: the weather data was available in hourly intervals).

2.2. Internal Surfaces Regeneration

During the night, air is passed through the channel and is dried by the desiccant. For the state of the dried air leaving the desiccant, the adsorption model is considered: the hysteresis is ignored such that the drying model was applied, albeit in reverse.

The specific humidity of air leaving the desiccant bed is expressed in Equation (9):

$$X_{ao} = max\left[\left(X_{ai} - \frac{m}{\dot{m}_a(1+X)}k(X - X_e)\right); X_a^*\right], \tag{9}$$

where X_a^* is the minimum possible specific humidity of the air at outlet temperature, this is the specific humidity where the air is assumed to have been dried to the point where it comes into equilibrium with desiccant and the desiccant is at its initial moisture content. Expressing X_a^* in terms of the moisture content [24] and substituting in Equation (9) results in Equation (10).

$$X_{ao} = max\left[\left(X_{ai} - \frac{m}{\dot{m}_a(1+X)}k(X - X_e)\right); 0.0061X^{0.4329}e^{0.062T}\right], \tag{10}$$

For the internal surfaces, the sorption rate and the equilibrium models are given by Equations (11) and (12) [23], where in Equation (11) t is in seconds:

$$\frac{dX}{dt} = 6 \times 10^{-5}(X_e - X)^{0.5286}, \tag{11}$$

$$X_e = 0.325\varnothing^{2.31}, \tag{12}$$

The temperature of the air leaving the desiccant is evaluated by applying Equation (13) (based on energy balance):

$$Q = m_a c_p\left(T_f - T_a\right) + mC_{sg}\left(T_{sg_i} - T_{sg_{i-1}}\right), \tag{13}$$

The dried air, as it passes through the room, picks up moisture from the moisture-laden surface and to obtain the state of the desiccant at any moment, a desorption model derived from data obtained from the literature through curve fitting was applied. Because the desorption process is endothermic, the interior surface of the room is assumed to be cooled in the process, and this is desirable for the cooling of the PCM. The outdoor air is expected to pass through the external desiccant bed and into the interior and then out through the ventilation.

2.3. Moisture Load Estimation Model

In an office space, the main contributors towards moisture build up are the occupants, infiltration and ventilation. The moisture load due to the respiration and perspiration of the occupants can be estimated using Equation (14):

$$\dot{m}_p = n_p F, \tag{14}$$

In Equation (14), $\dot{m}_p$ is moisture released into the occupied space by the occupants (people), n_p is the number of people and F is the moisture production rate per person. In their experimental work, Zemitis et al. [26] reported a moisture generation of 51 g/h per person as the generally approximated value for assumed office activities. Infiltration was considered to be largely through the door and the window, and an infiltration rate of 0.5 air changes (ACH) was considered [27].

2.4. Thermal Load Estimation Model

The thermal load of the office space consists of internal and external sources. Internal heat sources include lights, equipment and people, whereas external heat sources include solar irradiation and external convection. For the considered passive moisture buffering of applying a desiccant silica gel, there is need to also consider the heat of adsorption as an additional heat source in the thermal load model. The heat of adsorption is a function of the moisture adsorption rate and equations for this are taken from the literature [28]. The internal heat gains are shown in Table 1.

Table 1. Estimated Sensible Heat Gains.

Source	Estimated Heat Gain
Lights	20 W/m^2
People	65 W/person
Computers	60 W/computer
Printer	110 W/printer

External heat sources consist of solar radiation through fenestration, heat conduction across the envelope and air exchange with the ambient. For radiation heat that gains entry through fenestration, Equation (15) is applied:

$$Q_{sol} = AI(\tau + \alpha\frac{U}{h_o}), \tag{15}$$

A is the surface area of the transparent window, I is the total radiation on the window and τ and α are transmissivity and absorptivity, respectively. U is the overall heat transfer coefficient of the window, whereas h_o is the external convective heat transfer coefficient. For conductive heat gains through each envelope component, the sol-air temperatures (outside temperatures that consider the effect of the solar radiation) were evaluated and the difference between the sol-air and indoor air temperatures was applied to the conduction equation. The local hourly solar radiation data based on the available 1999 data and the hourly ambient temperatures from the meteorological department of Zimbabwe were employed. The sol-air temperature is defined as the outside air temperature that, in the absence of solar radiation, would give the same temperature distribution and rate of heat transfer through a wall (or roof) as exists due to the combined effects of the actual outdoor temperature distribution plus the incident solar radiation [29]. The sol-air temperature is given by Equation (16) [29]

$$T_{sol} = T_a + \frac{\alpha I - \epsilon \Delta R}{h_o}, \quad (16)$$

For a vertical wall, $\Delta R = 0$, whereas for a horizontal wall, $\Delta R = 63\ \text{W/m}^2$ [29].

The heat transfer coefficient of the building envelope was estimated based on a brick and plaster wall and a concrete roof slab. The estimated U-factors for the components are as given in Table 2. A spreadsheet calculator was prepared, such that each load component was estimated on an hourly basis.

Table 2. U-factor values.

Description of the Building Envelope Component	U-Factor Value (W/m^{2}K)
Wall: 12.5 mm cement plaster + 230 mm fired-clay brick + 12.5 mm PCM/cement plaster [30–33]	1.72
Roof: 15 mm cement layer + 20 mm waterproof layer + 200 mm concrete + 15 mm cement layer [34]	0.87
Window: Vacuum double glazing [35,36]	0.7

Heat that is transported by the air is estimated by:

$$Q_v = \rho \dot{V} c_p \Delta T, \quad (17)$$

where ρ is the mass density of air, $\dot{V}$ is volume flow rate of air exchanged between the indoor and outdoor environments, c_p is the specific heat capacity of air at constant pressure and ΔT is the temperature difference.

2.5. *Buffering Models*

2.5.1. Moisture Exchange Model

The moisture buffering potential depends on the rate of moisture removal, which depends on the surface conditions at occupation in the morning and the moisture entry into the occupied space. The change in specific humidity can be expressed by Equation (18):

$$\Delta X = \frac{(\dot{m}_p + \dot{m}_i + \dot{m}_s)\Delta\tau}{\rho V}, \quad (18)$$

$$\dot{m}_s = \dot{m}_w = \frac{0.216}{1+X} m (X_e - X)^{0.5286}, \quad (19)$$

$$\dot{m}_p = n_p F, \quad (20)$$

$$\dot{m}_i = \rho V X_a = \rho V (ACH)(0.003732 \varnothing e^{0.062T}), \quad (21)$$

In finite difference formulation we have Equation (23):

$$X_{(\tau+\Delta\tau)} = \frac{n_p F + (ACH)\rho V \times 0.003732 \varnothing e^{0.062T} - \frac{0.216m}{1+X}(X_e - X)^{0.5286}}{\rho V} + X_\tau, \quad (22)$$

2.5.2. Heat Exchange Model

The rate at which heat or moisture is exchanged between the room air and the envelope is crucial for the passive moderation of the room conditions. The moisture exchange between the air and the interior surfaces is considered to be characterized by the rate of moisture uptake by the silica-laden surface and the subsequent desorption. The rate is dependent on the moisture conditions of both the air and the surface. The thermal energy exchange between the indoor air and the interior surface is considered to be dependent on the perceived rate of heat transfer from the PCM to the air. The potential control will therefore depend on this rate of heat transfer and the thermal storage capacity of the PCM. According to Liu and Awbi [37], the equation for the convective heat transfer between a wall impregnated with a PCM and room air is given by the following equation:

$$h = 2.3 \times 1.82 \frac{\Delta T^{0.293}}{D^{0.121}}, \quad (23)$$

where D is the characteristic length, which is given by wall surface area divided by the perimeter and ΔT is the temperature difference. Equation (23) is applied in the simulation for the heat transfer between room air and the PCM/desiccant wall. The heat of adsorption was considered as part of the load. For the case under consideration, the value of D is approximately 0.8; this gives the following expression for the heat transfer coefficient:

$$h = 4.31 \Delta T^{0.293}, \quad (24)$$

Applying Newton's Law of cooling, the rate of heat transfer per unit area of wall was found to be given by Equation (25)

$$q = 4.31 \Delta T^{1.293}. \quad (25)$$

Conduction resistance was ignored: the Biot number was established to be plus or minus 0.1, which is a significant degree less than 1, hence the rate of thermal energy transfer was considered to be largely dependent on the convective rate.

$$\rho V C \frac{dT}{d\tau} = q_{load} - 4.31 A_{walls} \left(T_{room} - T_{pcm}\right)^{1.293}, \sum q_{load} \leq m H_f. \quad (26)$$

In Equation (28), q_{load} is the cooling load, m is mass of PCM in walls and H_f is the heat of fusion. A spreadsheet was applied to evaluate the temperature T_{room} at time τ using a finite difference formulation, Equation (27).

$$T_{(\tau+\Delta\tau)} = \frac{\Delta \tau}{\rho V C} \left(q_{load} - 4.31 A_{walls} \left(T_{\tau} - T_{pcm}\right)^{1.293}\right) + T_{\tau}, \quad (27)$$

The initial temperature at the start of the day is assumed to be equal to the temperature of the PCM, $T_{(\tau=0)} = T_{pcm}$. The room air and the PCM are assumed to be in thermal equilibrium. The heat of adsorption in kilojoules per kilogram of moisture adsorbed is dependent on the moisture content of the desiccant and, according to Ramzy et al. [28], this is given by Equation (28)

$$H_{ad} = 3500 - 13400X : X < 0.05 \text{ or } H_{ad} = 2950 - 1400X : X > 0.05, \quad (28)$$

2.5.3. Cost–Benefit Analysis

A concise cost–benefit analysis was carried out on the basis of investment and operating costs in comparison to the active air conditioning system. Table 3 below shows the costs involved for the passive and active systems.

Table 3. Estimated Costs for Passive System.

Material	Type	Cost (USD/Kg)
PCM	Paraffin	3.3–8.8 [38]
Silica gel		2.3 [39]
Glass	Building glass (6.38 mm safety glass)	11.0 [40]

2.6. Limitations

The moisture transfer models are premised on a thin-layer drying concept, such that their validity depends on the degree of acquiescence of the system to the thin-layer design limits. It takes the lumped parameter approach that ignores internal resistances and assumes a uniform state within the materials such that the interfacial vapor pressure difference creates the moisture exchange driving force. The mass transfer Biot number could be utilized to establish the validity of the model for any case application. The moisture exchange coefficient is considered constant; however, this assumption holds true only to a certain degree of accuracy within bounded thermal conditions. The curve-fitting-derived governing equations for the paint/desiccant sorption model need validation through empirical data for diverse real environmental conditions. Furthermore, the internal regeneration is largely driven by the vapor pressure difference and the kinetics rate to reach equilibrium is unclear. This determines the adequacy of the time steps valid for application. In the present study, hourly time steps are considered sufficient to assume equilibration. The heat exchange model also takes the lumped capacitance framework; thus, the model is applicable for low surface to air heat exchange rates that give the convective heat transfer dominancy. Therefore, the model is applicable to systems within a certain convective heat transfer range or design and can be characterized by the heat transfer Biot number. The Biot number is the ratio of thermal resistances within a material and at the surface. For Biot numbers considerably smaller than 1, the heat transfer is simplified by assuming uniform internal temperature, such that convection becomes the sole mode of heat transfer. The sole convective heat transfer formulation is a simplification, a more detailed and accurate approach to heat transfer maybe required for more accurate results.

The PCM and desiccant are treated as semi-independent individual layers in direct communication with the indoor air. The PCM's hygroscopic properties, the desiccant's thermal capacity and the participation of other materials have been ignored. The boundary conditions and the transport coefficient correlations and values are not easy to obtain or even estimate for real-life situations. This remains a limitation and leads to overreliance on general data; there is need for more reliable empirical data. Some data were limited, such as input weather data; however, effort was made to obtain existing data from the local meteorological department. Certain inputs are based on general estimates obtained from the literature due to the limited local empirical data. The model, although it cannot serve as a definitive design tool, provides a proof-of-concept confidence level for further developments towards more accurate design tools.

3. Results and Discussion

The question of providing thermal comfort passively in an office in a warm and humid climate was the focus of this study. A simulation was performed to evaluate the effectiveness of a novel passive humidity and temperature control strategy to provide thermal comfort for an office in a warm and humid climate. The strategy makes use of a PCM and a desiccant for temperature and humidity regulation, respectively, and employs a solar-assisted regeneration (SAR) approach. The degree to which the external silica gel bed is regenerated (dried) is simulated using January 1999 and June 1999 weather data; these were considered typical summer and winter months, respectively. The results presented are for a solar channel of 1 m width, 0.02 m in depth and 3 m in height. For the simulation, the channel is considered to be either east or west facing; these cardinal orientations receive significant amounts of solar radiation in both summer and winter. The regeneration (drying) curves for the external silica gel are presented in Figure 2. It can be observed from Figure 2 that the moisture content drops remarkably within the first 7 h of drying in both the summer and winter. Within the 7 h of regeneration, the moisture content dropped from the initial 0.35 kg/kg to approximately 0.07 kg/kg. In summer, notable drying carries on even towards end of day, but at a reduced rate. However, as the day progresses beyond 16:00, drying recedes and rewetting sets in. By the end of the day (18:00), the external silica gel bed had a moisture content of around 0.1 kg/kg for east orientation in winter and 0.06 kg/kg for west orientation in the summer and winter.

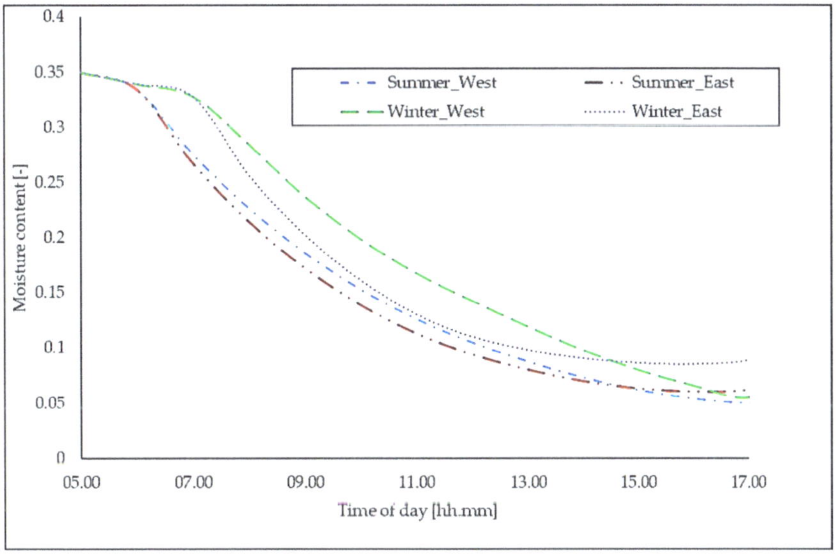

Figure 2. External Silica Gel Regeneration Curve, 15 kg Silica gel and b = 0.02 m (January and June 1999).

Considering the external bed's moisture content at the end of the day as the initial input, a simulation of night air conditions was carried out. Figure 3 shows the specific humidity of the air as it enters and leaves the regenerator (the external silica gel bed). The simulated results indicate that in the first 2 h, the air leaves the regenerator at a specific humidity of around 0.005 kg/kg. The specific humidity of the air leaving the regenerator then increases gradually to a maximum of around 0.006 kg/kg as the regeneration time approaches 8 h. An air specific humidity of 0.006 kg/kg corresponds to a relative humidity of 34% at a temperature of 25 °C, and the equilibrium moisture content of silica gel mixed with paint is approximately 0.02 kg/kg. The air, therefore, has some capacity to dry the interior surfaces that are painted with silica-gel-laden paint. Figure 4 shows the regeneration of the interior surfaces. From Figure 4, it can be observed that the interior silica gel surfaces are dried to approximately 0.06 kg/kg after 8 h of regeneration, after which rewetting begins. The result of this simulation implies that it is possible to regenerate the internal silica gel surface during the night to make it ready for the next day's dehumidification process. Figure 5 shows the simulated indoor relative humidity and temperature. A maximum relative humidity of around 60% is reached at around 13:00 and maintained until after 17:00 (the time the office is vacated). The temperature reaches a maximum of 27.5 °C in the afternoon at around 15:00.

Thermal comfort is a function of both the temperature and relative humidity in addition to other parameters such as activity level, clothing and local air velocity. Thermal comfort conditions are the conditions of the surroundings that allow optimal thermoregulation, which maintains the core body temperature at ±37 °C. The body can exchange heat with its surroundings due to the presence of a temperature difference or through sweat evaporation. For sweat evaporation to take place, the surrounding air must have the capacity to absorb moisture, and that capacity depends on the air relative humidity. Thus, whereas air temperature is the main parameter in determining thermal comfort, humidity plays a significant role in determining comfort outside the range 30–60%, within which humans do not perceive humidity variation [41]. High relative humidity inhibits sweat evaporation, so the perceived thermal sensation becomes warmer than if the humidity is relatively low.

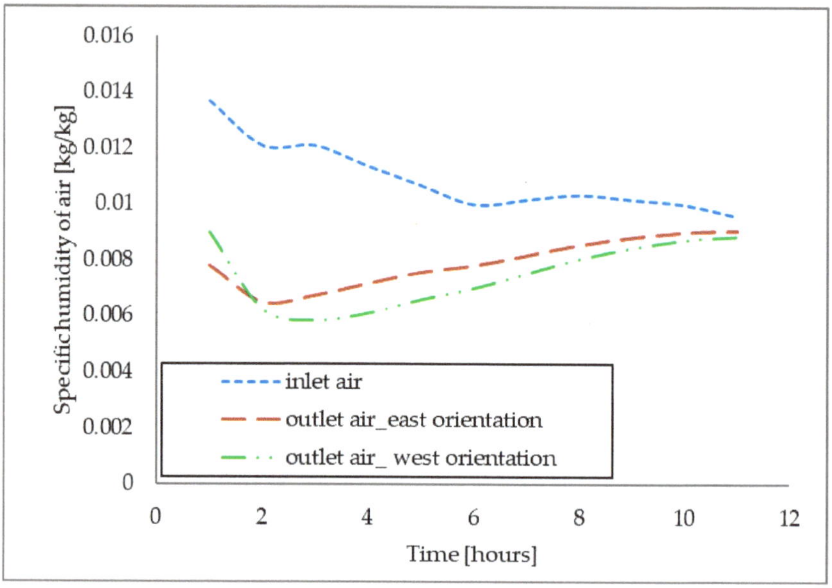

Figure 3. Simulated Specific Humidity of Air at Inlet and Outlet of the Solar Channel Regenerator (January 1999, typical summer day).

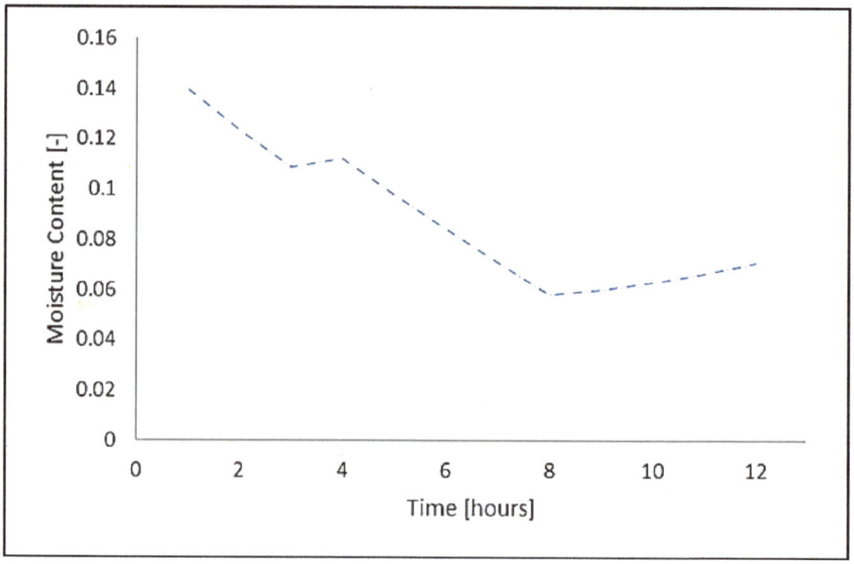

Figure 4. Simulated Interior Surfaces (with Silica gel) Moisture Content.

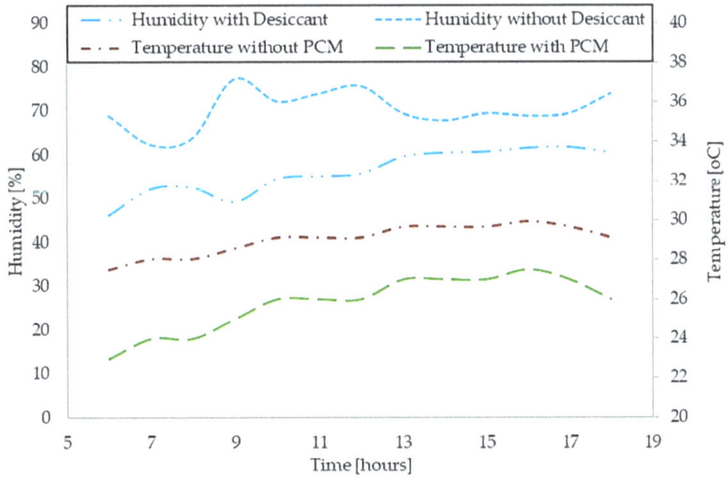

Figure 5. Simulated Indoor Humidity and Temperature (Typical Summer Day–17 January 1999).

The upper limits for temperature according to the ISO 7730 [42] and ASHRAE 55 [43] standards are set at a temperature of 26 °C [44]. The ASHRAE 55 standard upper limit of 26 °C assumes a relative humidity of 50%, whereas the ISO 7730 specifies relative humidity limits of 30–70%. According to the obtained results from 06:00 to about noon (about 6 h), the temperature is kept under and up to 26 °C, this is within the common two comfort standards' limits. Beyond noon, the temperature gradually rises, reaching a maximum of 27.5 °C towards end of day. The maximum temperature reached is 1.5 °C higher than the limits set by the ASHRAE/ISO standards, with an associated relative humidity around 60%, which is within the imperceptible range. The relatively low humidity ensures thermal comfort even at the higher temperature (Figure 6). Considering adaptation, the obtained results are considered to be within acceptable thermal comfort [45]. Thus, the proposed strategy can provide comfort to an office in a warm humid region.

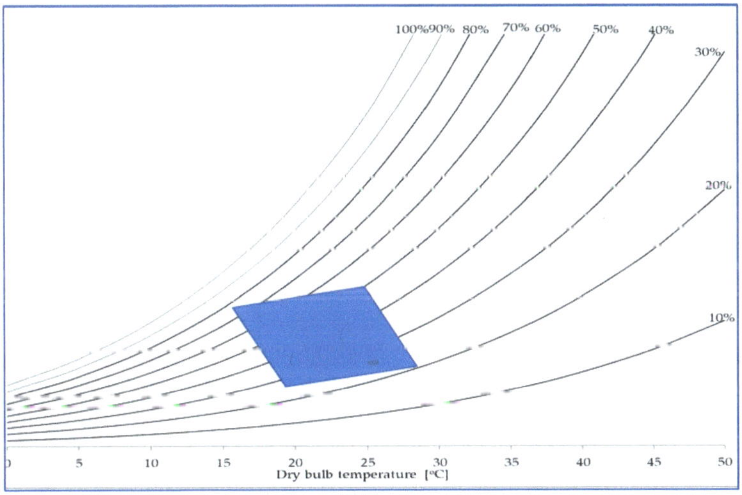

Figure 6. A Simple Psychrometric Chart showing some Comfort Zone temperature and humidity dependence.

The performance of such passive strategies is particularly dependent on the load regime; thus, this requires knowledge of the many parameters that characterize the load. An attempt to model the load regime has been performed and applied to the simulation process. The intermittency and indeterminant nature of the parameters involved present the biggest challenges. Due to the paucity of data, the study depended on the estimation of the input variables, which are spatially and temporally diverse in nature. For example, infiltration can have a significant impact; however, its variability is a challenge towards its estimation.

Critical parameters such as mass and heat transfer coefficients between the PCM/DESICCANT system and the air should be verified; these are quite important in the design and performance prediction of such a system. The values used were from the literature and were derived under similar but not identical conditions. Experimental determination of these parameters is therefore recommended. In addition, measurements on a full-scale model office will give even more reliable and conclusive results. Experimental determination of the dynamic parameters of silica gel–water vapor systems in various conditions is also proposed. This study applied simple algorithms for simulation and the results obtained call for more investigation to validate the performance.

The cost of implementing the passive system is estimated to be around USD 1000.00 based on cost of the materials given in Table 3. The capital cost of an air conditioning unit for the room is estimated to be around USD 750.00. The daily operational costs based on an energy tariff of USD 0.105/kWh, an energy requirement for powering the air conditioning unit of around 0.35 kW power input and 10 h of operation is USD 0.36. The passive system reaches a break-even point in comparison with the active system in year 3; thereafter, the passive system results in cost savings. In addition, the passive system has environmental benefits.

4. Conclusions

From the performed work, it can be concluded that with some load management interventions, systems utilizing PCM and desiccant combinations have the potential for moderating the temperature as well as the humidity in an office in Harare, Zimbabwe. The relative humidity in the room can be kept at around 60%, whereas the temperature can be maintained at around 27 °C or lower by utilizing 1.8 kg of desiccant and 14 kg of PCM per m^2 of floor area. Estimation of the indoor conditions without the PCM and desiccant indicates out of comfort conditions of up to around 80% and 30 °C for the relative humidity and temperature, respectively. Although many studies have been carried out on thermal storage in buildings using PCMs for temperature control, there have been very few studies on the possible application of desiccants within building components for moisture control. The present strategy brings the cooperative aspect of the desiccant to ensure comfort is achieved at the temperatures attainable for passive cooling using PCMs. Due to a number of limitations, although our results show some potential, it is important to note that more investigation is necessary. Furthermore, experimental investigation is necessary to validate and/or improve the model. The current study focuses on both temperature and humidity control in an office. The two parameters are critical for thermal comfort and health when considering warm and humid climates. The study is unique because of its application of both a PCM and a desiccant as part of the interior finish and the regeneration of both through night air drying and applying a solar-regenerated external desiccant as a dryer.

Author Contributions: Conceptualization, E.M.; methodology, E.M.; software, E.M.; formal analysis, E.M.; investigation, E.M.; data curation, E.M.; writing—original draft preparation, E.M.; writing—review and editing, J.N.C. and E.M.; supervision, J.N.C. and V.M. All authors have read and agreed to the published version of the manuscript.

Funding: This research received no external funding.

Data Availability Statement: The authors confirm that the data supporting the findings of this study are available within the article.

Acknowledgments: This work is carried out as part of duties of the first author at Chinhoyi University of Technology and as such he acknowledges the University for its support.

Conflicts of Interest: The authors declare no conflict of interest.

References

1. Frontck, M.; Wargocki, P. Literature survey on how different factors influence human comfort in indoor environments. *Build. Environ.* **2011**, *46*, 927–937.
2. Lan, L.; Wargocki, P.; Lian, Z. Optimal thermal environment improves performance of office work. *REHVA J.* **2012**, *49*, 12–17.
3. Gao, N.; Shao, W.; Rahaman, M.S.; Zhai, J.; David, K.; Salim, F.D. Transfer Learning for Thermal Comfort Prediction in Multiple Cities. In Proceedings of the IoTDI '21: The 6th ACM/IEEE Conference (2020), Sydney, Australia, 21–24 April 2020.
4. Alsaleem, F.; Tesfay, M.K.; Rafaie, M.; Sinker, K.; Besarla, D.; Arunasalam, P. An IoT Framework for Modeling and Controlling Thermal Comfort in Buildings. *Front. Built Environ.* **2020**, *6*, 87. [CrossRef]
5. Yang, L.; Yan, H.; Lam, J.C. Thermal comfort and building energy consumption implications—A review. *Appl. Energy* **2014**, *115*, 164–173. [CrossRef]
6. Nejat, P.; Jomehzadeh, F.; Taheri, M.M.; Gohari, M.; Majid, M.Z.A. A global review of energy consumption, CO_2 emissions and policy in 13 the residential sector (with an overview of the top ten CO_2 emitting countries). *Renew. Sustain. Energy Rev.* **2015**, *43*, 843–862. [CrossRef]
7. IEA. *The Future of Cooling Opportunities for Energy Efficient Air Conditioning*; IEA: Paris, France, 2018. [CrossRef]
8. Fanger, P.O. Assessment of man's thermal comfort in practice. *Br. J. Ind. Med.* **1973**, *30*, 313–324. [CrossRef] [PubMed]
9. Pham, H.; Sachs, H. Next Generations Refrigerants: Standards and Climate Policy Implications of Engineering Constraints. *ACEEE Summer Study Energy Effic. Build.* **2010**, 282–294. Available online: https://www.aceee.org/files/proceedings/2010/data/papers/1933.pdf (accessed on 10 July 2023).
10. McLinden, M.O.; Brown, J.S.; Brignoli, R.; Kazakov, A.F.; Domanski, P.A. Limited options for low-global-warming-potential refrigerants. *Nat. Commun.* **2017**, *8*, 14476. [CrossRef]
11. Belman-Flores, J.M.; Rangel-Hernandez, V.H.; Uson, S.; Rubio-Maya, C. Energy and exergy analysis of R1234yf as drop-in replacement for R134a in a domestic refrigeration system. *Energy* **2017**, *132*, 116–125. [CrossRef]
12. Nawaz, K.; Ally, M.R. Options for low-global-warming-potential and natural Refrigerants Part 2: Performance of refrigerants and systemic irreversibilities. *Int. J. Refrig.* **2019**, *106*, 213–224. [CrossRef]
13. Nair, V. HFO refrigerants: A review of its present status and future prospects. *Int. J. Refrig.* **2021**, *122*, 156–170. [CrossRef]
14. Li, G. Performance evaluation of low global warming potential working fluids as R134a alternatives for two-stage centrifugal chiller applications. *Korean J. Chem. Eng.* **2021**, *38*, 1438–1451. [CrossRef]
15. Waqas, A.; Ud Din, Z. Phase Change Material (PCM) Storage for Free Cooling of Buildings—A Review. *Renew. Sustain. Energy Rev.* **2013**, *18*, 607–625. [CrossRef]
16. Socaciu, L.; Plesa, A.; Unguresau, P.; Giurgiu, O. Review on phase change materials for building applications. *Leonardo El. J. Pract. Technol.* **2014**, *25*, 179–194.
17. Cui, Y.; Xie, J.; Liu, J.; Wang, J.; Chen, S. A review on phase change material in Building. *Adv. Mech. Eng.* **2017**, *9*, 1687814017700828. [CrossRef]
18. Al-Yasiri, Q.; Szabó, M. Paraffin as a Phase Change Material to Improve Building Performance: An Overview of Applications and Thermal Conductivity Enhancement Techniques. *Renew. Energy Environ. Sustain.* **2021**, *6*, 38. [CrossRef]
19. Yang, H.; Liu, Y.; Kong, X.-F.; Chen, W.-H.; Yao, C.-Q. Preparation and hygrothermal performance of composite phase change material wallboard with humidity control based on expanded perlite/diatomite/paraffin. *J. Cent. South Univ.* **2018**, *25*, 2387–2398. [CrossRef]
20. Hou, X.; Li, Q.; Yang, Z.; Zhang, Y.; Zhang, W., Wang, J. Temperature–humidity dual regulation of a single core–double-shell microcapsule fabricated by electrostatic assembly and chemical precipitation. *RSC Adv.* **2020**, *10*, 26494–26503. [CrossRef] [PubMed]
21. Padfield, T. The Role of Absorbent Building Materials in Moderating Changes of Relative Humidity. Ph.D. Thesis, Department of Structural Engineering and Materials, The Technical University of Denmark, Kongeno Lyngby, Denmark, 1998.
22. Rudd, A.F. Development of a Moisture Storage Coatings for Enthalpy Storage Wallboard. *ASHRAE Trans. Res.* **1994**, *100*, 84–90.
23. Available online: https://www.rubitherm.eu/en (accessed on 27 June 2023).
24. Manyumbu, E.; Martin, V. Towards Passive Humidity Control for an Office Building, Modeling and Spreadsheet Simulation of a Desiccant Regeneration Strategy. In Proceedings of the 2nd International Energy Storage Conference, Dublin, Ireland, 19–21 June 2013.
25. Jung, C.-Y.; Kim, J.-R.; Yi, S.-C. Two-dimensional simulation of silica gel drying using computational fluid dynamics. *J. Ceram. Process. Res.* **2008**, *9*, 184–188.
26. Zemitis, J.; Borodinecs, A.; Frolova, M. Measurements of moisture production caused by various sources. *Energy Build.* **2016**, *127*, 884–891. [CrossRef]
27. Tian, Z.; Yang, J.; Lei, Y.P.; Yang, L. Sensitivity Analysis of Infiltration Rates Impact on Office Building Energy Performance. *IOP Conf. Series Earth Environ. Sci.* **2019**, *238*, 012019. [CrossRef]

28. Ramzy, A.K.; Hamed, A.M.; Awad, M.M.; Bekheit, M.M. Theoretical investigation on the cyclic operation of radial flow desiccant bed dehumidifier. *J. Eng. Technol. Res.* **2010**, *2*, 96–110.
29. Kaska, Ö.; Yumrutas, R. Experimental investigation for total equivalent temperature difference (TETD) values of building walls and flat roofs. *Energy Convers. Manag.* **2009**, *50*, 2818–2825. [CrossRef]
30. Motawa, I.; Elsheikh, A.; Diab, E. Energy Performance Analysis of Building Envelopes. *J. Eng. Proj. Prod. Manag.* **2021**, *11*, 196–206.
31. Vijayan, D.S.; Mohan, A.; Revathy, J.; Parthiban, D.; Varatharajan, R. Evaluation of the impact of thermal performance on various building bricks and blocks: A review. *Environ. Technol. Innov.* **2021**, *23*, 101577. [CrossRef]
32. Lachheb, M.; Younsi, Z.; Naji, H.; Karkri, M.; Nasrallah, S.B. Thermal behavior of a hybrid PCM/plaster: A numerical and experimental investigation. *Appl. Therm. Eng.* **2017**, *111*, 49–59. [CrossRef]
33. Al-Absi, Z.A.; Hafizal, M.I.M.; Ismail, M.; Awang, H.; Al-Shwaiter, A. Properties of PCM-based composites developed for the exterior finishes of building walls. *Case Stud. Constr. Mater.* **2022**, *16*, e00960. [CrossRef]
34. Abdelraouf, Y.A.; El-desouky, A.A.; Moustafa, A.M. Passive House (PH) Standards for Achieving Energy-efficient Office Buildings in Egypt. *Eng. Res. J.* **2022**, *1*, 105–116. [CrossRef]
35. Fang, Y.; Arya, F. Evacuated glazing with tempered glass. *Sol. Energy* **2019**, *183*, 240–247. [CrossRef]
36. Aguilar-Santana, J.L.; Jarimi, H.; Velasco-Carrasco, M.; Riffat, S. Review on window-glazing technologies and future prospects. *Int. J. Low-Carbon Technol.* **2020**, *15*, 112–120. [CrossRef]
37. Liu, H.; Awbi, H. Performance of phase change material boards under natural convection. *Build. Environ.* **2009**, *44*, 1788–1793. [CrossRef]
38. Biswas, K.; Abhari, R. Low-cost phase change material as an energy storage medium in building envelopes: Experimental and numerical analyses. *Energy Convers. Manag.* **2014**, *88*, 1020–1031. [CrossRef]
39. Grande, C.A.; Morence, D.G.B.; Bouzga, A.M.; Andreassen, K.A. Silica Gel as a Selective Adsorbent for Biogas Drying and Upgrading. *Ind. Eng. Chem. Res.* **2020**, *59*, 10142–10149. [CrossRef]
40. Available online: https://www.elitesafetyglass.com/6mm-to-12mm-clear-laminated-glass-price/ (accessed on 28 June 2023).
41. Lstiburek, J. Relative Humidity. In Proceedings of the Indoor Air Conference, Austin, TX, USA, 23 April 2003.
42. *ISO 7730*; Standard Analytical Determination and Interpretation of Thermal Comfort Using Calculation of the PMV and PPD Indices and Local Thermal Comfort Criteria. AFNOR: Geneva, Switzerland, 2006.
43. *ASHRAE 55*; Thermal Environmental Conditions for Human Occupancy. ANSI/ASHRAE: Atlanta, GA, USA, 2017.
44. Kontes, G.D.; Giannakis, G.I.; Horn, P.; Steiger, S.; Rovas, D.V. Using Thermostats for Indoor Climate Control in Office Buildings: The Effect on Thermal Comfort. *Energies* **2017**, *10*, 1368. [CrossRef]
45. Singh, M.K.; Ooka, R.; Rijal, H.B.; Takasu, M. Adaptive thermal comfort in the offices of North-East India in autumn season. *Build. Environ.* **2017**, *124*, 14–30. [CrossRef]

energies

MDPI

Article

Achieving Net Zero Carbon Performance in a French Apartment Building?

Alpha Hamid Dicko, Charlotte Roux * and Bruno Peuportier

CES (Centre for Energy Efficiency of Systems), MINES Paris—PSL Research University, 75006 Paris, France; bruno.peuportier@mines-paristech.fr (B.P.)
* Correspondence: charlotte.roux@minesparis.psl.eu

Abstract: Containing global warming to 1.5 °C implies staying on a given carbon budget and therefore being able to design net zero carbon buildings by 2050. A case study corresponding to a French residential building is used to assess the feasibility of achieving this target. Starting from an actual construction built in 2016, various improvement measures are studied: lowering heating energy needs, implementing bio-sourced materials and renewable energy systems (geothermal heat pump, solar domestic hot water production, and photovoltaic electricity production). Dynamic thermal simulation is used to evaluate energy consumption and overheating risk in hot periods. Greenhouse gas emissions are quantified using a consequential life cycle assessment approach, considering that during a transition period, exporting electricity avoids impacts corresponding to marginal production on the grid. Avoided impacts decrease and become zero when the grid is ultimately "decarbonized". From this point, the building should be net zero emissions, but there remain unavoidable emissions. Residual GhG (greenhouse gas) emissions account for 5.6 $kgCO_2$ eq/m^2 annually. The possibility of offsetting these emissions is investigated, considering sequestration in forests or vegetation systems. A net zero emission level can be achieved, but on a national level, it would require that the whole sequestration potential of forest growth be devoted to offset emissions of new construction. A circular economy for construction products and equipment and considering water use will be needed to further decrease environmental impacts.

Keywords: life cycle assessment; energy simulation; carbon sequestration; zero carbon emission building

Citation: Dicko, A.H.; Roux, C.; Peuportier, B. Achieving Net Zero Carbon Performance in a French Apartment Building? *Energies* **2023**, *16*, 7608. https://doi.org/10.3390/en16227608

Academic Editor: Antonio Gagliano

Received: 27 October 2023
Revised: 7 November 2023
Accepted: 12 November 2023
Published: 16 November 2023

1. Introduction

The building sector accounts for 36% of the EU's final energy consumption and almost 40% of total direct and indirect greenhouse gas emissions [1]. Decarbonizing this sector is crucial to achieve the objectives set by international climate agreements [2] and to maintain the earth in a safe operating space [3–5]. This involves improving our construction standards to a net zero emission performance. However, analyzing the roadmaps for achieving climate targets in different regions of the world shows that achieving Zero Carbon and Energy Buildings (ZCEBs) by 2050 is still problematic [6,7]. These roadmaps rarely consider embodied emissions due to complexity, e.g., related to emissions outside national boundaries. Literature proposals for the Zero Energy Building definition also tend to focus only on operational energy use, see for instance [8].

At the EU level, where low emissions are targeted, the Energy Performance of Buildings Directive (EPBD) has defined a zero energy building target [1]. This is a positive initiative, though considering embodied carbon emissions remains important, as they can amount up to 75% of the total life cycle in net zero-energy buildings [9]. The concept of a zero emission building is still progressively becoming the target [9] and has even been extended to the neighborhood level [10,11]. Several definitions have been suggested for (net) zero carbon buildings and are thoroughly described and analyzed in [12]. The authors have identified large variations in methodological options (e.g., "system boundaries for both operational and embodied GhG emissions, the type of GhG emission factor for electricity use, the approach to the "time" aspect, and the possibilities of GhG emission compensation"). They finally acknowledge the unavoidable discrepancies among the ZCEB definitions across countries but urge the account of embodied carbon emissions and recommend the use of dynamic marginal electricity factors.

The design of ZCEBs remains highly dependent on the local context, e.g., availability of low impact materials, access to clean power or heat, and on-site renewable energy sources (RES). As a consequence, achieving a ZCEB could be close to impossible [13,14]. Aside from technical barriers, legislative, cultural and financial barriers have also been revealed in other countries, such as the UK [15]. Education and sensibilization aiming at applying Sustainable Development Goals in professional practice [16] are important, as well as combining qualitative and quantitative methodologies [17].

Life cycle assessment has been applied to buildings for a long time, and several reviews highlight the profusion of methods, data and accessible tools [18–21]. Some authors have even specifically reviewed consequential LCA in the building sector [21], which has been considered the more relevant methodological approach in an eco-design context [22]. The possibility of evaluating a consistent set of environmental indicators allows progress toward zero GhG emission building without degrading other environmental problems. It is mostly used in a comparative way, although recent efforts have been made to progress toward an "absolute" environmental evaluation [23,24], based upon the planetary boundary concept initially developed by Röckström and Steffen [3,4]. Combination with optimization strategies is recent and so is combination with the planetary boundary concept [25]. Zero emission buildings and districts are not always evaluated through a life cycle perspective, as explained by Brozovsy et al. [11].

Using wood or other bio-based materials is seen as one efficient solution to decrease embodied GhG emissions [26,27] and progress toward a circular economy [28]. Accounting for biogenic carbon is still a vivid debate among LCA researchers and practitioners, as various strategies coexist [29–31], and none are fully consensual. Some methods go up to complex modelling [32] integrating e.g., rotation period [33] or forestry carbon budget [34] but are not fully operational yet. Proper management of existing forests and forest landscape restoration (FLR) can be a relevant means for carbon storage and timber production [35].

Progress has also been made in decarbonizing building materials (e.g., cement, steel) through emission reduction and carbon capture technologies [36–38]. Despite higher costs, carbon capture can be made operational through economical circular CO_2 recovery [39], which would ease the achievement of zero carbon buildings.

Based upon previous works addressing zero carbon and energy efficiency objectives, assessment methods, design approaches and technical aspects, this paper attempts to answer the following research question: is it technically feasible to reach a net zero GhG emission balance in a building over its life cycle, and which techniques need to be implemented towards this objective? The available solutions and existing challenges are analyzed. The possibility of offsetting remaining emissions by carbon capture and storage (CCS) or soil and tree sequestration is explored. The method aims to pave the way towards planetary-boundary compliant buildings, starting with climate change and net zero emissions buildings in a case study. The order of magnitude of emissions offset in the case of a residential building, typical of new construction in France, is evaluated through an original prospective and consequential approach. Other types of buildings can be studied by applying the same methodology.

This article is structured as follows: first the method is presented, then the case study, including the improvement possibilities of the building envelope (insulation and windows), the choice of materials (structure, inertia and insulation), and the choice of equipment (heat pump, solar collectors, etc.). The results for the energy and environmental assessment of the actual vs improved building are then presented and discussed in a separate part. Description of the methods includes the energy simulation procedure as well as the life cycle assessment framework and hypothesis. The results include the analysis of the building's emissions as well as possible offsetting to achieve a net zero emissions balance at the building and further at the national scale. Sensitivity to data quality and uncertainty is explored in a specific discussion section.

2. Methods

2.1. Methodology Overview

The steps followed for the study are summed up in Figure 1 below. The study was carried out on a low energy gas-heated residential building that was built in 2016 in France. A first assessment is performed on the actual building. Then, alternative design options are studied using energy simulation and life cycle assessment (LCA) in order to evaluate the potential for reducing emissions by optimizing the building (architectural, technological and behavioral choices). The remaining GhG emissions to be offset are then quantified in order to derive the required amount of CO_2 to be captured and the feasibility of offsetting by, e.g., tree planting, as well as the possible obstacles to such implementation. Finally, a top-down approach is performed at the level of the French residential building stock to highlight the order of magnitude of emissions to be offset from a carbon-neutral perspective for the sector in 2050.

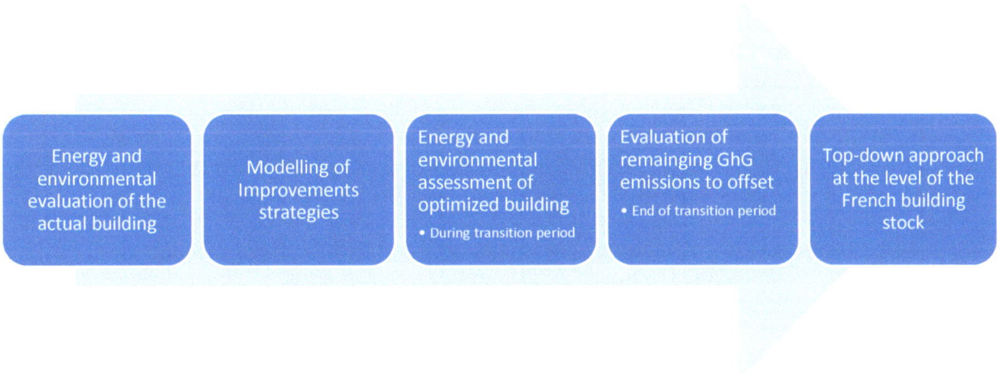

Figure 1. General overview of the methodology.

2.2. Building Energy Simulation

Energy performance is studied using the dynamic thermal simulation tool Pleiades STD Comfie [40]. Heating needs and consumption of the building are evaluated during a typical year with hourly resolution, based on thermal characteristics of envelope and systems, the site (climatic data, near and distant shading), occupancy scenarios (temperature set-point, internal heat gains corresponding to electricity consumption, domestic hot water (DHW), occupancy, etc.). The model is based on the concept of a thermal zone, a subset of the building considered with a homogeneous operating temperature. A finite volume discretization mesh is used. For each zone, the walls are divided into nodes that are sufficiently fine to be considered at a homogeneous temperature and an additional node corresponding to the air volume, furniture and light interior partitions. A heat balance is applied to each node, which can be represented at the zone level by a continuous and invariant linear system.

A modal reduction method is applied to each zone model to reduce the computation time. The reduced matrix systems of the zones are grouped by a coupling procedure. The outputs at each time step are calculated as a function of the indoor (heat gains from occupants and equipment) and outdoor (outdoor temperature, solar radiation) driving forces of the building. Non-linear phenomena (ventilation) or variable parameters (additional resistance due to shutters) are taken into account by correcting the driving force vector.

Model reliability was evaluated by comparison with real data [41] and by the international BESTEST procedure for numerical comparison of reference models [42,43].

2.3. *Life Cycle Assessment*

2.3.1. Tools and Database

Pleiades LCA Equer is used for the life cycle assessment according to the ISO 14040 and 14044 standards [44,45], allowing the quantification of the environmental impacts of a building over its life cycle according to multiple indicators. The Equer database provides information on the environmental impacts corresponding to a functional unit of a product, process or service according to several indicators. It is created using the Brightway2 framework [46] and the ecoinvent database [47,48] version 3.8 using a wide range of life cycle impact assessment methods. Unit process data are contextualized to the French context (e.g., regarding electricity production). The reliability of Pleiades LCA Equer has been studied by inter-comparison with other software in several research projects. The results showed good overall reproducibility, but discrepancies can arise from inventory data sources and methodological differences, e.g., allocation and accounting for biogenic carbon.

2.3.2. Main Assumptions

The functional unit considered for the case study is 1 m^2 of an apartment building housing 0.04 occupants per m^2 over one year, according to the occupancy scenarios shown in Table 1. A lifespan of 100 years is considered for the building (10 years for building finishes, 20 years for equipment, 25 years for PV modules and 30 years for windows). LCA is carried out under the conservative assumption of identical replacement of an element at the end of its lifespan.

An hourly resolution model is used for the electricity production mix, considering a consequential LCA approach. This approach is appropriate for buildings exporting electricity to the grid (photovoltaic generation) as it considers the complex interaction of the building with the grid, assuming that exported electricity avoids production by marginal generation technologies. Prospective scenarios from RTE (French electricity Transmission System Operator) and ADEME (French environmental agency) were considered for 2025, 2035 and 2050 [49]. To represent a 100-year life span, the 2025 mix is considered for 5 years, then 25 years for 2035 and 70 years for 2050. This calculation therefore corresponds to a transition period, and the indicators expressed per year correspond to a yearly average of the impacts over the building life cycle. The energy simulation results were used to evaluate the heating load and thermal comfort. In addition to the 40 L of hot water consumption, an average cold water consumption of 100 L/person/day is considered as well as wastewater treatment. The transport of occupants and domestic waste are not considered.

The end of life considered is the recycling of metals, photovoltaic systems and recyclable materials (e.g., concrete is crushed to produce aggregates). Plastics are incinerated and biobased materials are treated at the end of life so that biogenic carbon can be stored for a very long time. The rest are considered inert waste and sent to landfills.

2.3.3. Environmental Indicators

Because this article focuses on GhG emissions, the climate change indicator is the main focus. It is evaluated using the Environmental footprint v3.0 method developed by the JRC [50]. But damage indicators on human health, ecosystems and resources are also evaluated according to the Recipe 2016 method [51].

2.3.4. Consideration of Biogenic Carbon

In the EQUER method, negative biogenic CO_2 emissions are accounted for in the production stage if a new tree is growing, which is the case for wood from certified forests. But if the wood stems from non-certified forests, the same amount of carbon is stored in the building as if it were stored in the forest. Therefore, no carbon fixation is considered ("0" instead of "−1" according to the notation of EN 15804 standard [52]). At the end of life, the quantity of biogenic CO_2 is emitted if the wood is incinerated but not if the wood is landfilled or recycled (see Figure 2). Landfilling can delay emissions for a very long time, according to [53].

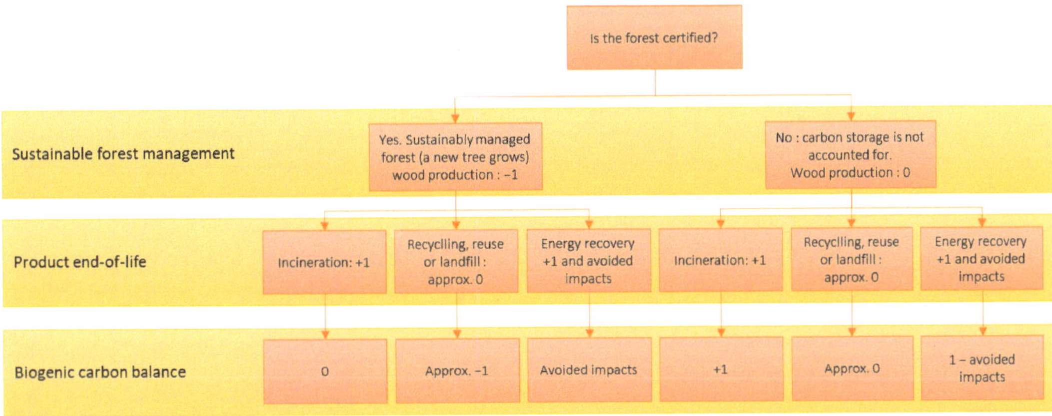

Figure 2. Biogenic carbon balance over the life cycle of wood (lab recherche environnement VINCI I PARISTECH).

2.4. Case Study Presentation

The residence Les roches blanches, located near Chambéry (Savoie, France), is composed of two low-energy apartment blocks built in 2016, each with 4 floors and 17 flats of different sizes (Figure 3). The total living area is 2414 m^2. The buildings have a concrete structure with external insulation (18 cm of rock wool on the walls, 30 cm on the sloped roofs and 30 cm of polyurethane on the flat roofs) and low emissivity double glazed windows. Space heating and domestic hot water (DHW) production are provided by a gas boiler. Ventilation is provided by a humidity-sensitive double flow ventilation system (exchanger efficiency: 80%). Climatic data correspond to a typical year in the region (Macon, France).

Figure 3. Residence Les Roches Blanches, source: Jean Paul Faure Architect.

The considered scenarios of temperature set points, occupancy, domestic hot water (DHW) consumption and heat gains corresponding to specific electricity consumption are defined in Table 1.

Table 1. Occupancy scenarios.

Category	Scenario
Heating temperature set point	20 °C (constant over the year)
Occupancy	Hourly scenario based on a stochastic model of occupancy developed by [54]
Internal gains	Hourly scenario based on a stochastic model of occupancy developed by [54]
Domestic hot water	40 L/day/person at 55 °C

2.5. *Improvement of the Building*

Starting from the actual building, an improved building model has been derived in order to evaluate a potential reduction of GhG emissions. Three main elements are considered: the structure and envelope of the building, heating and ventilation equipment, and the renewable energy system. The principle is first to decrease material and energy needs, then to improve energy efficiency, and finally to cover energy needs as much as possible through renewable production. Each improvement is evaluated using the energy simulation and life cycle assessment tools presented above.

The concrete structure of the actual building was replaced by timber frames (walls and roofs), and low carbon concrete was used for the foundation as well as the suspended floor. The intermediate floors remained in low-carbon concrete in order to add thermal mass to the wooden structure and improve summer comfort. A thin layer of raw earth was put on the walls and roofs for the same purpose. The insulation of the wooden walls and roof is made of 23.5 cm wood wool. The wood used in the construction is assumed to be grown in sustainably managed forests. The gas boiler for heating and domestic hot water (DHW) has been replaced by a geothermal heat pump (cop: 3.5 for heating; cop: 2.7 for DHW). Solar thermal collectors for DHW have been installed (140 m^2) providing most of the needs, complemented with the heat pump backup. The heat exchanger efficiency of the ventilation system has been increased from 80 to 85% in order to reduce heat losses.

Double glazing is replaced with triple glazing, except on the south facades in order to improve the insulation while providing high solar gains. Night ventilation by window opening is considered to improve summer comfort and blinds were installed with 80% reduction of solar factor during the summer on the parts most exposed to overheating. A 176 kWp photovoltaic system was set up on the roofs and southern external facades of the building in order to offset the carbon emissions of the electricity consumption (heating and DHW backup, lighting, ventilation and domestic appliances), taking into account the electricity production mix with higher emissions in winter than in summer.

3. Results

3.1. *Energy Simulation of the Actual Building*

The results are presented in Table 2. Areal ratios are provided per m^2 of net heated area.

Table 2. Actual building energy simulation results (annual balance).

Category	Equipment	Areal Ratio (kWh/m^2)
Heating needs	Gas	15
Energy needed for DHW	Gas	34
Electricity use	Grid	26
Electricity use for ventilation	CMV [1]	1

[1] Controlled Mechanical Ventilation of the building: CMV (0.45 ach) + air renewal (6 ach) by opening windows at night in summer if indoor temperature > 22 °C and outdoor temperature < indoor.

3.2. Life Cycle Assessment of the Actual Building

The environmental impacts in terms of greenhouse gas (GhG) emissions and damage indicators (human health and ecosystems) obtained for the base case (actual building) are given in Table 3, expressed per m^2 of net building area and per year so that they can be compared with benchmark references.

Table 3. GhG emissions and damage indicators of the actual building.

Impact	Unit	Construction	Use	Renovation	End-of-Life	Total
Climate change (EF v3.0)	$kgCO_2$ eq/m^2/yr	3.1×10^{0}	2.7×10^{1}	2.5×10^{0}	1.2×10^{-1}	3.3×10^{-1}
Damage to ecosystem (ReCiPe2016)	specie.yr/m^2/yr	1.6×10^{-8}	1.0×10^{-7}	2.5×10^{-8}	4.7×10^{-10}	1.4×10^{-7}
Damage to human health (ReCiPe2016)	DALY/m^2/yr	8.7×10^{-6}	3.9×10^{-5}	1.8×10^{-5}	1.7×10^{-7}	6.7×10^{-5}
Damage to resources (ReCiPe2016)	USB/m^2/yr	2.0×10^{-1}	2.8×10^{0}	2.8×10^{-1}	8.5×10^{-3}	3.3×10^{0}

The total climate change impact is around 33 kg CO_2 eq/m^2/year. By comparison, these emissions vary between 10 kg CO_2 eq/m^2/year (passive building with a photovoltaic system) and 160 kg CO_2 eq/m^2/year (uninsulated old building heated with gas) in a benchmark study performed in the frame of International Energy Agency Annex 72 [55]. The largest emissions correspond to the use stage, as can be seen in Figure 4.

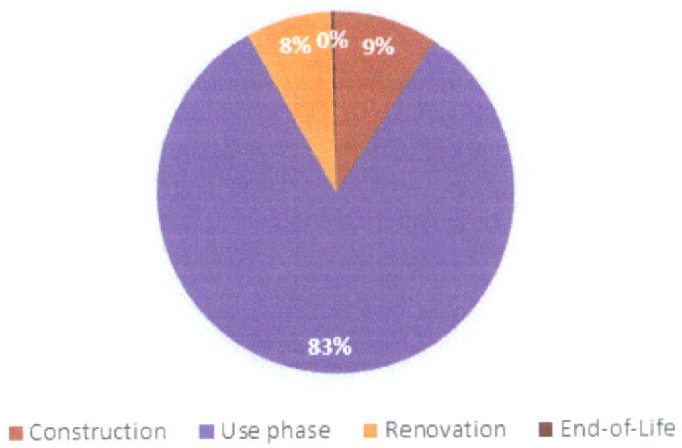

Figure 4. Contributions of the life cycle stages in GhG emissions (base case).

The objective of this case study is to investigate the feasibility of achieving net zero carbon emissions through eco-design measures, such as the use of bio-based materials, minimization of energy requirements and the use of low-impact energy sources, as well as the sequestration of the remaining emissions.

3.3. Energy Simulation of the Improved Building

Energy requirements were minimized, as shown in Table 4.

Summer comfort has also been studied. The increased thermal mass of the building, night ventilation and blinds have improved the comfort level, despite an overall lighter timber frame structure compared to the actual building. Aside from its importance for building quality, assessment of thermal comfort is crucial to prevent future usage of active cooling, which could downgrade the overall environmental performance of the building because of increased energy consumption and additional equipment. Figure 5 illustrates the effect of night ventilation, which allows, thanks to the thermal mass of the building and its good insulation to keep indoor temperature below the external one during hot periods. The choice of thicknesses of materials with high thermal mass (concrete floors, raw earth) in the improved building was made to maintain the annual temperature between 20 and 27 °C, with a maximum discomfort rate of 1% (percentage of hours above 27 °C or below 20 °C). According to these thermal simulation results, improvements proposed to reduce GhG emissions would not reduce the thermal comfort level of the building.

Table 4. Improved building energy simulation results (annual balance).

Category	Equipment	Areal Ratio (kWh/m^2)
Heating needs	Heat pump 30 kW	5
Energy needed for DHW	solar thermal collector (140 m^2) + electric back-up	13
Electricity use	Network	26
Electricity consumption for ventilation	CMV	1
Photovoltaic electricity production	PV panels	58

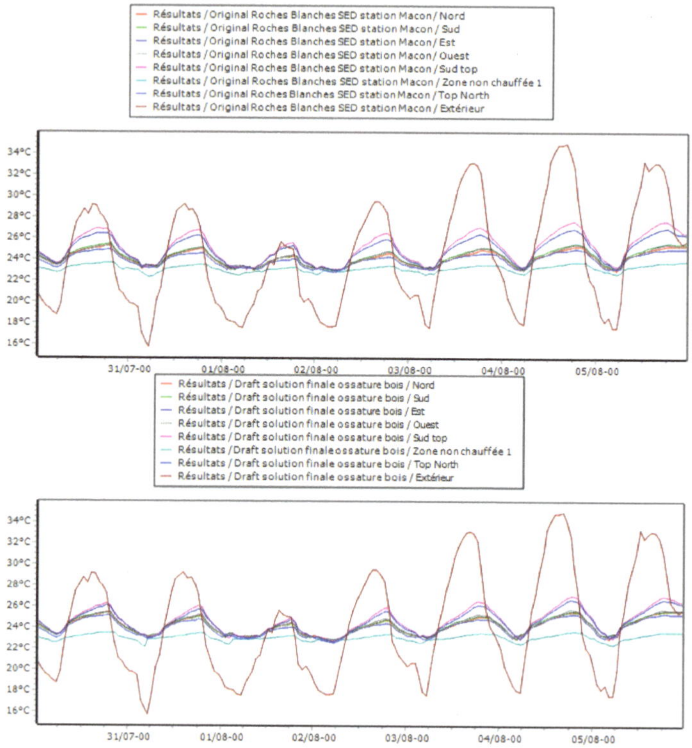

Figure 5. Temperature profiles during the hottest week (simulation results) for the actual building (**upper figure**) and the improved building (**bottom figure**).

3.4. *Life Cycle Assessment of the Improved Building*

The results of the LCA study show a potential GhG emission reduction of up to 97% (Table 5) using bio-based materials, minimizing heating needs and using low carbon energy sources through the implementation of appropriate equipment. The choice of a timber frame structure reduces construction emissions from 3.01 to −0.41 kg CO_2 eq/m^2/year.

Table 5. GhG emissions and damage indicators of the improved building.

Impact	Unit	Construction	Use	Renovation	End-of-Life	Total
Climate change (EF v3.0)	$kgCO_2$ eq/m^2/yr	-4.09×10^{-1}	-2.27×10^{0}	5.06×10^{0}	$1.49 \times 10-1$	2.53×10^{0}
Damage to ecosystem (ReCiPe2016)	specie.yr/m^2/yr	2.63×10^{-8}	-8.42×10^{-10}	3.97×10^{-8}	5.43×10^{-10}	6.57×10^{-8}
Damage to human health (ReCiPe2016)	DALY/m^2/yr	6.32×10^{-6}	-9.42×10^{-7}	2.65×10^{-5}	1.82×10^{-7}	3.20×10^{-5}
Damage to resources (ReCiPe2016)	USB/m^2/yr	1.94×10^{-1}	-2.18×10^{-1}	4.63×10^{-1}	7.23×10^{-3}	4.46×10^{-1}

A considerable reduction in operational emissions is achieved, made possible by replacing the gas boiler and using renewable energy sources (solar thermal and photovoltaic). The photovoltaic system is oversized in relation to the self-consumption needs to account for the difference between winter and summer grid emissions. Avoided impacts considering marginal production are accounted for, but they become zero when the national electricity grid mix is 100% renewable. The 100% renewable electricity mix considered is taken from the ADEME prospective study [56] and is composed of 63% wind power, 17% PV, 13% hydraulic and 7% thermal REN (waste incineration, biomass and biogas).

Emissions during renovation appear to be the most significant because of the plumbing, electricity cables and other equipment (ventilation, PV system, etc.) that is replaced several times over the lifetime of the building.

All the above measures have allowed a considerable reduction of the total GhG emissions of the building: more than 90% of the emissions have been cancelled compared to the actual building (see Figure 6). But this calculation corresponds to a transition period. It is also useful to evaluate building performance after this transition. In this case, when the electricity grid production is decarbonized, the reduction of emissions becomes 84% (see Figure 7) because there is no avoided impact from PV production anymore (a 100% renewable grid was considered in this scenario). However, GhG emissions due to construction products like plumbing, electrical installation, and equipment (solar collectors, heat pumps, ventilation, etc.) increase the emissions in renovation and make the total balance positive with a higher value than the actual building due to the effect of equipment replacement. The whole life cycle GhG emissions would then be around 5.6 kg CO_2 eq/m^2/year after the transition period.

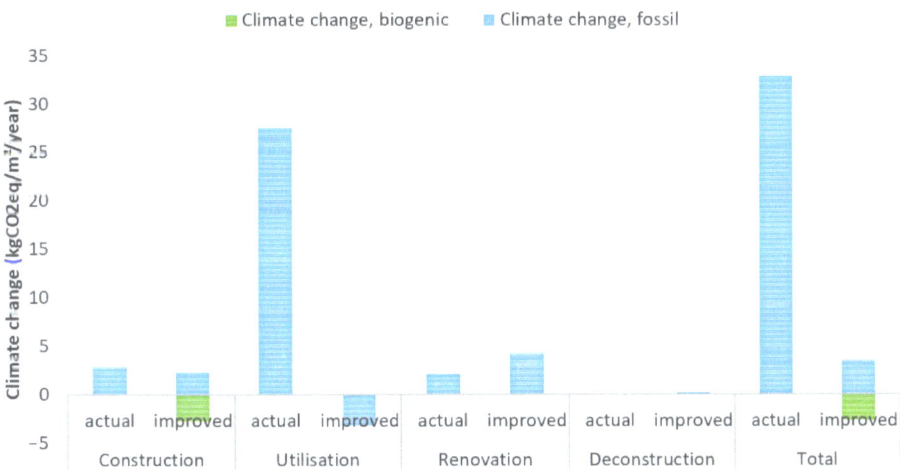

Figure 6. Yearly average life cycle GhG emissions of the actual and improved buildings during the transition period.

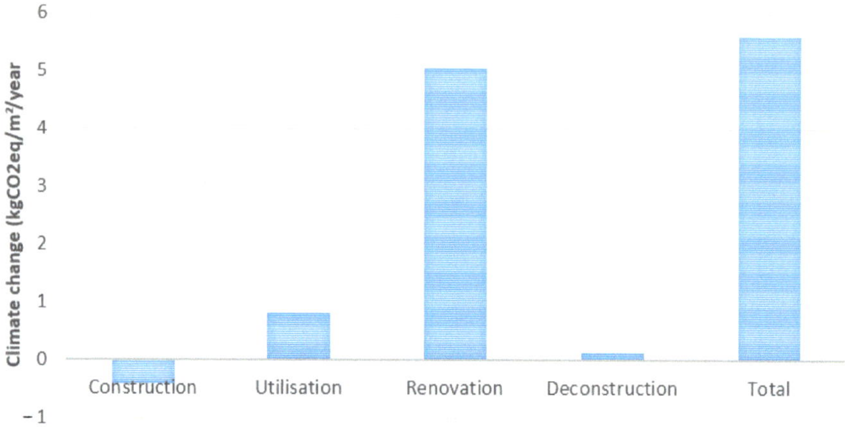

Figure 7. Life cycle GhG emissions of the improved building with a decarbonised grid scenario (after the transition period).

3.5. Compensation by Forest Sequestration and Extrapolation to the Dwelling Stock

In order to answer the research question regarding the feasibility of reaching a net zero GhG emission balance in a building, a top-down approach was performed. It consists of estimating a carbon budget corresponding to sequestration in forests, which can be expressed per m^2 considering the annual new construction area. The GhG emissions of the improved building can be compared to this carbon budget, allowing us to check if the climate planetary boundary is respected.

There are numerous possibilities for offsetting these emissions by sequestration, including storage in natural ecosystems (vegetation, soil, aquatic environments). Forest sequestration gives the possibility of replanting on the same surface and using wood as a low carbon construction material. Other means of in situ sequestration may also be of great interest, such as vertical vegetation systems (VGS) because of the limitation of external sequestration surfaces (forests, meadows, wetlands, etc.) and the possibility of optimizing the use of unused building surfaces (facades, roofs, etc.) allowing carbon sequestration while providing other positive externalities (e.g., well-being of occupants, cooling of contact surfaces, etc.). The corresponding biomass sequestration potential has been estimated by various studies [57–59] and varies between 0.44 and 3.18 kg CO_2 eq/m^2/year depending on climate, vegetation, life cycle treatment, etc.

From the estimate of the carbon sequestration capacity of European forests given by Lelarge and Birot [60], which is also found in the data of the National Forest Inventory in France [61], we deduce a storage of 1680 kg C/ha/year on average, which corresponds to 6160 kg CO_2 eq/ha/year in the biomass and forest soils. For this improved building with 2414 m^2 of living space, the balance to be compensated for after the transition period is 13.5 t CO_2 eq/year, which would correspond to the equivalent of 2.2 ha of European forest corresponding to around 9 m^2 of forest per m^2 of living space.

At the scale of the French territory, the number of dwellings built annually is estimated at 390,300 from 2000 to 2021, with an average living area of 90.9 m^2 (French Data and Statistical Studies Department [62]. The annual growth of French forests is estimated at 40,000 ha/year according to [63], which corresponds to a 246,000 ton CO_2 eq./year carbon budget considering the carbon sequestration estimate given above. If this whole budget could be allocated to compensate for new construction impacts, this would correspond to 7 kg CO_2 eq/m^2/year which is not much more than the improved building emissions. This means that our construction standards must be radically transformed, and that other compensation solutions must be found because new construction (which includes tertiary buildings) is yearly only 1% of the existing building stock, which produces much higher emissions.

4. Discussion

4.1. Vertical Vegetation Systems

Another way to further reduce emissions in buildings is to use vertical vegetation systems (VGS). Marchi et al. [58] described how they operate using a 5-step model to achieve real carbon sequestration in the soil. A potential of 0.44 to 3.18 kg CO_2 eq/m^2/year was obtained using this model. The process is as follows: plants growing in the VGS absorb CO_2 and use it to form biomass (step 1). Each year, a percentage of the plants in a VGS must be replaced (step 2). The removed plants are sent to a composting facility (step 3). There, some of the carbon is released in the form of CO_2 (step 4). The compost is then applied to agricultural soils, where some of the remaining carbon is absorbed by soil bacteria and eventually sequestered in the soil (step 5). The studied building has a total exterior opaque and unused facade area of 2069 m^2, which gives a maximum sequestration potential of 910 to 6580 kg CO_2 eq/year when fully vegetated. This sequestration does not take into account all the emissions related to the life cycle of the facility but only those from the biomass, for which it would be necessary to consider the emissions due to the fossil fuels and electricity needed to transport the plant residues to the composting facility, the management of the composting facility, and the transport and distribution of the compost produced to the agricultural soil.

Pulselli et al. [57] analyzed a case study considering the production chain up to the installation on a building facade as well as the maintenance of the VGS system and found that these emissions over the life cycle of the installation (here 25 years) can be equivalent to those sequestered by biomass according to the model of Marchi [58] and that it is necessary to take a local and responsible approach to the whole life cycle chain (emissions related to structure, transport, water and plant nutrients) in order not to release as much as the biomass sequestration of the VGS.

4.2. Forest Management

Sustainable forest management ensures a replanting of trees, but deforestation or overuse risk call for at least a national resource management plan of forests to ensure a sustainable use of the resources, improve ecosystem services and forest resilience [64]. Without biogenic carbon storage, the climate change impact of the improved building after the transition period almost reaches 8.5 kg CO_2 eq/m^2/yr, a 50% increase (see Figure 8, Optimized without biogenic carbon, noted wo Cbio). Moreover, the mitigation potential of forests can be hindered by climate change effects: increasing drought, fires, pest and disease outbreaks, wind storm [65].

4.3. Multi-Criteria Analysis

Beyond GhG emissions, it is important to also consider other environmental impacts in order to avoid impact shifting. Damage indicators were evaluated in this study, showing an important reduction by decarbonization measures for damage to health, damage to ecosystem and resource depletion (see Figure 8). These indicators are uncertain, and such an evaluation should be further improved by ongoing work regarding impact assessment models.

Damage to ecosystems, human health and resources are also decreasing but in a smaller proportion than climate change. Indeed, the climate contribution is only 20.6% and 14.2% for ecosystem and health, respectively, for the optimized building against 64.4% and 45.5% for the actual building. Resource impacts are largely dominated by fossil fuels (over 80% for all cases). Increasing the use of wood increases land use impacts and increasing the use of equipment increases mineral and metallic resource use. To prevent impact shifting, a multi-criteria analysis must be performed.

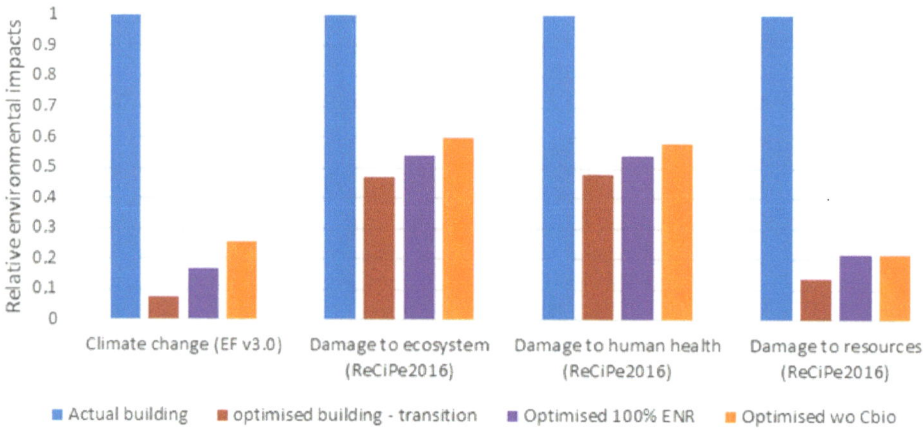

Figure 8. Environmental impacts of the optimized building relative to the actual building.

4.4. Circular Economy

Going further in reducing building impacts and easing the carbon offsetting effort would induce additional contributors that were previously considered minor contributors, such as equipment, electronics and material replacement. It would therefore be useful to further investigate reuse and reconditioning of old equipment, longer lifetime and other known circular economy levers. A first sensitivity analysis has been performed by increasing all lifetime of equipment and finishes to 30 years, and a second analysis considered French recycled copper for electronic manufacturing and plumbing. The results are presented in Figure 9 and show a great potential for continuing to decrease environmental impacts of the building sector. This also calls for a better evaluation of the composition and quantity of contributors such as equipment, electric and electronics, and plumbing. They have been proven to become an important contributor but were previously neglected [66] and as shown in this study, they will be the next lever to address in order to progress toward zero carbon buildings.

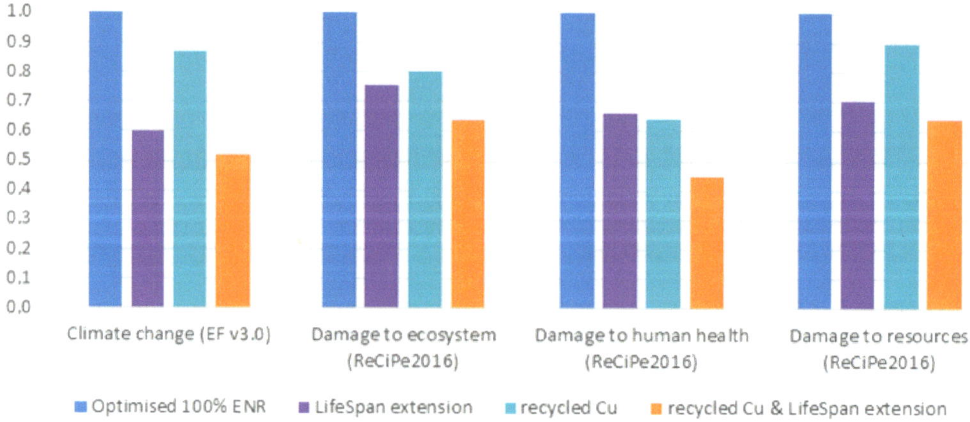

Figure 9. Environmental impacts of the optimized building, sensitivity to circular economy levers.

4.5. Behavioral Changes

Reducing water and electricity consumption through behavioral changes and efficient appliances would also further contribute to decreasing remaining GhG emissions and will have a positive effect on damage to health and ecosystems as well. It is outside the scope of this article, but the importance of behavior in building LCA has long been proven by previous studies, such as in Polster et al. in the 1990's [67].

Maybe in the future, suffering hazardous conditions, humanity should consider the decrease in life comfort, for instance:

- a radical reduction in the number of newly constructed buildings,
- increasing the density of people occupying buildings,
- accepting higher variability of air temperature indoors (using adaptive thermal comfort models in simulations).

4.6. Prospective Uncertainties

The study showed the possibility of reducing the emissions of new apartment buildings from 33 kg CO_2 eq/m^2/year to less than 6 kg CO_2 eq/m^2/year after a transition period. Assuming that the annual growth in forestry and housing stock follows the same trend as in previous years, emissions from new residential buildings would correspond nearly entirely to the possibility of sequestering carbon by forest growth. There is uncertainty about the evolution of these trends in the long term and, therefore, the possibility of long-term sequestration using this method. Annual CO_2 emissions from the existing housing stock were 98 Mt CO_2 eq in 2019, considering only energy related, no embodied emissions [68] for a total of 36.6 million dwellings. To sequester these emissions by forests, it would be necessary to cover about 16 million hectares, i.e., 95% of the French forest area in 2019. These emissions can be considerably reduced, as shown in the case study above, by minimizing heating needs and using RES.

Considering new dwellings to be built, it would be important to also consider the grid decarbonization effect on the manufacture of equipment and materials to be replaced along the life-cycle of the product. This has not been done in this project; thus, GhG emissions from renovation are potentially over-estimated. Such calculations have been made in the past for renewable energy technologies, showing a potentially significant effect on the results [69]. However, the order of magnitude given in this study would probably still hold as the French electricity grid is not carbon intensive. Such an integrated assessment is considered to be an interesting perspective of this work.

4.7. Temptation of Hasty Electrification

Electrification is seen in the building sector as a decarbonization lever, providing that clean power is made accessible. However, massive hasty electrification of uses, anticipating the future provisioning of low-impact electricity, could have an adverse effect and hinder the needed transition of the grid by unreasonably increasing the electricity demand. In this paper, the gas boiler is replaced by a heat pump, thus leading to an increase in electricity consumption. However, a significant effort is made to limit, at their minimum, the electricity needs for heating (additional insulation, heat recovery on ventilation) and DHW (solar panels). The improved building also provides renewable electricity to the grid thanks to the PV system. Even if its production is unlikely to coincide with its consumption, it takes part in the grid transition. This setup is thus considered to be consistent with an efficient decarbonization of the grid by 2050.

4.8. Generalisation of the Results

The case study is specific to the French context, and numerous assumptions and scenarios affect the results. Particularly, construction techniques, architecture, occupants' behavior and prospective aspects (e.g., regarding the evolution of electric system) may differ a lot in other contexts. The presented case study and corresponding results aim to show the possible application of the methodology, which could be used in other countries while adapting data and scenarios appropriately.

5. Conclusions and Perspectives

Limiting global warming to 1.5 °C implies not exceeding the remaining carbon budget, and therefore, designing net zero carbon buildings. An apartment building built in 2016 in France was redesigned in order to check the possibility of reaching this performance level by lowering heating energy needs and implementing bio-sourced materials and renewable energy systems (geothermal heat pump, solar domestic hot water production, and photovoltaic electricity production). GhG emissions were evaluated using life cycle assessment, integrating energy consumption calculated using a building energy simulation. During a transition period, exporting electricity avoids impacts corresponding to standard production on the grid. These avoided impacts decrease and become zero when the grid is decarbonized after the transition. At this date, the building should be net zero emissions, but there remain emissions related, e.g., to the replacement of construction products (e.g., equipment, windows, painting), drinking water production and wastewater treatment. More research is needed to better understand the amount of electric and electronic components in buildings depending on their uses and design options. Circular economy levers on such previously minor contributors (buildings equipment, electronics and plumbing) should be undertaken to further decrease the impacts of buildings.

The possibility of offsetting these emissions is therefore studied, considering sequestration in forests or vegetation systems. A net zero emission level can be achieved if the whole sequestration potential can be used to offset emissions by new construction. But emissions from existing buildings correspond to the potential of the whole French forest area, and the budget should also be shared with other sectors: transport, industry and agriculture. It is therefore needed to radically transform our construction standards, probably also our comfort and way of life standards, and to search for supplementary sequestration techniques.

In perspective, it would be useful to model the building stock using dynamic LCA and to allocate a part of the whole carbon budget to buildings in order to check if a net zero balance can be achieved on a national level.

Author Contributions: Conceptualization, A.H.D. and B.P.; Methodology, A.H.D.; Software, A.H.D. and C.R.; Validation, C.R.; Data curation, C.R.; Writing—original draft, A.H.D.; Writing—review & editing, C.R. and B.P.; Supervision, B.P. All authors have read and agreed to the published version of the manuscript.

Funding: This research received no external funding.

Data Availability Statement: Data are contained within the article.

Acknowledgments: This work was carried out during a research trimester of the engineering cycle at Ecole des Mines de Paris, using energy calculation and life cycle assessment tools that were previously developed in various research programs (European Commission, ADEME, Chair ParisTech VINCI Ecodesign of buildings and infrastructure).

Conflicts of Interest: The authors declare no conflict of interest.

References

1. European Parliament, Council of the European Union. EUR-Lex—32010L0031—EN—EUR-Lex. Available online: https://eur-lex.europa.eu/eli/dir/2010/31/oj (accessed on 9 May 2023).
2. United Nations Framework Convention on Climate Change. The Paris Agreement. 2015. Available online: https://unfccc.int/sites/default/files/english_paris_agreement.pdf (accessed on 11 May 2023).
3. Rockström, J.; Steffen, W.; Noone, K.; Persson, Å.; Chapin, F.S., III; Lambin, E.F.; Lenton, T.M.; Scheffer, M.; Folke, C.; Schellnhuber, H.J.; et al. A safe operating space for humanity. *Nature* **2009**, *461*, 472–475. [CrossRef]
4. Steffen, W.; Richardson, K.; Rockström, J.; Cornell, S.E.; Fetzer, I.; Bennett, E.M.; Biggs, R.; Carpenter, S.R.; De Vries, W.; De Wit, C.A.; et al. Planetary boundaries: Guiding human development on a changing planet. *Science* **2015**, *347*, 1259855. [CrossRef] [PubMed]

5. Richardson, K.; Steffen, W.; Lucht, W.; Bendtsen, J.; Cornell, S.E.; Donges, J.F.; Drüke, M.; Fetzer, I.; Bala, G.; von Bloh, W.; et al. Earth beyond six of nine planetary boundaries. *Sci. Adv.* **2023**, *9*, eadh2458. [CrossRef] [PubMed]
6. Mata, E.; Korpal, A.K.; Cheng, S.H.; Navarro, J.P.J.; Filippidou, F.; Reyna, J.; Wang, R. A map of roadmaps for zero and low energy and carbon buildings worldwide. *Environ. Res. Lett.* **2020**, *15*, 113003. [CrossRef]
7. Huovila, A.; Siikavirta, H.; Rozado, C.A.; Rökman, J.; Tuominen, P.; Paiho, S.; Hedman, A.; Ylén, P. Carbon-neutral cities: Critical review of theory and practice. *J. Clean. Prod.* **2022**, *341*, 130912. [CrossRef]
8. Sartori, I.; Napolitano, A.; Voss, K. Net zero energy buildings: A consistent definition framework. *Energy Build.* **2012**, *48*, 220–232. [CrossRef]
9. Kristjansdottir, T.F.; Heeren, N.; Andresen, I.; Brattebø, H. Comparative emission analysis of low-energy and zero-emission buildings. *Build. Res. Inf.* **2018**, *46*, 367–382. [CrossRef]
10. Lausselet, C.; Borgnes, V.; Brattebø, H. LCA modelling for Zero Emission Neighbourhoods in early stage planning. *Build. Environ.* **2019**, *149*, 379–389. [CrossRef]
11. Brozovsky, J.; Gustavsen, A.; Gaitani, N. Zero emission neighbourhoods and positive energy districts—A state-of-the-art review. *Sustain. Cities Soc.* **2021**, *72*, 103013. [CrossRef]
12. Satola, D.; Balouktsi, M.; Lützkendorf, T.; Wiberg, A.H.; Gustavsen, A. How to define (net) zero greenhouse gas emissions buildings: The results of an international survey as part of IEA EBC annex 72. *Build. Environ.* **2021**, *192*, 107619. [CrossRef]
13. Georges, L.; Haase, M.; Wiberg, A.H.; Kristjansdottir, T.; Risholt, B. Life cycle emissions analysis of two nZEB concepts. *Build. Res. Inf.* **2015**, *43*, 82–93. [CrossRef]
14. Moschetti, R.; Brattebø, H.; Sparrevik, M. Exploring the pathway from zero-energy to zero-emission building solutions: A case study of a Norwegian office building. *Energy Build.* **2019**, *188–189*, 84–97. [CrossRef]
15. Osmani, M.; O'Reilly, A. Feasibility of zero carbon homes in England by 2016: A house builder's perspective. *Build. Environ.* **2009**, *44*, 1917–1924. [CrossRef]
16. Oltra-Badenes, R.; Guerola-Navarro, V.; Gil-Gómez, J.-A.; Botella-Carrubi, D. Design and Implementation of Teaching–Learning Activities Focused on Improving the Knowledge, the Awareness and the Perception of the Relationship between the SDGs and the Future Profession of University Students. *Sustainability* **2023**, *15*, 5324. [CrossRef]
17. Ahmed, N. Developing a tailored tool to assess regenerative development and design within the built environment. *J. Green Build.* **2023**, *18*, 135–166. [CrossRef]
18. Sharma, A.; Saxena, A.; Sethi, M.; Shree, V. Varun Life cycle assessment of buildings: A review. *Renew. Sustain. Energy Rev.* **2011**, *15*, 871–875. [CrossRef]
19. Cabeza, L.F.; Rincón, L.; Vilariño, V.; Pérez, G.; Castell, A. Life cycle assessment (LCA) and life cycle energy analysis (LCEA) of buildings and the building sector: A review. *Renew. Sustain. Energy Rev.* **2014**, *29*, 394–416. [CrossRef]
20. Anand, C.K.; Amor, B. Recent developments, future challenges and new research directions in LCA of buildings: A critical review. *Renew. Sustain. Energy Rev.* **2017**, *67*, 408–416. [CrossRef]
21. Hansen, R.N.; Rasmussen, F.N.; Ryberg, M.; Birgisdóttir, H. A systematic review of consequential LCA on buildings: The perspectives and challenges of applications and inventory modelling. *Int. J. Life Cycle Assess.* **2022**, *28*, 131–145. [CrossRef]
22. Frischknecht, R.; Benetto, E.; Dandres, T.; Heijungs, R.; Roux, C.; Schrijvers, D.; Wernet, G.; Yang, Y.; Messmer, A.; Tschuemperlin, L. LCA and decision making: When and how to use consequential LCA; 62nd LCA forum, Swiss Federal Institute of Technology, Zürich, 9 September 2016. *Int. J. Life Cycle Assess.* **2016**, *22*, 296–301. [CrossRef]
23. Andersen, C.E.; Ohms, P.; Rasmussen, F.N.; Birgisdóttir, H.; Birkved, M.; Hauschild, M.; Ryberg, M. Assessment of absolute environmental sustainability in the built environment. *Build. Environ.* **2020**, *171*, 106633. [CrossRef]
24. Bjoern, A.; Chandrakumar, C.; Boulay, A.-M.; Doka, G.; Fang, K.; Gondran, N.; Hauschild, M.Z.; Kerkhof, A.; King, H.; Margni, M.; et al. Review of life-cycle based methods for absolute environmental sustainability assessment and their applications. *Environ. Res. Lett.* **2020**, *15*, 083001. [CrossRef]
25. Brejnrod, K.N.; Kalbar, P.; Petersen, S.; Birkved, M. The absolute environmental performance of buildings. *Build. Environ.* **2017**, *119*, 87–98. [CrossRef]
26. Ramage, M.H.; Burridge, H.; Busse-Wicher, M.; Fereday, G.; Reynolds, T.; Shah, D.U.; Wu, G.; Yu, L.; Fleming, P.; Densley-Tingley, D.; et al. The wood from the trees: The use of timber in construction. *Renew. Sustain. Energy Rev.* **2017**, *68*, 333–359. [CrossRef]
27. Carcassi, O.B.; Habert, G.; Malighetti, L.E.; Pittau, F. Material Diets for Climate-Neutral Construction. *Environ. Sci. Technol.* **2022**, *56*, 5213–5223. [CrossRef] [PubMed]
28. Dahiya, S.; Katakojwala, R.; Ramakrishna, S.; Venkata Mohan, S. Biobased products and life cycle assessment in the context of circular economy and sustainability. *Mat. Circ. Econ.* **2020**, *2*, 7. [CrossRef]
29. Matthews, R.; Sokka, L.; Soimakallio, S.; Mortimer, N.; Rix, J.; Schelhaas, M.; Jenkins, T.; Hogan, G.; Mackie, E.; Morris, A.; et al. Review of Literature on Biogenic Carbon and Life Cycle Assessment of Forest Bioenergy. Forest Research. 2014. Available online: http://www.energy-wsp.org/media/upload/veipraktiki34.pdf (accessed on 3 November 2015).
30. Hoxha, E.; Passer, A.; Saade, M.R.M.; Trigaux, D.; Shuttleworth, A.; Pittau, F.; Allacker, K.; Habert, G. Biogenic carbon in buildings: A critical overview of LCA methods. *Build. Cities* **2020**, *1*, 504–524. [CrossRef]
31. Andersen, C.E.; Rasmussen, F.N.; Habert, G.; Birgisdóttir, H. Embodied GHG Emissions of Wooden Buildings—Challenges of Biogenic Carbon Accounting in Current LCA Methods. *Front. Built Environ.* **2021**, *7*, 120. [CrossRef]

32. Tellnes, L.; Ganne-Chedeville, C.; Dias, A.; Dolezal, F.; Hill, C.; Escamilla, E.Z. Comparative assessment for biogenic carbon accounting methods in carbon footprint of products: A review study for construction materials based on forest products. *iForest Biogeosci. For.* **2017**, *10*, 815–823. [CrossRef]
33. Cherubini, F.; Guest, G.; Strømman, A.H. Application of probability distributions to the modeling of biogenic CO_2 fluxes in life cycle assessment. *GCB Bioenergy* **2012**, *4*, 784–798. [CrossRef]
34. Head, M.; Bernier, P.; Levasseur, A.; Beauregard, R.; Margni, M. Forestry carbon budget models to improve biogenic carbon accounting in life cycle assessment. *J. Clean. Prod.* **2019**, *213*, 289–299. [CrossRef]
35. Seymour, F. Seeing the Forests as well as the (Trillion) Trees in Corporate Climate Strategies. *One Earth* **2020**, *2*, 390–393. [CrossRef]
36. Habert, G.; Miller, S.A.; John, V.M.; Provis, J.L.; Favier, A.; Horvath, A.; Scrivener, K.L. Environmental impacts and decarbonization strategies in the cement and concrete industries. *Nat. Rev. Earth Environ.* **2020**, *1*, 559–573. [CrossRef]
37. Liu, X.; Peng, R.; Bai, C.; Chi, Y.; Li, H.; Guo, P. Technological roadmap towards optimal decarbonization development of China's iron and steel industry. *Sci. Total Environ.* **2022**, *850*, 157701. [CrossRef] [PubMed]
38. Tautorat, P.; Lalin, B.; Schmidt, T.S.; Steffen, B. Directions of innovation for the decarbonization of cement and steel production—A topic modeling-based analysis. *J. Clean. Prod.* **2023**, *407*, 137055. [CrossRef]
39. Chai, S.Y.W.; Ngu, L.H.; How, B.S.; Chin, M.Y.; Abdouka, K.; Adini, M.J.B.A.; Kassim, A.M. Review of CO_2 capture in construction-related industry and their utilization. *Int. J. Greenh. Gas Control.* **2022**, *119*, 103727. [CrossRef]
40. Peuportier, B.; Sommereux, I.B. Simulation tool with its expert interface for the thermal design of multizone buildings. *Int. J. Sol. Energy* **1990**, *8*, 109–120. [CrossRef]
41. Munaretto, F.; Recht, T.; Schalbart, P.; Peuportier, B. Empirical validation of different internal superficial heat transfer models on a full-scale passive house. *J. Build. Perform. Simul.* **2018**, *11*, 261–282. [CrossRef]
42. Judkoff, R.; Neymark, J. *International Energy Agency Building Energy Simulation Test (BESTEST) and Diagnostic Method*; NREL/TP-472-6231; National Renewable Energy Lab. (NREL): Golden, CO, USA, 1995. [CrossRef]
43. Judkoff, R.; Neymark, J. Twenty Years On!: Updating the IEA BESTEST Building Thermal Fabric Test Cases for ASHRAE Standard 140. In Proceedings of the BS 2013: 13th Conference of the International Building Performance Simulation Association, Chambery, France, 26–28 August 2013; pp. 63–70.
44. *ISO 14040:2006*; Environmental Management—Life Cycle Assessment—Principles and Framework. ISO: Geneva, Switzerland, 2006.
45. *ISO 14044:2006*; Environmental Management—Life Cycle Assessment—Requirements and Guidelines. ISO: Geneva, Switzerland, 2006.
46. Mutel, C. Brightway: An open source framework for Life Cycle Assessment. *J. Open Source Softw.* **2017**, *2*, 236. [CrossRef]
47. Frischknecht, R.; Jungbluth, N.; Althaus, H.-J.; Doka, G.; Dones, R.; Heck, T.; Hellweg, S.; Hischier, R.; Nemecek, T.; Rebitzer, G.; et al. Overview and Methodology. *Ecoinvent. Rep.* **2007**. Available online: http://www.ecoinvent.org/fileadmin/documents/en/01_OverviewAndMethodology.pdf (accessed on 20 June 2013).
48. Wernet, G.; Bauer, C.; Steubing, B.; Reinhard, J.; Moreno-Ruiz, E.; Weidema, B. The ecoinvent database version 3 (part I): Overview and methodology. *Int. J. Life Cycle Assess.* **2016**, *21*, 1218–1230. [CrossRef]
49. Frapin, M.; Roux, C.; Assoumou, E.; Peuportier, B. Modelling long-term and short-term temporal variation and uncertainty of electricity production in the life cycle assessment of buildings. *Appl. Energy* **2021**, *307*, 118141. [CrossRef]
50. European Commission, Joint Research Centre. *Environmental Footprint: Update of Life Cycle Impact Assessment Methods: Ecotoxicity Freshwater, Human Toxicity Cancer, and Non Cancer*; Publications Office of the European Union: Luxembourg, 2020; Available online: https://data.europa.eu/doi/10.2760/300987 (accessed on 3 September 2021).
51. Huijbregts, M.A.J.; Steinmann, Z.J.N.; Elshout, P.M.F.; Stam, G.; Verones, F.; Vieira, M.; Zijp, M.; Hollander, A.; van Zelm, R. ReCiPe2016: A harmonised life cycle impact assessment method at midpoint and endpoint level. *Int. J. Life Cycle Assess.* **2017**, *22*, 138–147. [CrossRef]
52. *EN 15804+A2*; Sustainability of Construction Works—Environmental Product Declarations—Core Rules for the Product Category of Construction Products. EU: Brussels, Belgium, 2019.
53. Levasseur, A.; Lesage, P.; Margni, M.; Samson, R. Biogenic Carbon and Temporary Storage Addressed with Dynamic Life Cycle Assessment. *J. Ind. Ecol.* **2013**, *17*, 117–128. [CrossRef]
54. Vorger, E.; Schalbart, P.; Peuportier, B. Integration of a comprehensive stochastic model of occupancy in building simulation to study how inhabitants influence energy performance. In Proceedings of the 30th International PLEA 2014 Conference, Ahmedabad, India, 16–18 December 2014; p. 8.
55. Wurtz, A.; Peuportier, B. Application of the Life Cycle Assessment to a Building Sample for in Order to Helping in Projects Evaluation. In Proceedings of the Climamed Conference, Lisbon, Portugal, 11 May 2021. Available online: https://hal.archives-ouvertes.fr/hal-03194021 (accessed on 8 March 2022).
56. ADEME. Mix Électrique 100% Renouvelable? Analyses et Optimisations. 2016. Available online: https://librairie.ademe.fr/cadic/2889/mix-electrique-rapport-2015.pdf (accessed on 14 June 2023).
57. Pulselli, R.M.; Saladini, F.; Neri, E.; Bastianoni, S. A comprehensive lifecycle evaluation of vertical greenery systems based on systemic indicators. In Proceedings of the Sustainable City 2014, 9th International Conference on Urban Regeneration and Sustainability is a Civil Engineering, Siena, Italy, 23–25 September 2014; pp. 1017–1024.
58. Marchi, M.; Pulselli, R.M.; Marchettini, N.; Pulselli, F.M.; Bastianoni, S. Carbon dioxide sequestration model of a vertical greenery system. *Ecol. Model.* **2015**, *306*, 46–56. [CrossRef]

59. Rowe, T.; Poppe, J.; Buyle, M.; Belmans, B.; Audenaert, A. Is the sustainability potential of vertical greening systems deeply rooted? Establishing uniform outlines for environmental impact assessment of VGS. *Renew. Sustain. Energy Rev.* **2022**, *162*, 112414. [CrossRef]
60. Lelarge, K.; Birot, J. *Etude de la Séquestration de Carbone Par Les éCosystèmes de la Réserve Naturelle du Pinail*; GEREPI: Vouneuil-sur-Vienne, France, 2021; 26p.
61. IFN, Inventaire Forestier National (IFN). La Forêt Française: Un Puit de Carbone? Son Role Dans la Limitations des Changements Climatiques. 2005. Available online: https://inventaire-forestier.ign.fr/IMG/pdf/L_IF_no07_carbone.pdf (accessed on 11 May 2023).
62. French Data and Statistical Studies Department (SDES). Key Figures on Housing—2022 Edition. 2022. Available online: https://www.statistiques.developpement-durable.gouv.fr/edition-numerique/chiffres-cles-du-logement-2022/pdf/Chiffres-cles-logement-2022.pdf (accessed on 28 June 2023). (In French)
63. French Ministry of Agriculture and Food Security. Forest—Wood. Ministère de l'Agriculture et de la Souveraineté Alimentaire. Available online: https://agriculture.gouv.fr/foret-bois (accessed on 28 June 2023). (In French)
64. Huuskonen, S.; Domisch, T.; Finér, L.; Hantula, J.; Hynynen, J.; Matala, J.; Miina, J.; Neuvonen, S.; Nevalainen, S.; Niemistö, P.; et al. What is the potential for replacing monocultures with mixed-species stands to enhance ecosystem services in boreal forests in Fennoscandia? *For. Ecol. Manag.* **2021**, *479*, 118558. [CrossRef]
65. Lindner, M.; Maroschek, M.; Netherer, S.; Kremer, A.; Barbati, A.; Garcia-Gonzalo, J.; Seidl, R.; Delzon, S.; Corona, P.; Kolström, M.; et al. Climate change impacts, adaptive capacity, and vulnerability of European forest ecosystems. *For. Ecol. Manag.* **2010**, *259*, 698–709. [CrossRef]
66. Hoxha, E.; Maierhofer, D.; Saade, M.; Passer, A. Influence of technical and electrical equipment in life cycle assessments of buildings: Case of a laboratory and research building. *Int. J. Life Cycle Assess.* **2021**, *26*, 852–863. [CrossRef]
67. Polster, B.; Peuportier, B.; Sommereux, I.B.; Pedregal, P.D.; Gobin, C.; Durand, E. Evaluation of the environmental quality of buildings towards a more environmentally conscious design. *Sol. Energy* **1996**, *57*, 219–230. [CrossRef]
68. French Ministry of Ecological Transition and Territorial Cohesion. Reporting on French Low Carbon Strategy. 2021. Available online: https://www.ecologie.gouv.fr/sites/default/files/2021_Indicateurs%20de%20r%C3%A9sultats_SNBC-vF.pdf (accessed on 28 June 2023). (In French)
69. Hertwich, E. Understanding the Climate Mitigation Benefits of Product Systems: Comment on "Using Attributional Life Cycle Assessment to Estimate Climate-Change Mitigation...". *J. Ind. Ecol.* **2014**, *18*, 464–465. [CrossRef]

energies

Article

Theoretical Framework and Research Proposal for Energy Utilization, Conservation, Production, and Intelligent Systems in Tropical Island Zero-Carbon Building

Qiankun Wang [1,2], Ke Zhu [1,2] and Peiwen Guo [3,*]

1 Sanya Science and Education Innovation Park, Wuhan University of Technology, Sanya 572000, China; wangqk@whut.edu.cn (Q.W.); zhuke2018@whut.edu.cn (K.Z.)
2 School of Civil Engineering and Architecture, Wuhan University of Technology, Wuhan 430070, China
3 School of Real Estate and Management Science, Chongqing University, Chongqing 400044, China
* Correspondence: guopeiwen@whut.edu.cn

Abstract: This study aims to theoretically explore the technological systems of tropical island zero-carbon building (TIZCB) to scientifically understand the characteristics of these buildings in terms of energy utilization, energy conservation, energy production, and intelligent system mechanisms. The purpose is to address the inefficiencies and resource wastage caused by the traditional segmented approach to building energy consumption management. Thus, it seeks to achieve a comprehensive understanding and application of the zero-carbon building (ZCB) technology system. This article focuses on the demands for energy-efficient comfort and innovative industrialization in construction. Through an analysis of the characteristics of TIZCB and an explanation of their concepts, it establishes a theoretical framework for examining the system mechanisms of these buildings. Additionally, it delves into the energy utilization, energy conservation, energy production, and intelligent system from macro, meso, and micro perspectives. This approach results in the development of an implementation strategy for studying the mechanisms of energy usage, conservation, and intelligent production systems in TIZCB. The results show that: (1) this study delves into the theoretical underpinnings of TIZCB, emphasizing their evolution from a foundation of low-carbon and near-zero energy consumption. The primary goal is to achieve zero carbon emissions during building operation, with reliance on renewable energy sources. Design considerations prioritize adaptation to high-temperature and high-humidity conditions, integrating regional culture along with the utilization of new materials and technologies. (2) A comprehensive technical framework for TIZCB is proposed, encompassing energy utilization, conservation, production capacity, and intelligent systems. Drawing from systems theory, control theory, and synergy theory, the research employs a macro–meso–micro analytical framework, offering extensive theoretical support for the practical aspects of design and optimization. (3) The research implementation plan establishes parameterized models, unveiling the intricate relationships with building performance. It provides optimized intelligent system design parameters for economically viable zero-carbon operations. This study contributes theoretical and practical support for the sustainable development of TIZCB and aligns with the dual carbon strategy in China and the clean energy free trade zone construction in Hainan.

Keywords: tropical island zero-carbon building (TIZCB); energy utilization–conservation–production; theoretical basis; research framework

Citation: Wang, Q.; Zhu, K.; Guo, P. Theoretical Framework and Research Proposal for Energy Utilization, Conservation, Production, and Intelligent Systems in Tropical Island Zero-Carbon Building. *Energies* **2024**, *17*, 1339. https://doi.org/10.3390/en17061339

Academic Editors: Vincenzo Costanzo, Zsuzsa Szalay and Bruno Peuportier

Received: 31 December 2023
Revised: 27 February 2024
Accepted: 29 February 2024
Published: 11 March 2024

1. Introduction

Reducing greenhouse gas emissions is a crucial step in addressing global climate change [1]. The Paris Agreement is a landmark international accord aimed at addressing climate change and its adverse impacts. The main goal of the agreement is to limit global warming to well below 2 °C above pre-industrial levels, with efforts to limit the temperature increase to 1.5 °C [2]. The Paris Agreement sets a net-zero carbon emission target for the world, leading the Chinese government to establish a system of achieving peak carbon dioxide emissions by 2030 and carbon neutrality by 2060 [3]. According to the International Energy Agency (IEA), buildings' lifecycle emissions constitute about 39% of global CO_2 emissions, significantly impacting global greenhouse gas contributions. The "14th Five-Year Plan" period is a critical time to implement these dual carbon goals. According to the 2022 China Urban and Rural Construction Carbon Emission Series Research Report [4], China's total building carbon emissions in 2020 were 508 million tCO_2, accounting for 50.9% of the country's total carbon emissions. From 2005 to 2020, national building carbon emissions increased from 223 million tCO_2 to 508 million tCO_2, a 2.3-fold increase with an average annual growth rate of 5.6%. This indicates that energy conservation and carbon reduction in the construction sector are essential for achieving China's dual carbon goals, as continuously growing building carbon emissions pose significant challenges to these targets. China has long promoted building energy efficiency and green buildings, but there is a lack of on-site energy supply methods. The Ministry of Housing and Urban-Rural Development's "the 14th Five Year Plan for Building Energy Conservation and Green Building Development" upgrades traditional "energy conservation and emission reduction" to "energy conservation-production capacity-carbon reduction", providing new technical ideas for ZCBs [5]. The Ministry of Science and Technology and Nine Other Departments' "Implementation Plan for Science and Technology to Support Carbon Peaking and Carbon Neutrality (2022–2030)" calls for focusing on decarbonization, emission reduction, and energy efficiency improvement to promote building energy conservation, emission reduction standards, and whole-process carbon reduction [6]. Tropical island regions in China have long summers without winters and small temperature differences between day and night, resulting in high cooling energy demand for buildings. At the same time, these areas have strong solar radiation capabilities and abundant renewable solar thermal resources. Theoretical exploration of the energy conservation, production capacity, and intelligent system mechanisms of TIZCB has become a key issue supporting the sustainable development of such buildings.

ZCBs, as a novel technological paradigm, have been extensively studied by numerous scholars from diverse perspectives. Moghayedi et al. (2023) [7] evaluated the impact of factors influencing the achievement of net-zero carbon goals on the circular economy based on obstacles and driving forces in employing green technology methods in construction. Scherz et al. (2020) [8] outlined the implementation of system design models during the early design stages of buildings to address the challenges of achieving net-zero carbon building environments. Leung et al. (2018) [9] conducted a study on a ZCB in Hong Kong, analyzing the dissemination of bacterial communities in green building air through the investigation of bacteria's 16S rRNA genes. Bui et al. (2022) [10] comprehensively surveyed the field of zero-carbon renovation in existing buildings, employing a mixed-methods approach to data analysis, revealing evolving hotspots and core research themes. Zhao et al. (2017) [11] developed a conceptual model covering project and organizational aspects to explore the theoretical relationship between business models and ZCBs, aiming to drive sustainable development in the ZCB market. Pan et al. (2014) [12], through a review of zero-carbon village policies and support literature, established a social–technical transformation framework providing information for discussions on current and future zero-carbon emission policies. Girard et al. (2012) [13] argued that the shift towards low-carbon or ZCBs introduces a paradigm shift in energy supply infrastructure, emphasizing the need to understand buildings not only from the demand side but also from the perspective of electricity. Bui et al. (2022) [14] conducted an exploratory study through semi-structured interviews with New Zealand building experts to investigate challenges and issues faced by the New Zealand construction industry in ZCBs. Nidhin et al. (2023) [15] surveyed building stakeholders involved in New Zealand architectural design and construction to understand their awareness of zero-carbon initiatives, providing a practical foundation for the market promotion and policy implementation of ZCBs. Hasan et al. (2023) [16] discussed the legitimacy and legality of implementing net-zero carbon building standards by examining policy documents from the World Green Building Council. Torriti et al. (2015) [17], drawing on policy framework theory and sociological knowledge, systematically explored the significant challenges in the implementation of zero-carbon residential policies. Walker et al. (2016) [18] reflected on the relationship between politics and governance revealed in the mainstreaming process of abandoning zero-carbon housing in the UK, emphasizing the foundation and positioning of carbon responsibility, and the essential role of the state in achieving daily reproduction of low-carbon living. Lees et al. (2014) [19] investigated the actual use of low-carbon and zero-carbon technologies by housing developers to understand the main drivers and significant roles of ZCB technology choices. Steijger et al. (2013) [20] discussed the constraints of compact urban housing on achieving zero-carbon performance in apartment buildings in the context of the Sustainable Homes Code, highlighting the need for off-site net import of electricity and/or heat. McLeod et al. (2012) [21], through comparative analysis, elucidated the revised definition of zero-carbon housing in the UK and the approaches advocated by the Zero Carbon Hub for policy implementation. In summary, these studies lay the foundation for the theoretical development and practical application of TIZCB. However, these studies primarily focus on macro-level policy implementation, involve fewer specific implementation technical measures, and do not consider the building's intrinsic production capacity from the perspective of the resource endowment of tropical island regions. They lack a comprehensive exploration of the building energy flow system from the perspective of the building integrated system, making it challenging to adapt to industrialized, modular, and integrated pattern innovations and integrated research. Furthermore, they struggle to achieve an effective balance between "zero carbon emissions" and "energy-efficient comfort".

In summary, there is currently no uniform definition of a ZCB [22]. For example, in the United States, the "National Definition of Zero Emission Buildings: Part 1 Operational Emissions (Version 1.00), Draft Standard" defines a ZCB as highly energy efficient, having no on-site energy use emissions, and being entirely powered by clean energy. In the European Union's "Energy Performance of Buildings Directive", a ZCB is defined as having very high energy performance, with its very low energy needs fully covered by renewable sources, without any on-site fossil fuel carbon emissions. In the United Kingdom's "Net Zero Carbon Building Framework Definition", a ZCB is described as high-efficiency, producing or procuring carbon-free renewable energy or high-quality carbon offsets in quantities sufficient to offset annual carbon emissions from building materials and operations. The traditional design approach in tropical island regions emphasizes passive energy conservation as the primary focus and active energy conservation as a secondary consideration [23]. However, TIZCB lack an integrated design that incorporates precise regulation of energy utilization for carbon reduction, energy-efficient building envelopes [24], high-efficiency multi-source energy production [25], and intelligent system integration for carbon reduction [26].

The research gaps mainly manifest as: (1) a lack of unified standard definitions for TIZCB, incomplete technical systems, and immature integrated applications [27]; (2) difficulty in achieving coordinated operation among energy use, energy saving, production capacity, and intelligent systems; and (3) this resulting in the insufficient comprehensive efficiency of zero-carbon operations in tropical island buildings [28].

In light of this, the research objectives of this paper are: (1) to construct a theoretical analysis model based on systems theory, synergy, and control theory by defining the inherent characteristics and technical system of TIZCB. This model will provide theoretical support for the study and optimization of mechanisms and systems related to energy utilization, energy conservation, energy production, and intelligent systems in TIZCB. (2) To establish a quantitative analysis framework for energy utilization, energy conservation, energy production, and intelligent systems from macro, meso, and micro perspectives. In this framework, the macro perspective refers to system decomposition and integration from the perspective of the entire building system of TIZCB; the meso perspective refers to system interaction and coupling from the perspective of energy utilization, energy conservation, energy production, and intelligent systems in TIZCB; and the micro perspective refers to quantification and regulation from the perspective of design parameters of energy utilization, energy conservation, energy production, and intelligent systems in TIZCB. (3) To form the research logic and implementation plan for the system mechanism of energy utilization, energy conservation, energy production, and intelligent systems in TIZCB. This implementation plan will provide technical support for the practical application and verification analysis of the mechanisms and systems related to energy utilization, energy conservation, energy production, and intelligent systems in TIZCB.

The rest of this paper is organized as follows: in Section 2, based on the introduction of zero carbon architecture and tropical island architecture, TIZCB and their technical system are defined; in Section 3, research methods such as systems theory, control theory, and synergy theory are summarized, and their main roles in this article are introduced; Section 4 develops the theoretical framework for researching integrated energy utilization, conservation, production, and intelligent systems in TIZCB; Section 5 outlines the logical thinking of the research on integrated energy utilization, conservation, production, and intelligent systems in TIZCB; Section 6, in conjunction with the research objectives and major research content, constructs an implementation plan for the study of integrated energy utilization, conservation, production, and intelligent systems in TIZCB; and Section 7 summarizes the research conclusions of this paper.

2. Conceptual Definition

2.1. Zero-Carbon Buildings

ZCBs are developed based on low-carbon buildings and near-ZCBs are developed based on low-carbon buildings, near-zero energy consumption buildings, and zero energy consumption buildings. According to the World Green Building Council [29], a ZCB refers to a building whose carbon emissions during the operation phase are ≤ 0, and the operating energy comes entirely from renewable sources. Improving building energy efficiency (reducing energy consumption) and optimizing building energy structure (a reduction in fossil fuels and strengthening the application of renewable energy) are important ways to reduce building carbon emissions. According to the announcement of the General Office of the Ministry of Housing and Urban–Rural Development on soliciting public opinions on the draft national standard "Zero Carbon Building Technology Standard (Draft for Comments)", a ZCB is one that adapts to climate characteristics and site conditions, reduces building energy demand through optimized architectural design while meeting indoor environmental parameters, improves the efficiency of energy equipment and systems, fully utilizes renewable energy and building energy storage, and can combine carbon offset methods such as carbon emission rights trading and green electricity trading on the basis of achieving near-zero carbon buildings, in accordance with Article 3.2.5 or 8.4.7 of this standard [30]. Combining domestic and foreign core definitions of ZCBs, it can be seen that ZCBs advocate the technical principle of "passive priority, active optimization, and balancing of renewable energy", which aims to eliminate the fossil fuel component in building energy consumption on the basis of reducing energy demand, and can be achieved through the use of local and surrounding renewable energy applications exclusively for their use, as well as financial mechanisms such as carbon trading and green electricity.

2.2. Tropical Island Architecture

Tropical island architecture design aims to adapt to the hot, humid, high-salt, and high-UV tropical island climate environment, reflecting the regional culture and tropical island cultural characteristics of Hainan architecture [31]. It emphasizes the integration of architecture with the ecological environment and urban design, creating open spaces that are accessible to the public through natural ventilation and lighting. At the same time, it actively adopts new materials, technologies, and processes to design and construct outstanding buildings with profound connotations, good functionality, and unique shapes. In terms of China, the biggest difference between tropical island architecture and other architectural systems lies in the tropical island climate environment. To create assembly-style buildings suitable for tropical island characteristics, the Hainan Provincial Department of Housing and Urban–Rural Development and other departments jointly issued the "Three Year Action Plan for the Development and Improvement of Prefabricated Buildings (Green Buildings) in Hainan Province (2023–2025)", which clearly proposes to build good houses that are suitable for Hainan's tropical climate and marine island characteristics [32]. It continuously carries out research on key technologies related to assembly-style buildings, green buildings, ultra-low (near zero) energy consumption buildings, low-carbon (zero-carbon) buildings, and other tropical building sciences under Hainan's tropical marine island environmental conditions. Therefore, this article is based on the tropical island climate environment, rationally plans the comfortable and healthy energy demand of tropical island architecture, actively adopts structural optimization design using enclosure structure materials to promote energy conservation and carbon reduction, focuses on utilizing Hainan's abundant renewable energy sources such as solar heat to support the building's energy production capacity, and combines machine learning and multi-objective optimization algorithms to enable continuous optimization during the operation phase of tropical island architecture.

2.3. *Tropical Island Zero-Carbon Buildings*

According to the national dual carbon strategy and the requirements of Hainan's clean energy free trade island construction, it is still necessary to vigorously promote energy conservation and carbon reduction in the construction industry for a considerable period until reaching the peak carbon dioxide emissions by 2060. ZCBs, which have zero carbon emissions during their operation, are considered as the key focus for achieving the dual carbon goals in the construction industry. By combining the concepts of ZCBs and tropical island buildings, this article defines TIZCB as follows: adapting to the climatic environmental characteristics of tropical islands, reflecting the regional culture of Hainan's architecture, and on the basis of maintaining indoor healthy and comfortable environment energy demand, achieving building energy conservation through optimizing the design of enclosure structure systems, making full use of renewable energy for building body production capacity, and adopting integrated building design and intelligent control methods to make TIZCB technically feasible and economically reasonable. Among them, energy consumption is the foundation for creating a comfortable building environment; energy conservation is the prerequisite for new building energy-conservation design, and ZCBs are first and foremost green, low-carbon, and energy-conservation buildings; production capacity is the key to achieving ZCBs; and intelligence is the means of using intelligent methods for parameterized multi-objective regulation in integrated design, improving the efficiency of energy equipment and systems.

2.4. *TIZCB Technology System*

A TIZCB is a new technology system. In the past, for new building systems, emphasis was placed on technological research and innovation, while existing technology integration and innovation were neglected. This approach failed to balance the principles of technological advancement, feasibility, and reasonable economic costs, hindering the sustainable development of ZCBs. Analysis of ZCBs on tropical islands shows that they emphasize energy-conservation and production capacity technology system integration, as well as the prominent role of intelligent means in energy-conservation and carbon reduction during the design phase. Therefore, under the premise that there is no uniform definition standard for ZCBs at home and abroad, how to learn from foreign beneficial experiences, integrate key technologies such as environmental control energy-conservation and carbon reduction, passive energy-conservation and carbon reduction, skin production capacity carbon reduction, and system intelligence carbon reduction through quantitative evaluation and design optimization during the design phase, so that the implementation plan of ZCBs is technologically advanced and feasible, economically cost-effective, and thus establish a

ZCB technology system suitable for the climatic environment characteristics and economic development features of tropical islands, has become the key to the sustainable development of TIZCB. On the basis of meeting the technical indicators of ZCBs, by adopting low-carbon building materials, low-carbon structural forms, and material reduction design, it is possible to achieve total carbon emissions of not more than zero during the building operation process. The TIZCB technology system is shown in Table 1.

Table 1. TIZCB technology system.

Number	Classification	Main Technologies
1	Energy Utilization Technology	Ground (water) source heat pump system, displacement ventilation system, radiant cooling system, room personnel density and occupancy rate, electrical equipment power density and utilization rate, lighting schedule, etc.
2	Energy Conservation Technology	Exterior Wall Energy-Conservation Technologies: Wall composite technologies include an internal insulation layer, an external insulation layer, and a sandwich insulation layer. Door and Window Energy-Conservation Technologies: Double-glazed windows, multi-layer glass, coated glass (including reflective glass, absorbent glass), high-strength LOW2E fire-resistant glass (high-strength low-emissivity coated fire-resistant glass), and glass with a metallized silver layer. The airtightness of the building envelope. Roof Energy-Conservation Technologies: Solar heat collecting roofs and controllable ventilation roofs, etc.
3	Energy Production Technology	Development and Utilization of New Energy: Solar water heaters, photovoltaic roof panels, photovoltaic exterior wall panels, photovoltaic sun-shading panels, photovoltaic window walls, photovoltaic skylights, photovoltaic glass curtain walls, etc.
4	Intelligent Technology	Machine Learning Prediction Technology: AdaBoost Regressor, Bagging Regressor, CAT Boost Regressor, Decision Tree Regressor, Extral Tree Regressor, GBDT Regressor, KNeighbors Regressor, Lasso Regressor, LGBM Regressor, Linear Regressor, LSTM Regressor, Multilayer Perceptron Regressor, Random Forest Regressor, Support Vector Machine Regressor, XGBoost Regressor, etc. Intelligent Algorithm Optimization Technology: GA, NSGA-II, NSGA-III, etc. Techno-economic evaluation: Cost–benefit analysis, cost-effectiveness analysis, return on investment, net present value, internal rate of return, sensitivity analysis, risk analysis, life cycle cost analysis, multi-criteria decision analysis, etc.

3. Theoretical Methodology

3.1. System Theory

A system is an organic whole composed of several interacting, interdependent, and interconnected components that have specific functions. The system itself is also a part of a larger system to which it belongs. System theory aims to explore and understand the nature, structure, and behavior of systems, emphasizing the use of holistic thinking to investigate the laws of things and using mathematical models to define, express, and quantify the characteristics and functions of systems.

In the early 20th century, Ludwig von Bertalanffy, an Austrian biologist of American nationality, was the founder of General Systems Theory (GST, 1968) [33]. He defined a system as "a complex composed of several interacting elements", thus defining the three basic characteristics of purposefulness, dynamics, and orderliness of systems theory and emphasizing the importance of system thinking [34]. Bertalanffy's work inspired other scholars, including Bertalanffy's work inspired other scholars, including Talcott Parsons and Niklas Luhmann, who applied systems theory to the study of social systems [35]. In the mid-20th century, ecologists Howard T. Odum and Eugene P. Odum introduced system theory into the field of ecology, pioneering research in ecosystem ecology, energy ecology, etc., and proposing the ecosystem theory [36]. Fritjof Capra combined systems theory with ecology, physics, and philosophy at the end of the 20th century, proposing the connection between ecology and science [37]. Peter Senge, on the other hand, proposed a theory of organizational learning in the fields of organizational science and management, applying system thinking to organizational management [38]. Modern systems science has promoted interdisciplinary research and the spread of systemic thinking in various fields, while also facilitating interdisciplinary collaboration among different domains, including engineering, ecology, sociology, management, and environmental science [39]. This helps in understanding and addressing complex issues [40].

Overall, systems theory underscores the interdependent, interactive, and constraining relationships between the whole and parts, among parts, and between the system itself and its external environment. This facilitates the analysis of complex system behaviors, optimization of system performance, and support for decision-making. Regarding the technological system of ZCB systems theory allows for the building to be viewed at a macro-level as an entity comprising distinct systems with different functional characteristics. This approach involves two fundamental steps: system deconstruction and system integration. System deconstruction entails dividing the building into various technical systems based on different target functions, each characterized by distinct functionalities. System integration then combines these technical systems into a building system that is highly cohesive and loosely coupled.

3.2. Cybernetics

The basic idea of cybernetics is to achieve effective control and management of a system by understanding the interactions between its various components. Its core concept is circular causality or feedback, where the result of an action serves as input for further actions. Cybernetics studies the state, function, behavior, and changing trends of systems, revealing common control laws for different systems through stable control of systems.

Control theory originated in the 1940s, with early contributors including Norbert Wiener, Warren McCulloch, Arturo Rosenblueth, etc. They promoted the formation of control theory at events such as Macy Conferences. Control theory broke free from the constraints of Newtonian classical mechanics and Laplacian mechanical determinism, using new statistical theories to study various possibilities for system motion states, behavior patterns, and changing trends [41]. The work of scholars like Jay Wright Forrester in the field of system dynamics has provided significant support for the development of cybernetics [42]. The concept of intelligent control emerged in the 1960s. With the increasing uncertainty, high dimensionality, and nonlinearity of things [43], new methods and technologies such as uncertain mathematical models and high-dimensional nonlinear intelligent means have gradually gained prominence [44]. In the 1970s, new cybernetic researchers such as Humberto Maturana proposed a new cybernetics that is more adaptable to biological systems [45], guiding the evolution of the field. The application of economists like Oskar Ryszard Lange in economic systems also provides a new perspective for the expansion of cybernetics [46]. Control theory has a wide range of concerns [47], including environmental, technological, biological, cognitive, and social systems [48], as well as practical activities such as design, learning, management, and dialogue [49].

Overall, cybernetics provides a systematic approach to thinking and methodology that aids in understanding and optimizing the comprehensive performance of building systems. It allows for perception, learning, and adjustment at the micro-level, quantifying the parametric design model of TIZCB systems. This forms the basis for the intelligent control of building systems, aiming to achieve multi-objective optimization of various systems, thereby promoting innovation and optimization in architectural design. This involves two fundamental steps: system quantification and system control. System quantification involves characterizing and modeling the system parameters obtained from system deconstruction to reveal the quantitative laws of each system. System control, based on understanding the operational laws of each system, adjusts the quantified parameters within the system to achieve the optimization of system objectives.

3.3. Synergetics

Synergetics is a discipline that studies the interaction and synergistic effects of systems. Synergy refers to the collective effect produced by the interaction of subsystems in complex systems. Synergetics believes that things are unified entities organized by many systems, with complementary, collaborative, and coordinated relationships between subsystems. It points out that the coupled evolution of natural systems and social systems from disorder to order is the result of mutual influence, interaction, and constraint among the elements of the system.

Synergetics, proposed by West German scientist Hermann Haken in the 1970s, emphasizes the generation of ordered structures and the issue of maintaining a constant supply of energy within a system. It further addresses the challenge of how complex systems transition from a state of disorder to order, facilitating the self-organizing evolution of systems from disorder to order. This transformation allows for the realization of system efficiency where 1 + 1 becomes greater than 2, transcending the conventional understanding of $1 + 1 \leq 2$ [50]. After decades of integrated development, modern synergetics has evolved from the "old trinity" of system theory, information theory, and dissipative structure theory within general systems theory to the "new trinity" of self-organizing system analysis. This progression has significantly advanced the theoretical foundations of self-organizing systems. Architects and structural engineers, such as Richard Buckminster Fuller and Marshall Applewhite, noted in "Synergetics: Explorations in the Geometry of Thinking" that synergetics reveals the logical principles governing natural laws, and they applied synergetic concepts in the field of architecture. The theoretical framework of synergetics combines system dynamics with statistics, providing essential conceptual tools for the study of the dynamic evolution of complex systems from micro- to macro-levels [51]. This integration has accelerated the development of modern nonlinear science and systems theory, highlighting synergetics as an intrinsic and essential force driving the internal evolution and formation of new ordered structures within systems [52]. Synergetics has laid the foundation for the study of ordered structures within various interdisciplinary fields. It finds extensive application in the control processes of engineering and technological systems [53], socio-economic systems [54], and ecological and environmental systems, among other domains [55].

Overall, systems theory emphasizes the interactive relationships of mutual action, influence, and constraint among systems, highlighting how coordinated coupling can generate new system structures, offering an important cognitive tool for studying complex system patterns. For the TIZCB systems, synergy theory explores the internal laws of several building systems from both qualitative and quantitative perspectives at the meso-level. This process involves two basic steps: system interaction and system coupling. System interaction involves comparative analysis and preparatory integration of concepts and definitions among systems, representing a qualitative level of system study; system coupling involves quantitative analysis and integrated optimization of models and relationships following system interaction, representing a quantitative level of system exploration.

3.4. Overview of Theoretical Application

The method system and technical approach of system theory, synergy theory, and control theory provide a new perspective for the deconstruction, integration, coupling, interaction, and intelligent control of the enclosure structure system in TIZCB, as illustrated in Figure 1.

Analyzing Figure 1, the interrelationship among these three methods is as follows: (1) systems theory divides and integrates TIZCB into four systems; (2) synergetics facilitates the interactive coupling of the four systems in TIZCB, which is crucial for technological updates and cost reduction in TIZCB; and (3) cybernetics, through "Intelligent Technology", implements "Energy Utilization Technology", "Energy Conservation Technology", and "Energy Production Technology" in TIZCB. Lastly, supporting the rest of the study with these three theoretical methods indeed provides a solid theoretical foundation and practical significance, especially in offering significant guidance for the rational division of building systems.

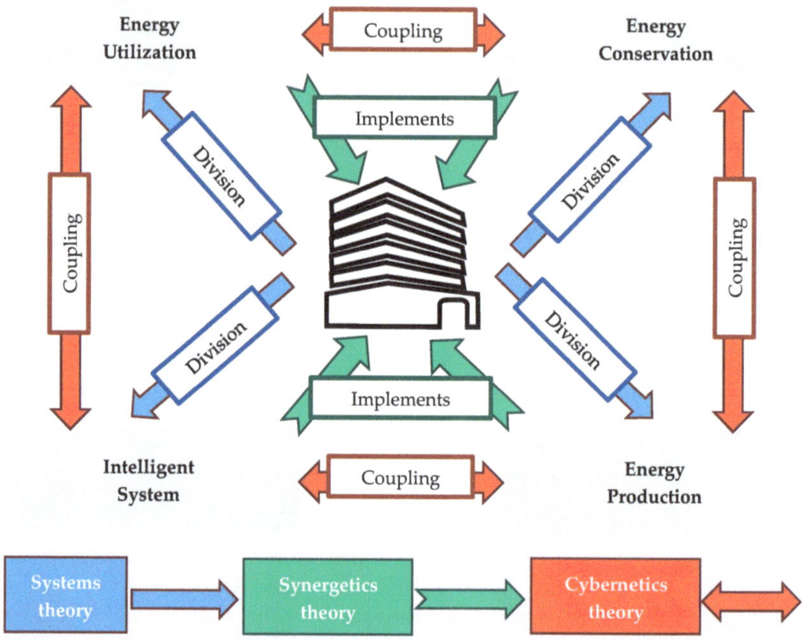

Figure 1. Overview of theoretical application.

4. Establishing a Theoretical Framework

4.1. Macroscopic System Decomposition and Integration

System theory, operating at the macro-level, conducts goal decomposition and system integration for the technical system of ZCBs. Its objective is to approach the energy conversion aspects such as energy utilization, conservation, and production, focusing on energy-efficient comfort and the goals of new industrialization in building construction. System theory perceives TIZCB as complex systems and systematically organizes the energy utilization, conservation, production, and intelligent technology systems in these buildings. This is achieved through defining functional objectives, analyzing complex systems, optimizing system feedback, and ultimately forming the macro-system decomposition and integration of TIZCB, as illustrated in Figure 2.

Analyzing Figure 2, it is evident that the key to the decomposition and integration of the macro-level technical system lies in goal identification and system integration. In accordance with tropical island building regulations, climate characteristics, and constituent elements, applicable technical systems for TIZCB are systematically outlined. Focusing on the requirements of zero-carbon construction, energy-efficient comfort, and the goals of new industrialization, the technical system for TIZCB is defined from the perspectives of building energy flow and intelligent control. The identified four major goals for the technical system of TIZCB are "precise energy utilization and carbon reduction at the environmental control endpoints, passive energy conservation in the building envelope, efficient energy production in the building shell, and intelligent coupling and control for carbon reduction". These goals are integrated into the four major systems of the technical system for TIZCB, namely, the energy utilization system, energy conservation system, energy production system, and intelligent system. Specifically, the energy utilization system describes the basic energy consumption in creating a comfortable building environment, focusing on the energy planning of the building's environmental control system. The energy conservation system characterizes the integrated optimization design through the building envelope, focusing on energy-conservation components such as walls, doors, windows, and roofs. The energy production system depicts the utilization of renewable energy technologies, particularly photovoltaic and solar-thermal integration, focusing on the integrated energy-producing surface of walls, doors, windows, and roofs. The intelligent system represents the use of intelligent methods for integrated design with parameterized multi-objective control, achieving technically advanced and economically reasonable integrated designs for energy planning, energy-conservation components, and energy-producing surfaces.

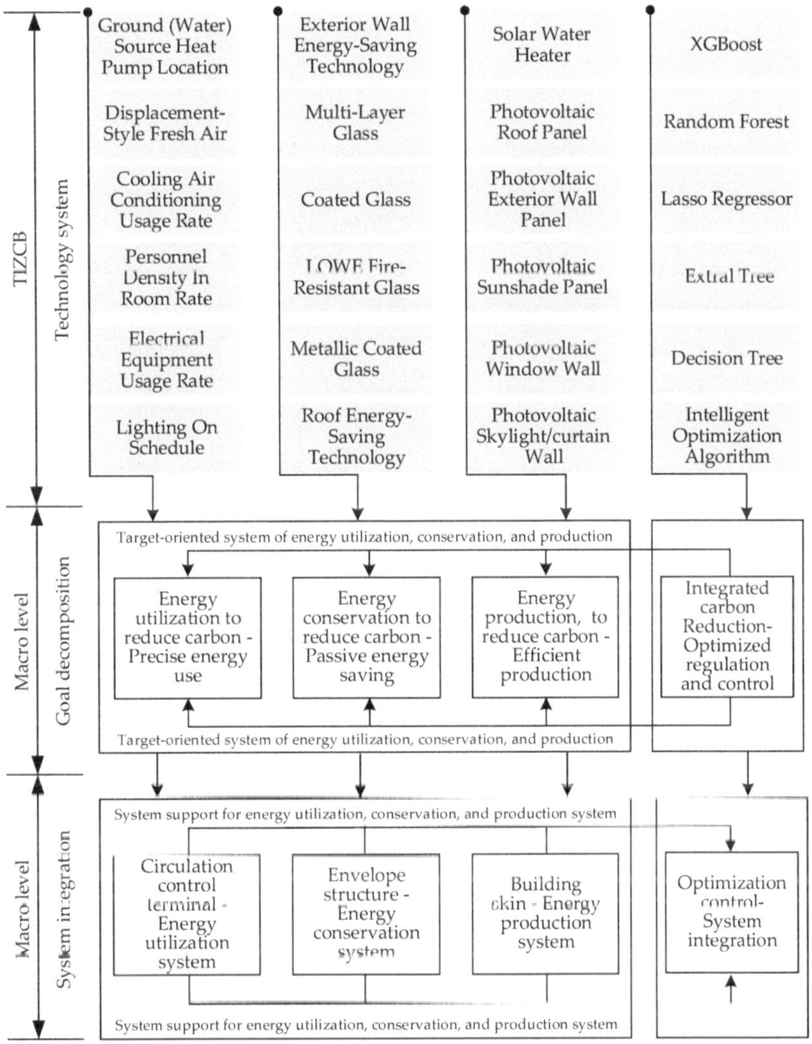

Figure 2. Macro system decomposition and integration of TIZCB based on system theory.

Overall, decomposition and integration involve defining functional modules, analyzing complex systems, and optimizing system feedback to achieve energy-conservation and green comfort in buildings. Through the analytical process of decomposition and integration based on system theory, the technical system of TIZCB is viewed as a complex system. The functional goals of the TIZCB technical system are decomposed, and the four major systems of "energy utilization, energy conservation, energy production, and intelligent systems" are integrated. Therefore, it is crucial to recognize that while macro-level decomposition and integration of the four systems of energy utilization, conservation, production, and intelligent systems are essential, a micro-level understanding of the internal influencing mechanisms of these systems is equally important. This micro-level analysis forms the basis for exploring the mechanisms of TIZCB systems and is critical for achieving the goals of energy-conservation comfort and new industrialization in TIZCB.

4.2. Microparameter Quantification and Regulation

Control theory, operating at the micro-level, involves quantifying and intelligently controlling design parameters for ZCBs. This process constitutes the micro-level quantification and regulation of parameters in TIZCB, as illustrated in Figure 3.

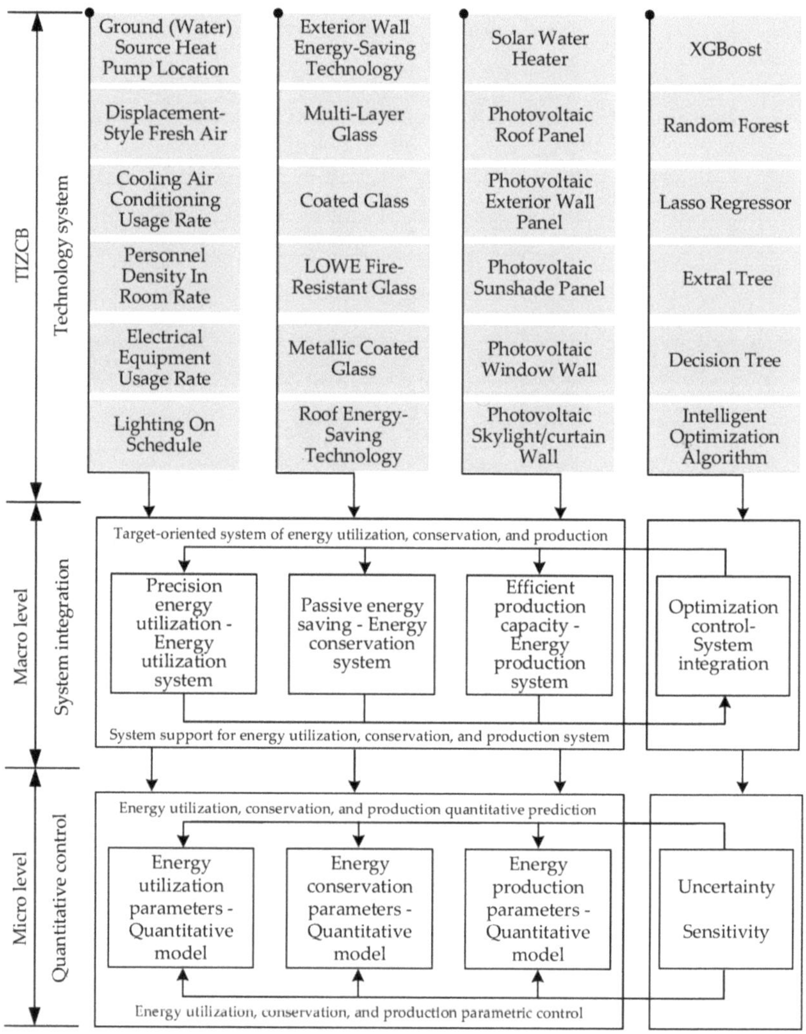

Figure 3. Microscopic parameter quantification and regulation of TIZCB based on cybernetics.

From the analysis of Figure 3, it is evident that the key to micro-level quantification and regulation lies in quantitative prediction and intelligent control. Its objective is to quantitatively predict and intelligently control the design parameters of the energy utilization, conservation, and production systems in TIZCB. Using machine learning models from intelligent systems, the theory aims to reveal the impact mechanisms of variations in design parameter combinations on building performance through quantitative simulation results. Intelligent control of the energy utilization, conservation, and production systems is achieved through multi-objective optimization algorithms. Based on the functional goals of the energy utilization, conservation, and production intelligent system in TIZCB, a model for quantifying design parameters of the energy utilization, conservation, and production systems is constructed. Intelligent systems incorporating machine learning and optimization algorithms are developed. Building upon this foundation, through quantitative simulation, performance prediction, and multi-objective optimization, the study reveals the impact mechanisms of variations in design parameter combinations on building performance. It explores the optimal combinations of design parameters for the energy utilization, conservation, and production systems in TIZCB at the micro-level. The micro-level quantification and regulation of design parameters determine the direction of modular, parameterized, and intelligent optimization design for the energy utilization, conservation, and production systems in TIZCB. This establishes the basic premise for studying the mechanism of energy utilization, conservation, and production intelligent systems and provides specific implementation plans. Quantitative prediction focuses on predictive analysis of the energy utilization, conservation, and production systems in TIZCB using machine learning integrated models built by intelligent systems. Intelligent control, based on predictive analysis, utilizes multi-objective intelligent optimization algorithms to optimize the energy utilization, conservation, and production systems, achieving optimal performance in the integrated design of energy planning, conservation components, and production surfaces in TIZCB. Overall, quantification and regulation constitute the process of revealing system regularities, exploring system mechanisms, and achieving optimal design parameter combinations for building performance. The process of quantification and regulation based on control theory proposes a conceptual framework for quantifying design parameters of the energy utilization, conservation, and production systems in TIZCB and constructs an intelligent system prediction optimization method based on data-driven approaches. Therefore, it is essential to recognize that, while decomposing and integrating the four major systems at the macro-level and quantifying and regulating the impact mechanisms at the micro-level, achieving the interaction and coupling of building systems and design parameters at the meso-level is crucial. This is a prerequisite for the integration of energy planning, conservation components, and production surface technologies and a key factor in realizing the integrated design of TIZCB.

4.3. *Interaction and Coupling of Mesoscopic Systems*

Synergetics, operating at the mid-level, engages in model interaction and goal coupling of TIZCB systems. Its aim is to, on the basis of model interaction in the energy utilization, conservation, and production systems of TIZCB, employ intelligent systems for multi-system quantified prediction and multi-objective control decision-making of the coupled systems. This process utilizes the mid-level building system interaction and coupling to link the macro-level technical systems and micro-level design parameters. Figure 4 illustrates the interaction and coupling of the mid-level system in TIZCB. The mid-level interaction coupling generates a multi-system coupled quantification model, laying the foundation for achieving the overall performance optimization goals of energy utilization planning, energy-conservation components, and capacity-building skin at the macro-level. This illustrates that mid-level interaction fusion of energy utilization, conservation, and production systems has become a crucial link in connecting the macro and micro-levels.

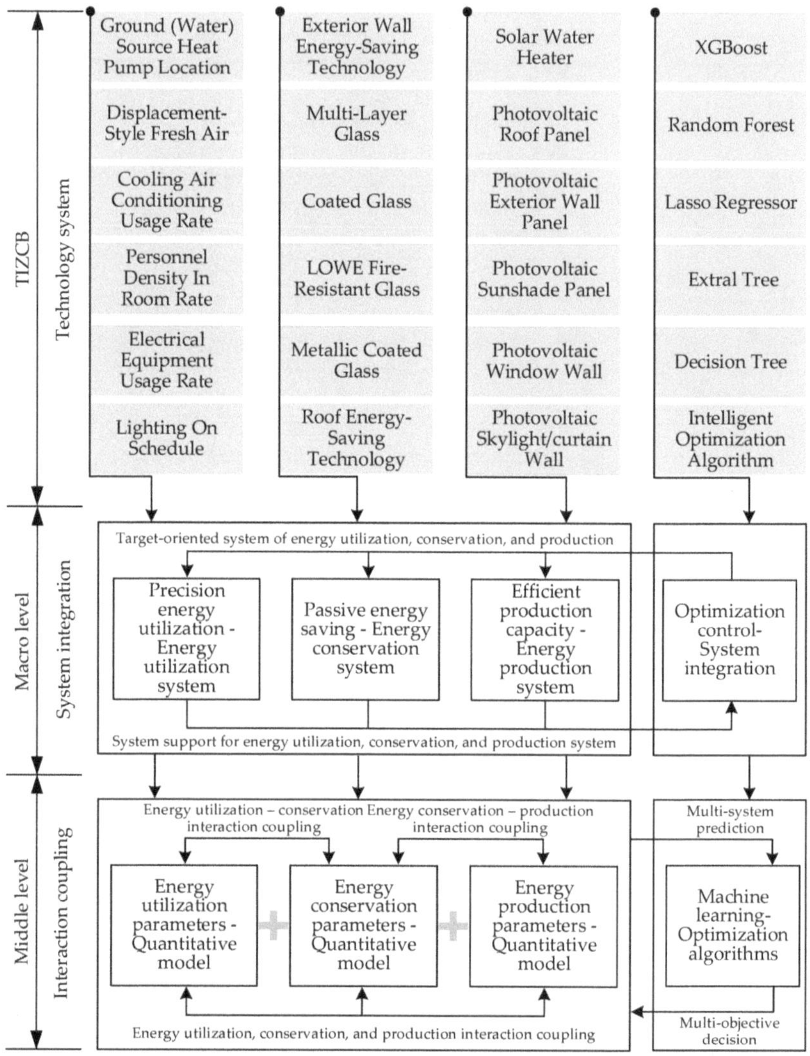

Figure 4. Intermediate interaction and coupling in TIZCB systems based on synergetics.

Analyzing Figure 4, it is evident that the key to mid-level interaction and coupling lies in multi-system quantified prediction and multi-objective control decisions. Building upon the macro-level decomposition of the four major goals, namely "precise energy reduction in environmental control, passive energy reduction in envelope structure, efficient energy reduction in building skin, and intelligent reduction in carbon through integrated control of systems" and the integration of the four major systems, namely "energy utilization system, energy-conservation system, capacity-building system, and intelligent system", intelligent systems with integrated learning models and optimization algorithms are utilized. This is carried out for the quantified prediction and intelligent control of design parameters at the micro-level of the energy utilization, conservation, and production system. Further, the coupled systems of energy utilization, conservation, and production undergo multi-system quantified prediction and multi-objective control decisions, thereby achieving the organic unity of the technical system, building system, and design parameters in the intelligent system of TIZCB. The macro-level derivation and integration of the four major systems provide a fundamental guideline for micro-level quantification and control.

Overall, interaction and coupling involve the interplay of system models and the coupling of system goals, realizing the optimization of design parameter combinations for the overall performance of TIZCB. Based on the analysis of the interaction and coupling process using synergetics and building on research at the macro- and micro-levels, a research path is proposed for the quantified prediction of multi-systems and the multi-objective control decision-making of energy utilization, conservation, and production in TIZCB. This forms the basis for the evaluation of intelligent system technology and economic analysis in TIZCB. It is crucial to recognize that the overall performance of TIZCB is not simply the sum of energy utilization, conservation, production, and intelligent systems. Instead, it is the result of the mutual influence, interaction, and constraint of each system. By exploring the optimal levels of each system, understanding the coupling effects and principles of each system's complex relationships, the overall energy and carbon reduction in TIZCB can be effectively optimized.

4.4. Analysis of Intelligent Systems Based on Macro–Medium–Micro

To theoretically explore the mechanisms of energy utilization, conservation, production, and intelligent systems in TIZCB and to summarize the research focus while elucidating the research approach, this section employs system engineering theories such as system theory, synergetics, and control theory. The analysis is conducted through processes that involve macro-level goal decomposition and system integration, micro-level model quantification and parameter control, and meso-level parameter interaction and system coupling. The resulting framework, as illustrated in Figure 5, establishes a comprehensive analysis structure for understanding the mechanisms of energy utilization, conservation, production, and intelligent systems in TIZCB based on macro–meso–micro perspectives.

From the analysis of Figure 5, a horizontal perspective reveals the following: At the macro-level, the four major goals of "precision energy utilization reduction, passive energy conservation of enclosure structure, efficient energy production of building envelope, and intelligent regulation for integrated control to reduce carbon emissions" are decomposed. This results in the integration of the TIZCB technical system into four major systems: "energy utilization system, energy conservation system, production system, and intelligent system". At the micro-level, a quantitative model of design parameters for the "energy utilization system, energy conservation system, and production system" is established. Machine learning and optimization algorithms from the "intelligent system" are then applied to quantitatively predict and intelligently control the three major quantitative models of energy utilization, conservation, and production systems. At the meso-level, interactions occur among the "energy utilization system, energy conservation system, and production system", and coupling takes place among the "energy parameter, conservation parameter, and production parameter" quantitative models. This achieves system quantification prediction and multi-objective intelligent control of the coupled model of energy utilization, conservation, and production through the "intelligent system" for technical evaluation and economic analysis. From a vertical perspective, building upon the analysis of technical system goals and system integration at the macro-level, the micro-level design parameter quantification prediction and intelligent control are carried out. This achieves a direct mapping from the macro to the micro-level, laying the foundation for the micro-level quantification of macro goals. At the meso-level, using system interaction and goal coupling as intermediaries, macro-level technical systems and micro-level design parameters are interactively coupled and organically linked. This results in multi-system quantification prediction and multi-objective intelligent control of energy utilization, conservation, and production. Thus, the framework for analyzing TIZCB systems is established at multiple levels, including macro, meso, and micro, encompassing multiple systems, and addressing multiple objectives in terms of technical evaluation and economic analysis.

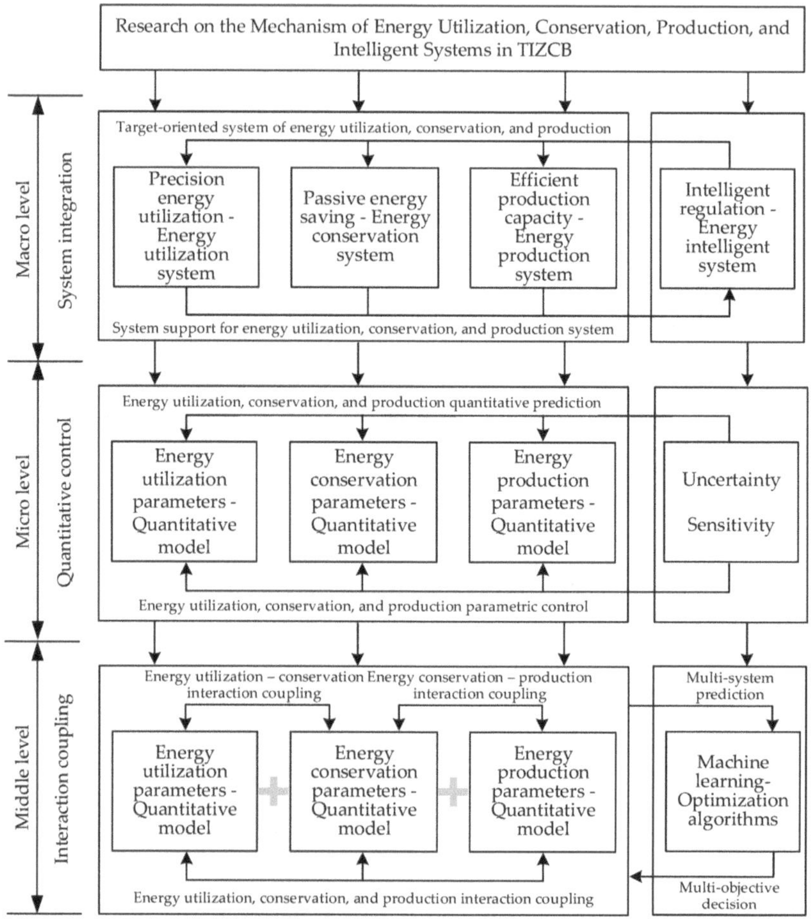

Figure 5. Analysis framework of TIZCB systems based on macro–meso–micro perspectives.

Overall, the TIZCB system analysis framework based on macro–meso–micro perspectives aims to provide a holistic, multi-level, multi-system, and multi-objective research perspective from global to local and macro to micro. This framework assists in exploring and understanding the parameterized model construction, influencing relationships, and optimization analysis of energy utilization, conservation, production, and intelligent systems in TIZCB from various perspectives. It establishes a theoretical foundation for the study of mechanisms in the energy utilization, conservation, production, and intelligent systems of TIZCB.

5. Research Logical Framework

To support China's dual-carbon strategy in the construction industry and the establishment of Hainan as a clean energy free trade island, this study addresses the current challenges in TIZCB standards, including inconsistent definitions, incomplete technical systems, and immature integrated applications. Focusing on the demands for energy conservation, comfort, and the new industrialization of buildings and considering the unique climate characteristics of Hainan's tropical island environment, this paper explores the energy utilization, conservation, production, and intelligent systems in TIZCB. It employs methods from systems engineering and parametric design technology to summarize and address common issues such as "how to accurately establish quantitative models for system design parameters and building performance", "how to finely investigate the relationships between system design parameters and building performance", and "how to dynamically optimize system design parameters and building performance for comprehensive objectives". The study systematically distills the key scientific and technological problem of "mechanisms and intelligent optimization methods for the energy utilization, conservation, production, and intelligent systems in TIZCB". The framework for theoretical analysis and research entitled "Mechanisms of the Energy Utilization, Conservation, Production, and Intelligent System in TIZCB" is established as a crucial component. Corresponding research contents include "Mechanism of the Energy Utilization System in TIZCB", "Mechanism of the Energy Conservation System in TIZCB", "Mechanism of the Production System in TIZCB", and "Mechanism of the Intelligent System in TIZCB". These components collectively contribute to the objectives of "establishing a quantitative model for the design parameters of the energy utilization, conservation, production, and intelligent system in TIZCB", "achieving a quantitative analysis of the energy utilization, conservation, production, and intelligent system in TIZCB", and "optimizing the comprehensive objectives of the energy utilization, conservation, production, and intelligent system in TIZCB". The overall research framework for the energy utilization, conservation, production, and intelligent system in TIZCB is illustrated in Figure 6.

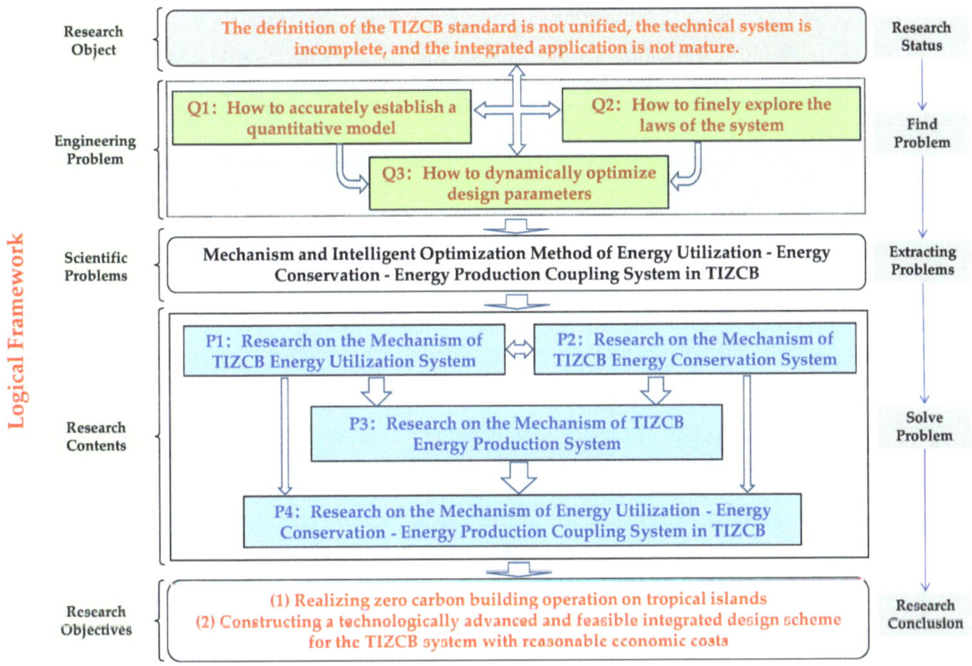

Figure 6. Comprehensive research approach for TIZCB.

6. Research Implementation Plan

6.1. TIZCB Energy Utilization System Mechanism Research Plan

6.1.1. Research Objectives

To survey and analyze the distribution characteristics of energy utilization parameters in tropical island buildings and establish a parametric design reference model based on the goals of TIZCB. Through quantitative simulation and sensitivity analysis of the energy utilization system, reveal the impact relationships between variations in design parameter combinations of the TIZCB energy utilization system islands and building performance. Develop data-driven models for building performance prediction and multi-objective optimization analysis, forming optimized combinations of design parameters for the energy utilization system in TIZCB.

6.1.2. Main Contents

(1) Construction of a Parametric Model for the Energy Utilization System in TIZCB

First, based on tropical island building design standards and climate characteristics, conduct research on design parameters such as personnel, equipment, HVAC systems, lighting control, and ventilation strategies in the energy utilization system of TIZCB. Then, utilize statistical analysis software to process research data, perform reliability and validity tests, and employ methods such as descriptive statistics, correlation analysis, and regression analysis to reveal the distribution patterns of design parameters in the TIZCB energy utilization system. Finally, based on the characteristics of energy utilization parameter distribution, select typical design parameters for the energy utilization system and, with reference to tropical island building design standards, establish a parametric design reference model based on the goals of TIZCB.

(2) Impact Relationships between the Design Parameters of the Energy Utilization System and Building Performance in TIZCB

First, based on the distribution characteristics of energy utilization system design parameters, use the Latin hypercube sampling (LHS) method to sample design parameters for the energy utilization system in TIZCB. Generate parameterized design schemes suitable for the batch input of IDF scripts in EnergyPlus and simulate the energy utilization system in batches using the Jeplus plugin. Then, using quantitative tools such as histograms, scatter plots, and standard regression coefficients, conduct correlation, determinacy, and uncertainty analyses of energy utilization system design parameters. Finally, perform local and global sensitivity analyses of energy utilization system design parameters based on the R language, revealing the impact relationships between variations in design parameter combinations of the TIZCB energy utilization system and building performance.

(3) Optimization Analysis of the Impact of the Design Parameters of the Energy Utilization System on Building Performance in TIZCB

First, based on the characteristics of energy utilization parameters and building performance indicators, select machine learning prediction models such as Bayesian regression, ridge regression, linear regression, support vector machine regression, random forest regression, and artificial neural networks. Construct a data-driven prediction and analysis model for the TIZCB energy utilization system. Then, use metrics such as RMSE, MSE, SE, MAPE, MAE, and NMSE to evaluate the model. Build multiple integrated models with excellent performance and conduct predictive analyses of energy utilization parameters and building performance. Finally, based on the predictive analysis, conduct multi-objective analyses of building performance indicators using intelligent optimization algorithms, forming optimized combinations of design parameters for the energy utilization system in TIZCB.

6.2. TIZCB Energy Conservation System Mechanism Research Plan

6.2.1. Research Objectives

To survey and analyze the thermal parameters of building energy conservation systems on tropical islands, establish a parametric design reference model based on the goals of TIZCB. Through quantitative simulation and sensitivity analysis of the thermal parameters of the energy conservation system, reveal the impact relationships between variations in design parameter combinations of the TIZCB energy conservation system and building performance. Under the guidance of thermal parameters, establish a material-structure parametric design model, develop data-driven models for building performance prediction, and form optimized combinations of design parameters for the energy conservation system in TIZCB.

6.2.2. Main Contents

(1) Construction of a Parametric Model for the Energy Conservation System in TIZCB

First, based on building design standards, climate characteristics, and energy utilization habits on tropical islands, research thermal performance parameters such as window-wall ratio, solar heat gain coefficient, and heat transfer coefficient. Then, use statistical analysis software to process research data, perform reliability and validity tests, and employ methods such as descriptive statistics, correlation analysis, and regression analysis to reveal the distribution patterns of thermal performance design parameters in the energy conservation system. Finally, based on the characteristics of thermal parameter distribution, select typical design parameters for the energy conservation system and, with reference to tropical island building design standards, establish a parametric design reference model based on the goals of TIZCB.

(2) Impact Relationships between Design Parameters of the Energy Conservation System and Building Performance in TIZCB

First, based on the distribution characteristics of thermal parameters in the energy conservation system, use the Latin hypercube sampling (LHS) method for sampling. Generate parameterized design schemes suitable for the batch input of IDF scripts in EnergyPlus and simulate the thermal parameters of the energy conservation system in batches using the Jeplus plugin. Then, using quantitative tools such as histograms, scatter plots, and standard regression coefficients, conduct correlation, determinacy, and uncertainty analyses of thermal design parameters in the energy conservation system. Finally, perform local and global sensitivity analyses of thermal design parameters based on the R language, revealing the impact relationships between variations in design parameter combinations of the TIZCB energy conservation system and building performance.

(3) Optimization Analysis of the Impact of the Design Parameters of the Energy Conservation System on Building Performance in TIZCB

First, under the guidance of thermal parameters, conduct batch simulations of material performance (thermal conductivity, density, and specific heat) and structural system (wall orientation, construction hierarchy, and material thickness) combination design parameters. Then, select machine learning prediction models such as Bayesian regression, ridge regression, linear regression, support vector machine regression, and random forest regression. Construct a data-driven prediction and analysis model for the TIZCB energy conservation system. Use metrics such as RMSE, MSE, SE, MAPE, MAE, and NMSE to evaluate the model. Build multiple integrated models with excellent performance and conduct predictive analyses of thermal parameters and building performance. Finally, based on the predictive analysis, conduct multi-objective analyses of building performance indicators using intelligent optimization algorithms, forming optimized combinations of design parameters for the energy conservation system in TIZCB.

6.3. TIZCB Production System Mechanism Research Plan

6.3.1. Research Objectives

To survey and analyze the distribution characteristics of building production parameters on tropical islands, establish a parametric design reference model based on the goals of TIZCB. Through quantitative simulation and sensitivity analysis of the production system, reveal the impact relationships between variations in design parameter combinations of the TIZCB production system and building performance. Develop data-driven models for building performance prediction, and form optimized combinations of design parameters for the production system in TIZCB.

6.3.2. Main Contents

(1) Construction of a Parametric Model for the Production System in TIZCB

First, based on building design standards, climate characteristics, and energy utilization habits on tropical islands, research design parameters such as photovoltaic–thermal integrated walls, multifunctional photovoltaic windows, and photovoltaic–thermal integrated roofs. Then, use statistical analysis software to process research data, perform reliability and validity tests, and employ methods such as descriptive statistics, correlation analysis, and regression analysis to reveal the distribution patterns of design parameters in the production system. Finally, based on the characteristics of production parameter distribution, select typical design parameters for the production system and, with reference to tropical island building design standards, establish a parametric design reference model based on the goals of TIZCB.

(2) Impact Relationships between the Design Parameters of the Production System and Building Performance in TIZCB

First, based on the distribution characteristics of design parameters in the production system, use the Latin hypercube sampling (LHS) method for sampling. Generate parameterized design schemes suitable for the batch input of IDF scripts in EnergyPlus and simulate the production system in batches using the Jeplus plugin. Then, using quantitative tools such as histograms, scatter plots, and standard regression coefficients, conduct correlation, determinacy, and uncertainty analyses of design parameters in the production system. Finally, perform local and global sensitivity analyses of design parameters based on the R language, revealing the impact relationships between variations in design parameter combinations of the TIZCB production system and building performance.

(3) Optimization Analysis of the Impact of the Design Parameters of the Production System on Building Performance in TIZCB

First, based on the characteristics of production parameters and building performance indicators, select machine learning prediction models such as Bayesian regression, ridge regression, linear regression, support vector machine regression, random forest regression, and artificial neural networks. Construct a data-driven prediction and analysis model for the TIZCB production system. Then, use metrics such as RMSE, MSE, SE, MAPE, MAE, and NMSE to evaluate the model. Build multiple integrated models with excellent performance and conduct predictive analyses of production parameters and building performance. Finally, based on the predictive analysis, conduct multi-objective analyses of building performance indicators using intelligent optimization algorithms, forming optimized combinations of design parameters for the production system in TIZCB.

6.4. TIZCB Intelligent System Mechanism Research Plan

6.4.1. Research Objectives

To interact and couple the design parameters of energy utilization, conservation, and production systems to obtain the distribution characteristics of the intelligent system and design parameters in TIZCB. Establish a parametric design reference model based on the goals of TIZCB. Develop data-driven models for building performance prediction and multi-objective optimization analysis, forming optimized combinations of design parameters for the intelligent system in TIZCB. Under the premise of achieving the zero-carbon operation goal of tropical island buildings, conduct an economic evaluation of the optimized combinations of design parameters for the intelligent system, obtaining economically reasonable design parameter combinations for the intelligent system in TIZCB.

6.4.2. Main Contents

(1) System-Coupled Parametric Modeling of the Intelligent System in TIZCB

First, interact the combined design schemes of the energy utilization, conservation, and production systems in TIZCB. Form the intelligent system of TIZCB through the coupling of the energy utilization—conservation—production system. Then, outline the technical solutions for the energy utilization planning, conservation components, and production system integrated design technologies of the intelligent system. Obtain the distribution characteristics of the design parameters of the intelligent system in TIZCB. Finally, based on the distribution characteristics of intelligent system parameters, select typical design parameters for the intelligent system and, with reference to tropical island building design standards, establish a parametric design reference model based on the goals of TIZCB.

(2) Technical Evaluation-Based Optimization and Regulation of the Intelligent System in TIZCB

First, based on the distribution characteristics of intelligent system design parameters, use the Latin hypercube sampling (LHS) method to sample design parameters for the intelligent system in TIZCB. Generate parameterized design schemes suitable for the batch input of IDF scripts in EnergyPlus and simulate the intelligent system in batches using the Jeplus plugin. Then, based on the distribution characteristics of intelligent system parameters, select machine learning prediction models such as Bayesian regression, ridge regression, linear regression, support vector machine regression, random forest regression, and artificial neural networks. Construct a data-driven prediction and analysis model for the TIZCB intelligent system. Use metrics such as RMSE, MSE, SE, MAPE, MAE, and NMSE to evaluate the model. Build multiple integrated models with excellent performance and conduct predictive analyses of intelligent system parameters and building performance. Finally, based on the predictive analysis, conduct multi-objective analyses of building performance indicators using intelligent optimization algorithms, forming optimized combinations of design parameters for the intelligent system in TIZCB.

(3) Decision Assessment of the TIZCB Intelligent System Based on Economic Evaluation

Firstly, based on the optimization and control results of the intelligent system in TIZCB, analyze the numerical relationship between energy utilization, energy conservation, and production systems to obtain an integrated optimization scheme for achieving zero-carbon operation of the building. Subsequently, apply the life cycle cost analysis method to evaluate the initial investment, operational maintenance, energy conservation, and equipment replacement costs of the integrated optimization scheme. Establish an economic decision assessment model for the integrated optimization scheme. Finally, through sensitivity analysis, identify and quantify uncertainties in economic evaluations, and formulate a technically advanced, feasible, and economically reasonable integrated optimization scheme for the intelligent system design in TIZCB.

7. Conclusions

(1) This study defines TIZCB, achieving ZCBs during building operation by relying entirely on renewable energy sources. It emphasizes the need for architectural design to adapt to tropical climate conditions, integrate local culture, and utilize innovative technologies and materials. A technical framework is proposed, focusing on energy utilization, energy conservation, energy production, and intelligent technologies, grounded in theories including system theory, control theory, and synergy theory.

(2) Using a macro–meso–micro analytical framework for TIZCB, this study outlines the macro objectives and micro parameter controls of such systems. System theory is employed for goal setting, control theory for parameter prediction, and synergy theory for establishing system interactions, providing a foundation for practical design and optimization.

(3) By integrating systems engineering theory and parametric design technology, this research investigates the impact of design parameter models on the performance of ZCBs. A theoretical framework is established covering energy utilization, energy conservation, energy production, and intelligent systems, offering clear research strategies for implementation.

(4) Through meticulous planning, this study develops parameter models and data-driven analysis for TIZCB, ensuring both zero-carbon operation and economic feasibility. It provides a systematic framework and practical guidance to advance clean energy development in Hainan and China's dual-carbon strategy.
(5) The TIZCB proposed in this study achieves zero carbon emissions only during the building operation phase. The next step is to advance TIZCB to further reduce carbon emissions from a whole lifecycle perspective; on the other hand, the framework and technical solutions proposed in this study still need to be validated in actual cases.

Author Contributions: Conceptualization, Q.W., K.Z. and P.G.; methodology, K.Z. and P.G.; Software, K.Z. and P.G.; validation, K.Z. and P.G.; formal analysis, K.Z. and P.G.; investigation, K.Z. and P.G.; resources, K.Z. and P.G.; data curation, K.Z. and P.G.; writing—original draft, K.Z. and P.G.; writing—review and editing, Q.W., K.Z. and P.G.; visualization, K.Z. and P.G.; supervision, Q.W.; project administration, Q.W.; funding acquisition, Q.W. All authors have read and agreed to the published version of the manuscript.

Funding: This research was funded by Key Research and Development Program of the People's Republic of China, grant number 2023YFC3106605, Hainan Province Major Science and Technology Plan Project, grant number ZDKJ2021024, the Project of Sanya Yazhou Bay Science and Technology City, grant number SKJC-2022-PTDX-021, the Wuhan Key R&D Plan, grant number 2023020402010590, Wuhan University of Technology Sanya Science and Education Innovation Park Open Fund Project, grant number 2022KF0030, the PhD Scientific Research and Innovation Foundation of Sanya Yazhou Bay Science and Technology City, grant number HSPHDSRF-2022-03-001, the PhD Scientific Research and Innovation Foundation of Sanya Yazhou Bay Science and Technology City, grant number HSPHDSRF-2022-03-002, the PhD Scientific Research and Innovation Foundation of Sanya Yazhou Bay Science and Technology City, grant number HSPHDSRF-2023-03-001.

Data Availability Statement: The sample data is derived from research interviews conducted by our research team, and due to confidentiality reasons, the data cannot be disclosed.

Conflicts of Interest: The authors declare no conflict of interest.

References

1. Kongboon, R.; Gheewala, S.H.; Sampattagul, S. Greenhouse gas emissions inventory data acquisition and analytics for low carbon cities. *J. Clean. Prod.* **2022**, *343*, 130711. [CrossRef]
2. Too, J.; Ejohwomu, O.A.; Hui, F.K.P.; Duffield, C.; Bukoye, O.T.; Edwards, D.J. Framework for standardising carbon neutrality in building projects. *J. Clean. Prod.* **2022**, *373*, 133858. [CrossRef]
3. Hao, J.; Gao, F.; Fang, X.; Nong, X.; Zhang, Y.; Hong, F. Multi-factor decomposition and multi-scenario prediction decoupling analysis of China's carbon emission under dual carbon goal. *Sci. Total Environ.* **2022**, *841*, 156788. [CrossRef] [PubMed]
4. China Building Energy Conservation Association. 2022 China Urban and Rural Construction Carbon Emission Series Research Report. 2023. Available online: https://www.cabee.org/site/content/24420.html (accessed on 7 October 2023).
5. The Central People's Government of the People's Republic of China. Notice of the Ministry of Housing and Urban Rural Development on Issuing the 14th Five Year Plan for Building Energy Conservation and Green Building Development. 2022. Available online: https://www.gov.cn/zhengce/zhengceku/2022-03/12/content_5678698.htm (accessed on 7 October 2023).
6. Notice of the Central People's Government of the People's Republic of China. Ministry of Science and Technology and Nine Other Departments on Issuing the Implementation Plan for Science and Technology to Support Carbon Peaking and Carbon Neutrality (2022–2030). 2022. Available online: https://www.gov.cn/zhengce/zhengceku/2022-08/18/content_5705865.htm (accessed on 7 October 2023).
7. Moghayedi, A.; Michell, K.; Hübner, D.; Jeune, K.L.; Massyn, M. Examine the impact of green methods and technologies on the environmental sustainability of supportive education buildings, perspectives of circular economy and net-zero carbon operation. *Facilities* **2023**, *42*, 201–222. [CrossRef]
8. Scherz, M.; Passer, A.; Kreiner, H. Challenges in the achievement of a net zero carbon built environment—A systemic approach to support the decision-aiding process in the design stage of buildings. *IOP Conf. Ser. Earth Environ. Sci.* **2020**, *588*, 032034. [CrossRef]
9. Leung, M.; Tong, X.; Tong, J.; Lee, P. Airborne bacterial assemblage in a zero carbon building: A case study. *Indoor Air* **2018**, *28*, 40–50. [CrossRef]
10. Bui, T.T.P.; Domingo, N.; Macgregor, C.; Wilkinson, S. Zero carbon refurbishment for existing buildings: A literature review. *IOP Conf. Ser. Earth Environ. Sci.* **2022**, *1101*, 022017. [CrossRef]

11. Zhao, X.; Pan, W. Co-productive interrelations between business model and zero carbon building: A conceptual framework. *Built Environ. Proj. Asset Manag.* **2017**, *7*, 353–365. [CrossRef]
12. Pan, W.; Ning, Y. A socio-technical framework of zero carbon building policies. *Build. Res. Inf.* **2014**, *43*, 94–110. [CrossRef]
13. Girard, S.; Mckoen, K.; Koch, A. Towards a neighbourhood scale for low- or zero-carbon building projects. *Build. Res. Inf.* **2012**, *40*, 527–537. [CrossRef]
14. Bui, T.T.P.; Macgregor, C.; Wilkinson, S.; Domingo, N. Towards zero carbon buildings: Issues and challenges in the New Zealand construction sector. *Int. J. Constr. Manag.* **2022**, *23*, 2709–2716. [CrossRef]
15. Nidhin, B.K.S.N.; Domingo, N.; Bui, T.T.P.; Wilkinson, S. Construction stakeholders' knowledge on zero carbon initiatives in New Zealand. *Int. J. Build. Pathol. Adapt.* **2023**. *ahead of print*. [CrossRef]
16. Hasan, L.; Lizarralde, G.; Lachapelle, E. The problem with net zero building policy: Rreflecting on the making, and future, of a global discourse. *IOP Conf. Ser. Earth Environ. Sci.* **2023**, *1196*, 012002. [CrossRef]
17. Torriti, J.; Lees, T.; Schweber, L. Framing evidence: Policy design for the zero-carbon home. *Build. Res. Inf.* **2015**, *43*, 420–434. [CrossRef]
18. Walker, G.; Karvonen, A.; Guy, S. Reflections on a policy denouement: The politics of mainstreaming zero-carbon housing. *Trans. Inst. Br. Geogr.* **2016**, *41*, 104–106. [CrossRef]
19. Lees, T.; Sexton, M. An evolutionary innovation perspective on the selection of low and zero-carbon technologies in new housing. *Build. Res. Inf.* **2014**, *42*, 276–287. [CrossRef]
20. Steijger, L.A.; Buswell, R.A.; Smedley, V.A.; Firth, S.K.; Rowley, P. Establishing the zero-carbon performance of compact urban dwellings. *J. Build. Perform. Simul.* **2013**, *6*, 319–334. [CrossRef]
21. Mcleod, R.S.; Hopfe, C.J.; Rezgui, Y. An investigation into recent proposals for a revised definition of zero carbon homes in the UK. *Energy Policy* **2012**, *46*, 25–35. [CrossRef]
22. Liu, Y.; Xue, S.; Guo, X.; Zhang, B.; Sun, X.; Zhang, Q.; Wang, Y.; Dong, Y. Towards the goal of zero-carbon building retrofitting with variant application degrees of low-carbon technologies: Mitigation potential and cost-benefit analysis for a kindergarten in Beijing. *J. Clean. Prod.* **2023**, *393*, 136316. [CrossRef]
23. Kalbasi, R.; Afrand, M. Which one is more effective to add to building envelope: Phase change material, thermal insulation, or their combination to meet zero-carbon-ready buildings? *J. Clean. Prod.* **2022**, *367*, 133032. [CrossRef]
24. Manyumbu, E.; Martin, V.; Chiu, J.N. Prospective PCM–desiccant combination with solar-assisted regeneration for the indoor comfort control of an office in a warm and humid climate—A numerical study. *Energies* **2023**, *16*, 5391. [CrossRef]
25. Matias, M.L.; Carlos, E.; Branquinho, R.; Valle, H.D.; Marcelino, J.; Morais, M.; Pimentel, A.; Rodrigues, J.; Monteiro, T.; Fortunato, E.; et al. A comparison between solution-based synthesis methods of ZrO_2 nanomaterials for energy storage applications. *Energies* **2022**, *15*, 6452. [CrossRef]
26. Pei, L.; Schalbart, P.; Peuportier, B. Quantitative evaluation of the effects of heat island on building energy simulation: A case study in Wuhan, China. *Energies* **2023**, *16*, 3032. [CrossRef]
27. Dicko, A.H.; Roux, C.; Peuportier, B. Achieving net zero carbon performance in a French apartment building? *Energies* **2023**, *16*, 7608. [CrossRef]
28. Gao, H.; Wang, L.; He, S.; Liu, J. Zero-carbon-driven multi-energy coordinated sharing model for building cluster. *J. Clean. Prod.* **2023**, *430*, 139658. [CrossRef]
29. World Green Building Council. The Net Zero Carbon Buildings Commitment. 2022. Available online: https://worldgbc.org/thecommitment/?_sfm_signatory_type=gbc (accessed on 7 October 2023).
30. Ministry of Housing and Urban Rural Development of the People's Republic of China. Notice of the General Office of the Ministry of Housing and Urban Rural Development on Public Solicitation of Comments on the National Standard "Zero Carbon Building Technology Standard (Draft for Comments)". 2023. Available online: https://www.mohurd.gov.cn/gongkai/zhengce/zhengcefilelib/202307/20230724_773337.html (accessed on 7 October 2023).
31. Ministry of Housing and Urban Rural Development of the People's Republic of China. Hainan Building Prefabricated Buildings Suitable for Tropical Island Characteristics. 2023. Available online: https://www.mohurd.gov.cn/xinwen/dfxx/202312/20231207_775564.html (accessed on 27 January 2024).
32. Hainan Provincial Department of Housing and Urban Rural Development. Notice on Issuing the Three Year Action Plan for the Development and Improvement of Prefabricated Buildings (Green Buildings) in Hainan Province (2023–2025). 2023. Available online: https://zjt.hainan.gov.cn/szjt/0404/202311/ee84db1de1e844bba234994c958c9579.shtml?ddtab=true (accessed on 7 October 2023).
33. Ceylan, B.O.; Karatuğ, Ç.; Akyuz, E.; Arslanoğlu, Y.; Boustras, G. A system theory (STAMP) based quantitative accident analysis model for complex engineering systems. *Saf. Sci.* **2023**, *166*, 106232. [CrossRef]
34. Poller, A. Exploring and managing the complexity of large infrastructure projects with network theory and model-based systems engineering—The example of radioactive waste disposal. *Syst. Eng.* **2020**, *23*, 443–459. [CrossRef]
35. Lo, S.-H.; Liu, C.-D. Furthering systems theory towards social theory by revisiting critical systems thinking: Connecting problem-solving and theory-building. *Kybernetes* **2023**, *52*, 182–206. [CrossRef]
36. Corning, P.A. A systems theory of biological evolution. *Biosystems* **2022**, *214*, 104630. [CrossRef]
37. Delcea, C.; Javed, S.A.; Florescu, M.-S.; Ioanas, C.; Cotfas, L.-A. 35 years of grey system theory in economics and education. *Kybernetes* **2023**. *ahead of print*. [CrossRef]

38. Yonat, N.; Isaac, S.; Shohet, I.M. Complex infrastructure systems analysis and management: The theory of faults. *Smart Sustain. Built Environ.* **2023**. *ahead of print*. [CrossRef]
39. Hazelrigg, G.A. Toward a theory of systems engineering. *J. Mech. Des. Trans. ASME* **2022**, *144*, 011402. [CrossRef]
40. Xing, J.; Meng, H.; Meng, X. An urban pipeline accident model based on system engineering and game theory. *J. Loss Prev. Process Ind.* **2020**, *64*, 104062. [CrossRef]
41. Alvarez, J.T.; Ramírez-Correa, P. A brief review of systems, cybernetics, and complexity. *Complexity* **2023**, *2023*, 8205320. [CrossRef]
42. Thomakos, D. The origins of forward-looking decision making: Cybernetics, operational research, and the foundations of forecasting. *Decis. Anal. J.* **2023**, *8*, 100284. [CrossRef]
43. Scholte, T.; Sweeting, B. Possibilities for a critical cybernetics. *Syst. Res. Behav. Sci.* **2022**, *39*, 986–989. [CrossRef]
44. Burbano, A.; Reyes, E. Capsaicin and cybernetics: Mexican intellectual networks in the foundation of cybernetics. *AI Soc.* **2022**, *37*, 1013–1025. [CrossRef]
45. Aboulmouna, L.; Raja, R.; Khanum, S.; Gupta, S.; Maurya, M.R.; Grama, A.; Subramaniam, S.; Ramkrishna, D. Cybernetic modeling of biological processes in mammalian systems. *Curr. Opin. Chem. Eng.* **2020**, *30*, 120–127. [CrossRef]
46. Nagel, L.; Roth, S.; Kleve, H. Theory of the third: A cybernetic approach to escalative conflict dynamics in business families. *Syst. Res. Behav. Sci.* **2023**, *1*. [CrossRef]
47. Wu, C.; Gu, W.; Luo, E.; Chen, X.; Lu, H.; Yi, Z. An economic cybernetic model for electricity market operation coupled with physical system dynamics. *Appl. Energy* **2023**, *335*, 120764. [CrossRef]
48. Yang, B.; Serrano, J.V.; Launer, M.A.; Wang, L.; Rabiei, K. A comprehensive and systematic study on the cybernetics management systems. *Syst. Pract. Action Res.* **2023**, *36*, 479–504. [CrossRef]
49. Lavanderos, L. From cybersin to cybernet. Considerations for a cybernetics design thinking in the socialism of the XXI century. *AI Soc.* **2022**, *37*, 1279–1292. [CrossRef]
50. Bruun, O.; Rubin, O. Authoritarian environmentalism—Captured collaboration in Vietnamese water management. *Environ. Manag.* **2023**, *71*, 538–550. [CrossRef] [PubMed]
51. Quader, M.A.; Khan, A.U.; Malak, M.A.; Kervyn, M. Mainstreaming decentralization and collaboration in disaster risk management: Insights from coastal Bangladesh. *Int. J. Disaster Risk Sci.* **2023**, *14*, 382–397. [CrossRef]
52. Ulibarri, N.; Imperial, M.T.; Siddiki, S.; Henderson, H. Drivers and dynamics of collaborative governance in environmental management. *Environ. Manag.* **2023**, *71*, 495–504. [CrossRef] [PubMed]
53. Prabhakaran, K.K.; Saravanakumar, R. Maximum power extraction from a wind turbine using terminal synergetic control. *IETE J. Res.* **2023**, *69*, 1152–1159. [CrossRef]
54. Olsén, M.; Oskarsson, P.-A.; Hallberg, N.; Granåsen, M.; Nordström, J. Exploring collaborative crisis management: A model of essential capabilities. *Saf. Sci.* **2023**, *162*, 106092. [CrossRef]
55. Dong, Z.; Xing, J.; Zhang, F.; Wang, S.; Ding, D.; Wang, H.; Huang, C.; Zhen, H.; Jiang, Y.; Hao, J. Synergetic $PM_{2.5}$ and O_3 control strategy for the Yangtze River Delta, China. *J. Environ. Sci.* **2023**, *123*, 281–291. [CrossRef]

energies

MDPI

Article

Heat Pumps with Smart Control in Managing Australian Residential Electrical Load during Transition to Net Zero Emissions

Adrian Rapucha [1], Ramadas Narayanan [2,*] and Meena Jha [1]

1 School of Engineering and Technology, Central Queensland University, Sydney, NSW 2000, Australia; adrian.rapucha@cqumail.com (A.R.); m.jha@cqu.edu.au (M.J.)
2 School of Engineering and Technology, Central Queensland University, University Drive, Bundaberg, QLD 4670, Australia
* Correspondence: r.narayanan@cqu.edu.au

Abstract: Australia, like many other countries around the world, is undergoing a transition toward net zero emissions. It requires changes and development in many sectors, which not only bring benefits but also challenges. The rapid growth in renewable energy sources (RESs) is necessary to decarbonise electricity generation but negatively affects grid stability. Residential buildings also contribute to this issue through specific load profiles and the high penetration of rooftop photovoltaic (PV) installations. Maintaining grid balance will be crucial for further emissions reductions. One of the potential solutions can be the replacement of conventional heating and cooling systems in houses with solutions capable of storing energy and shifting the electrical load. As presented in this paper, heat pumps and hydronic systems can significantly improve the electrical load of a typical South Australian household when they are controlled by algorithms reacting to the current grid conditions and household-generated electricity compared to conventional solutions. TRNSYS 18 simulations of air source and ground source heat pump systems with smart control based on measured electricity consumption and domestic hot water usage data showed the possibility of total energy consumption reduction, shifting the load from peak periods towards periods of excessive RES generation and increasing self-consumption of rooftop PV electricity. These improvements reduce the amount of emissions generated by such a household and allow for further development of other sectors.

Keywords: net zero emissions; renewable energy; heat pump; hydronic system; smart grid; peak demand; grid stability

Citation: Rapucha, A.; Narayanan, R.; Jha, M. Heat Pumps with Smart Control in Managing Australian Residential Electrical Load during Transition to Net Zero Emissions. *Energies* **2024**, *17*, 2977. https://doi.org/10.3390/en17122977

Academic Editors: Zsuzsa Szalay, Francesco Minichiello and Bruno Peuportier

Received: 7 May 2024
Revised: 11 June 2024
Accepted: 13 June 2024
Published: 17 June 2024

1. Introduction

Australia fulfilled its 2020 carbon emissions plans and is on track to meet the 2050 goal of reaching net zero emissions [1]. However, it requires multifield improvements. Among the affected sectors is electricity generation, with a modelled reduction in emissions of up to 97% compared to 2005 [1]. This is to be achieved mainly by an enormous increase in the contribution of RESs. Another affected sector is Australian buildings, including residential ones, which account for over 10% of Australia's current carbon emissions [2]. These two sectors are connected by the electrical grid, which also needs to undergo changes to adapt to new, less carbon-emission-dependent realities.

Most Australian residences are connected to the grid operated by the Australian Energy Market Operator (AEMO). The Australian National Electricity Market (NEM) is one of the longest interconnected electricity grids in the world (about 5000 km), covering six out of eight Australian states (Queensland, New South Wales, Australian Capital Territory, Victoria, South Australia and Tasmania) [3]. AEMO allows for trading and exchanging energy between regions to meet the demand in every part of the grid. AEMO is also responsible for demand prediction and maintaining the necessary level of reserves [4].

However, these tasks are becoming more and more difficult due to reasons such as the increasing penetration of RESs and the peak demand phenomenon. According to ENERGEIA forecasts, issues such as instabilities in demand and complexity of electricity generation control will become more challenging in the following decades [5].

Grid stability is a cornerstone of a well-functioning electrical power system, which is indispensable in modern society. An imbalance between electricity generation and demand can lead to voltage and frequency deviations, potentially causing equipment damage and disruptions in the power supply. Moreover, grid instability can result in blackouts or brownouts, negatively impacting both residential and industrial consumers. Australia is currently confronted with issues related to grid instability, and the problem is likely to worsen [6]. This contradicts one of the key principles of Australia's Long-Term Emissions Reduction Plan, which mentions providing reliable power [1]. Therefore, the implementation of increased grid-preserving technologies is required.

1.1. Residential Electrical Load

The residential sector is a significant contributor to grid stress [7]. Specific consumption profiles and increasing decentralised electricity generation cause demand and supply instability. These problems also concern the Australian National Grid. One of the contributors to grid imbalance is the high penetration of rooftop PV systems, which causes excess and uncontrollable generation. Australia is the world leader in rooftop PV systems, and around 35% of Australian households have installed a solar PV system, with states such as Queensland and South Australia reaching the level of about 45% [8]. The other contributor is a fluctuating load of heating and cooling devices due to weather conditions and residents' behaviour patterns. The number of small and medium reverse-cycle air conditioners, which are typical for the residential market, is increasing [9]. This creates additional load, especially during extreme weather events. This trend is likely to continue due to decarbonisation and electrification of heating [10] and increasing cooling and dehumidification demand [11].

Therefore, many research considerations are focused on improvements in energy consumption in the housing sector. Most proposed solutions are concerned with increasing rooftop PV generation self-consumption and demand-response programs. The desirable result would be shifting the load from peak hours to periods with the highest PV generation. This would help to utilise carbon emission-free energy instead of other energy sources, including fossil-based generators. As the major contributors to household energy consumption, space conditioning and water heating [8] are the main factors affecting the residential electricity load profile. Therefore, many researchers have investigated the possibility of adapting air conditioners and domestic hot water (DHW) heaters in the smart grid to address the demand response in Australia.

1.2. Current Solutions

The factors mentioned above necessitate the introduction of new solutions, such as smart management. The smart grid allows for detecting and reacting to changes in electricity demand and fluctuations in RES output using a digital communication network [12]. All the receivers and generators are connected to the network and monitored in real time, and if necessary, the grid operator can introduce appropriate actions. Smart management is based on the Internet of Things (IoT) concept, which relies on interconnections of various devices and objects over the global network to allow them to exchange data and interact all the time [13]. Therefore, the IoT opens new possibilities for measuring, understanding, and controlling more and more complex demand and supply phenomena.

Numerous research programs have proved that IoT has great potential in managing residential peak demand [13–15]. Australia has already implemented or is still implementing IoT and smart management solutions, beginning from the most basic, such as controlled load. This is a separate circuit powering large receivers, such as hot water heaters or pool filtration and heat pump systems, which is only activated during off-peak hours [16]. Another more complex and flexible solution is remote limiting or turning off large devices if the grid struggles with too high demand. An example of this system is the PeakSmart air conditioning program, where consumers agree to add a remote control to their air conditioners in exchange for a reward. If the grid experiences unusually high peak demand, the AC unit output can be limited, or the device can be completely turned off until the peak drops [17]. Finally, some programs analyse the entire household's energy consumption and production and suggest when is the best time to use electric appliances or when it is desired not to use them. An example of this solution is granted by the SA government EmberPuls program [18]. It offers energy usage and production analysis, operation cost optimisation suggestions, and remote control of various electric circuits (appliances) in a household.

Unfortunately, most of these solutions have a major disadvantage. They require changes in the behaviour patterns of users and affect their comfort. This is especially noticeable when controlling AC load. Residential AC is one of the major contributors to peak demand events, particularly on extremely hot or cold days [19]. Therefore, the demand-response programs usually limit AC load when it is the most needed. Due to the low heat capacity of air, the lack of air conditioning is quickly apparent, causing aversion to these types of programs among the population.

Another approach to stabilise residential electrical load is transitioning towards the self-consumption of energy produced from a rooftop PV system proposed in many studies [20–22]. The considerations contain conclusions indicating financial (reducing bills), technical (decreasing grid load), and environmental (reduced carbon emission) benefits. However, they also mention the need to have batteries to store energy. Due to variability in PV generation that does not coincide with residential electricity demand, it is necessary to store exceeded energy to be able to use it later. Therefore, instead of pushing electricity to the grid during midday and contributing to afternoon peak demand, charging batteries while the generation is too high to use stored energy later would be more desirable.

Nevertheless, some researchers warn against the limited lifetime, dangerous materials used for production, energy-consuming production, and costly recycling of batteries, and they suggest considering other solutions to store energy [22]. This fact should be taken into account to mitigate environmental or health problems. Another issue causing a hurdle for battery storage adoption is the cost of the system, which is still high.

1.3. Heat Pumps in Grid Stabilisation

Heat pumps are also refrigeration devices, such as AC. However, they require a hydronic system to transfer heat energy. Hydronic systems use water as an additional medium to distribute thermal energy in the building. There are two types of residential heat pumps. An air source heat pump (ASHP), similar to AC, uses outside air as one of the reservoirs, whereas a ground source heat pump (GSHP) uses the ground to collect or dump heat [23]. Hydronic systems are also found in many variants. The most common solutions for heating are hydronic underfloor heating (HUFH) and radiators, whereas a hydronic fan coil unit allows for both heating and cooling. The hydronic system can also prepare DHW. An additional heat exchanger is necessary for this task because a hydronic system is a closed circuit and cannot mix with drinking water. For domestic applications, this is usually a DHW tank with a coil inside, which is heated by circulating hydronic system water. Another common part of the system is a buffer tank, which, among other tasks, allows for the accumulation of thermal energy. Therefore, it is often named as thermal energy storage (TES). It also allows for the separation of a heat pump from the space conditioning system. Thanks to that, both parts of the system can work independently, as long as the water temperature in the buffer tank is in the operating range of a heat pump and still allows for conditioning (is colder or warmer than the conditioned space).

Storage ability and independent heat pump and distribution system operation are desirable for smart grid management [24]. It allows for controlling the heat pump output without disturbing the indoor temperature, maintaining constant thermal comfort [24].

According to simulations from Spain [25], heat pumps can provide heat reserve during electricity overproduction time, consuming problematic surplus. A hydronic system can accumulate it, delivering only the required heat to the conditioned rooms. Moreover, a heat pump can be turned off, or its output can be tempered, during peak hours, reducing the grid load while the building is still conditioned by energy stored in water. Therefore, hydronic systems with heat pumps would be much more suitable for stabilising the residential load than conventional solutions. Heat pumps with TESs tend to affect users' comfort much less than AC when their load is adjusted to the current grid needs. They also do not use as many dangerous and environmentally affecting materials as battery systems.

The importance of heat pumps in smart grid management has been studied in numerous research works, which can be categorised into three groups regarding the approach: grid-focused, price-focused, and renewable-energy-focused [24]. An example of grid-focused research is presented in a study from Germany that explores the high penetration of PV with and without heat pump integration [26]. The results show fewer critical grid parameters under high penetration of heat pumps. A study from Denmark, in turn, aims at the dynamic pricing of electricity, which corresponds to the grid's conditions [27]. This type monitors the current electricity price, which reflects the supply–demand ratio. When there is too much supply, the price is low to encourage customers to use more energy. When the grid suffers from insufficient generation, the price rises to promote a reduction in consumption. Heat pumps under price-reacting management helped to keep the grid under grid load limits even during critical scenarios. Continuing, the problem with increasing fluctuations in Ireland's grid caused by the growing penetration of wind power resulted in an investigation of electrification of residential heating as a solution [28]. According to the results, 20% of national domestic heat demand delivered by heat pumps with thermal storage would add only 3% CO_2 emissions by the electricity sector, simultaneously decreasing much more pollution from fuel burning. Moreover, storage allows for reducing peak demand and influences the flattening load profile. These research projects, mainly from Europe and the U.S., demonstrate that hydronic heating and cooling systems based on heat pumps have a huge potential for application as a tool in smart grid management.

Unfortunately, heat pumps and hydronic systems are not popular in Australia due to the lack of awareness among builders and house owners, the complexity of the system, and higher than conventional system investment costs. Therefore, their potential for stabilising the Australian grid has not been adequately investigated. One of the few studies was conducted at RMIT University in Melbourne, where the authors analysed the usage of an air source heat pump and thermal storage to deliver DHW, heating, and cooling while utilising rooftop PV energy and decreasing peak loads [29,30]. The simulations conducted for a representative Australian house located in Brisbane and equipped with a 5 kW PV installation showed 76% potential to reduce annual grid load compared with a system without thermal storage. Moreover, the results showed the potential to decrease peak load by approximately 45% and increase solar energy self-consumption to 56%. Nevertheless, the study had some limitations. First of all, the model only took rooftop PV generation into consideration and did not react to grid conditions. Secondly, the model used very simplified occupancy and behaviour patterns. Lastly, the air source heat pump used in their model is typical for commercial installations, not residential systems.

Therefore, it has been decided to investigate the potential of hydronic heat pump systems with smart management in flattening the residential electrical demand and generation, which will be crucial in the ongoing transition towards net zero carbon emission. The study presented in this paper focused on analysing the impact of such systems in a typical South Australian household compared to conventional solutions.

2. Study

The major part of the study was designing and conducting TRNSYS simulations. Their goal was to allow a comparison of electricity consumption profiles in a typical South Australian household between a conventional system with reverse-cycle air conditioning coupled with an electric water heater and a hydronic heat pump system for space conditioning and DHW preparation. Two heat pump types were investigated: ASHP and GSHP. Heat pump operation was controlled by a standard or a smart control strategy. The second one focused on adjusting heat pump output to the current grid condition and surplus PV generation. The current grid conditions were determined based on the wholesale electricity price. The simulations were performed for the Adelaide location for a year's duration. The TRNSYS timestep was set to 1 min to ensure high accuracy of the calculations. However, input and result data timesteps were set to 30 min. Any later-mentioned timesteps refer to 30 min intervals.

2.1. TRNSYS Model and Input Data

The study was based on the electricity consumption and PV generation load of dozens of Adelaide households measured by the Commonwealth Scientific and Industrial Research Organisation (CSIRO) [31]. Monitoring was conducted from 2012 to 2017. The averaged data from 39 households was collected from various circuits with a 30 min resolution. Due to the significant drop in the number of households participating in the program in the second half of 2017, it was decided to use data from 2016. Table 1 presents the electricity consumption and generation used in the study. Due to the size of the data, the table contains only total annual values for PV generation, AC load, and remaining load. The remaining load is a sum of all other circuits, excluding AC and DHW. DHW load was excluded due to the fact that some of the houses were equipped with a gas DHW heater, which could distort the results.

Table 1. Total annual load of a typical South Australian household from CSIRO study.

Load Type	Usage
PV Load	1264.7 kWh
AC Load	835.5 kWh
Remaining Load	2524.6 kWh

An AC electricity load constituted the consumption of the conventional space conditioning system for the rest of the study. Similar to the previously presented case for Queensland [32], the electric load was used to calculate a typical household space conditioning thermal load. The electricity consumption of the reverse-cycle air conditioning unit and its thermal energy capacity are relative to indoor and outdoor conditions. Knowing the electricity power P and using a coefficient of performance at specific outdoor temperatures COP_T of the popular middle-range split air conditioner series [33], it was possible to estimate the thermal load Q produced by the AC using the equation below:

$$Q = COP_T \cdot P \quad (1)$$

The required weather data for the analysed year were acquired from Meteostat [34], while indoor conditions were assumed to be constant: 24 °C and 55% relative humidity for the cooling season and 21 °C for the heating season. Another assumption was that days with a mean ambient temperature <18 °C were heating days, and ≥18 °C were cooling days. This process allowed for the creation of house heating and cooling loads, which are affected not only by weather conditions and building properties but also by behaviour patterns and user preferences. The peak cooling load reached 5 kW, while the peak heating load was 2.8 kW. The loads are relatively low. However, it needs to be remembered that this is an average load of 39 households.

As explained earlier, a DHW load for the conventional solution had to be produced. The measured data from another study allowed for the preparation of a daily DHW usage profile [35] to represent typical household usage accurately. This load was then loaded to a simple TRNSYS simulation of a 200l electric water heater to create an electricity consumption profile of conventional DHW heating.

The TRNSYS model of a hydronic system was composed of the following components:

- ASHP with implemented weather data or GSHP with implemented GHX model;
- 200l DHW tank with outlet equipped with a tempering valve set to 50 °C;
- 350l TES for cooling;
- 250l TES for heating;
- Pumps, piping, and valves;
- DHW, heating and cooling circuits imposing load;
- Control system components.

Performance data of the heat pumps were modelled based on real devices available locally or in nearby markets. The GSHP was modelled according to Ecoforest ecoGEO+ 1–9 [36] performance. The TRNSYS ground source heat pump component is single-speed. Therefore, only data for the maximum output was considered. However, due to the capacity of GSHP significantly exceeding the peak loads, 60% of heating and cooling outputs were assumed, reaching 6.54 kW for both. Electricity power was accordingly adjusted. The GSHP was connected to a 110 m vertical ground heat exchanger. The model of the ASHP was based on the performance of a Mitsubishi Electric PUHZ-SHW 80VAA/YAA [37]. Similar to the GSHP, the TRNSYS component from the ASHP is single-speed. However, the performance of the ASHP is highly dependent on the outdoor temperature, and therefore, most typical residential devices are variable-speed. To mirror this for cooling, the performance data for various outdoor temperatures are modelled for different compressor speeds; the hotter the ambient temperature, the higher the compressor speed used. Due to the heating load being significantly lower than the cooling, only the middle compressor speed was used. As previously for the GSHP, the output was scaled to match the household space conditioning load. This time, it was 80% of the capacity, which reached 5.12 kW for heating and 5.68 kW for cooling. TES tank sizes were determined using Equation (2), recommended by various manufacturers [38], and rounded to the nearest commonly available sizes.

$$V_{TES} = 81.54 + 53.8 \cdot Q_{max} \quad (2)$$

The model's operation principle is that the heat pump provides heating or cooling to one of the tanks. For space conditioning, it is a TES heating or a TES cooling tank, while the hot water system is a DHW tank. The tanks' temperatures are maintained according to the strategy further described and kept within the hysteresis sets. Discharging of the tanks is then conducted by another circuit connected to each of them. This allows for the independent operation of the heat pump and DHW, as well as heating and cooling circuits. All of the main components are presented in Figure 1.

2.2. *Electricity Trading Price as a Grid Condition Indicator*

The model required an indicator of whether the grid is well balanced or if it suffers from overgeneration or insufficient generation. Electricity trading prices are a good index of the grid condition. The price drops when the electricity demand is low, or generation is too high. Meanwhile, high demand or insufficient generation makes the prices rise. The Australian Energy Market Operator (AEMO) shares and archives the electricity trading prices and total demand on its website [39]. The data were acquired and processed to meet the model's needs. It was necessary for the model to use the price for 2016 to ensure that all of the inputs (energy consumption and generation, weather data, and electricity price) aligned and presented the same events, such as peak load periods during hot days. The price was divided into five categories, with the lowest referring to extreme overgeneration and the highest referring to extreme peak demand. The price ranges for each category are presented in Table 2.

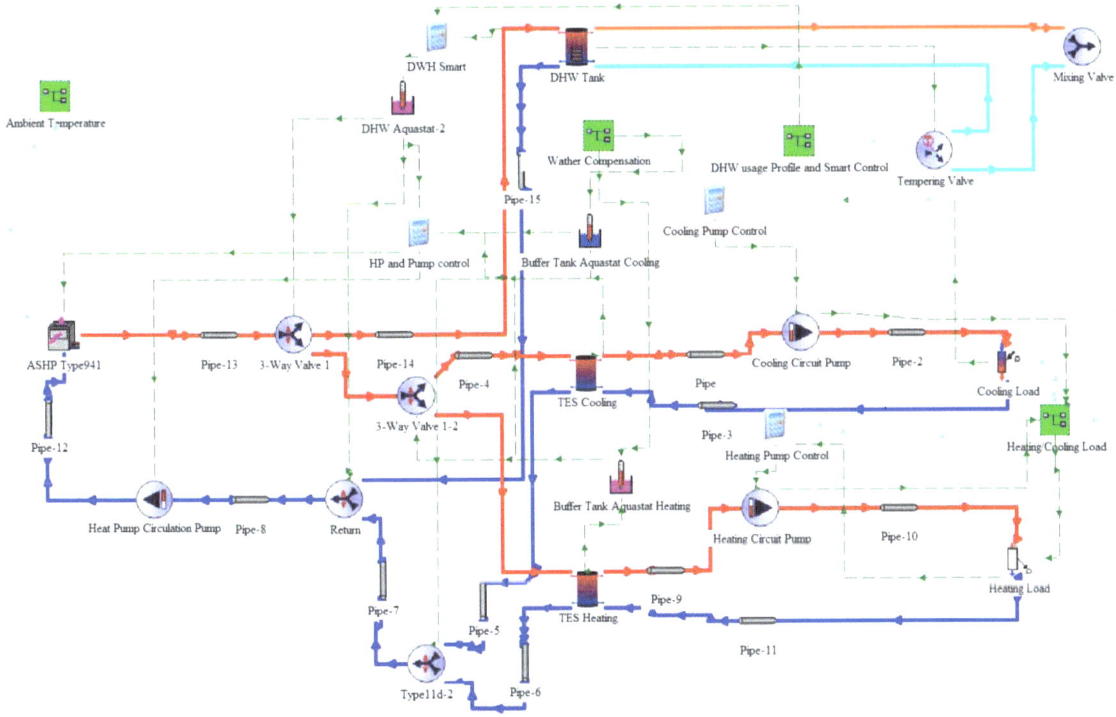

Figure 1. TRNSYS model of the hydronic system based on ASHP.

Table 2. SA electricity price in AUD/MWh divided into five categories

Category	1	2	3	4	5
Mean Δ% Price	<−50% AUD 40.29	<−25% AUD 60.44	Mean AUD 80.59	≥25% AUD 100.74	≥50% AUD 120.88

2.3. Control Strategies

Control strategies were based on the systems typical in residential heat pumps. The heat pump prioritises DHW preparation over space conditioning. Therefore, whenever DHW aquastat calls DHW tank heating, the heat pump is turned on or switched from another mode and starts domestic hot water tank heating until the aquastat is satisfied. When the DHW tank temperature is satisfied, the heat pump can switch to supply TES heating or TES cooling tanks. Both TES tanks' temperatures are maintained according to the weather compensation (heating and cooling curves). This allows for adjustment of the temperature in the TES tanks according to the outdoor temperature. Weather compensation is commonly used in this type of system to increase efficiency [40].

2.3.1. Standard Control Strategy

A standard control strategy was created to allow a direct comparison between a conventional and a hydronic system. This strategy does not include any components focused on matching the grid needs or increasing the self-consumption of household-generated PV electricity. This step also allowed for the determination of whether any positive or negative impact on the grid was due to hydronic system properties or smart strategies.

This control strategy assumes a fixed DHW tank temperature of 57 °C with 6 °C hysteresis. The weather compensation curve for cooling was created in such a way that for the hottest timestep of the analysed year, the TES cooling temperature was 7 °C. When the outdoor temperature reaches 18 °C, which is the cooling and heating mode boundary value, the TES cooling temperature is 18 °C. It was assumed that the system was equipped with a hydronic fan coil with a cooling capacity for 7 °C entering water temperature, which exceeds the peak cooling load by 25%. It turned out that a linear characteristic was not sufficient to provide enough cooling capacity in the middle of the range. Therefore, an additional point was added to create a quadratic characteristic.

The heating curve was created using the same methodology. For the coldest timestep, the TES heating tank is heated to 45 °C, whereas for 18 °C outdoor, the tank temperature set is 30 °C. For simplification purposes, it was assumed that the heating system uses the same hydronic fan coil as for cooling. In this case, a linear characteristic turned out to be sufficient. Weather compensation for heating and cooling are presented in Figure 2. Both aquastats for TES tanks were also programmed to work with 2 °C hysteresis.

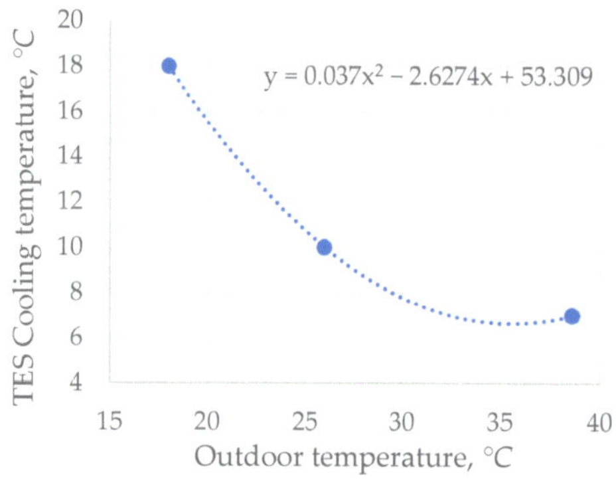

(**a**)

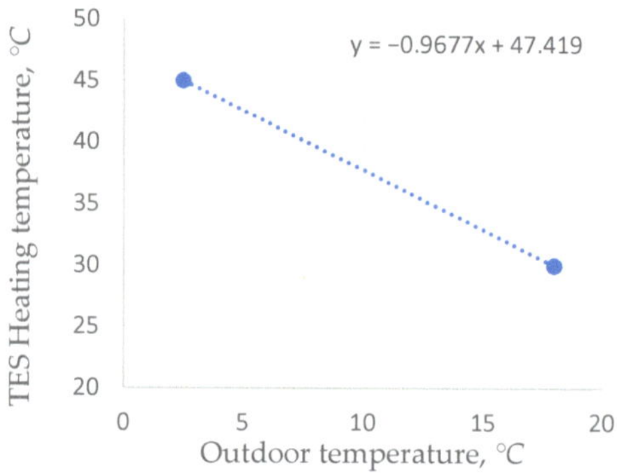

(**b**)

Figure 2. Weather compensation for (**a**) cooling and (**b**) heating.

2.3.2. Smart Control Strategy (Price Only)

A smart control strategy was created based on a standard control strategy. The first version of smart control used only price categories to shift the electricity load from periods with high prices to periods with low prices. This was achieved thanks to an algorithm that adjusted weather compensation sets according to the current price category. For categories 1 and 2 (low price), the standard TES cooling temperature (TES_{CST}) was replaced by a new, lower smart TES cooling set (TES_{CSMART}). Accordingly, a similar procedure was conducted for the standard TES heating temperature (TES_{HST}), which was raised to the new smart TES heating (TES_{HSMART}). Thanks to these actions, the heat pump increases energy usage and charges storage tanks. Opposite operations are conducted for timesteps with categories 4 and 5. Lifted sets for cooling and dropped sets for heating cause the heat pump to work less or not at all during these periods, resulting in space conditioning being supplied from TES tanks. For category 3 the algorithm checks four previous timesteps (2 h). If there is at least one timestep with category 1 or 2, the raised (for heating) or lowered (for cooling) temperatures in TES tanks are maintained. If the lowest price category is category 3, the sets are returned to values from the standard control strategy. If tanks have any surplus thermal energy stored, return to standard sets is conducted slowly. For any other scenario, the standard sets remain unchanged. Hysteresis for both TES tank aquastats was changed to 1 °C to ensure a quick reaction to the changes. Figure 3 presents the entire algorithm with details on values and limits.

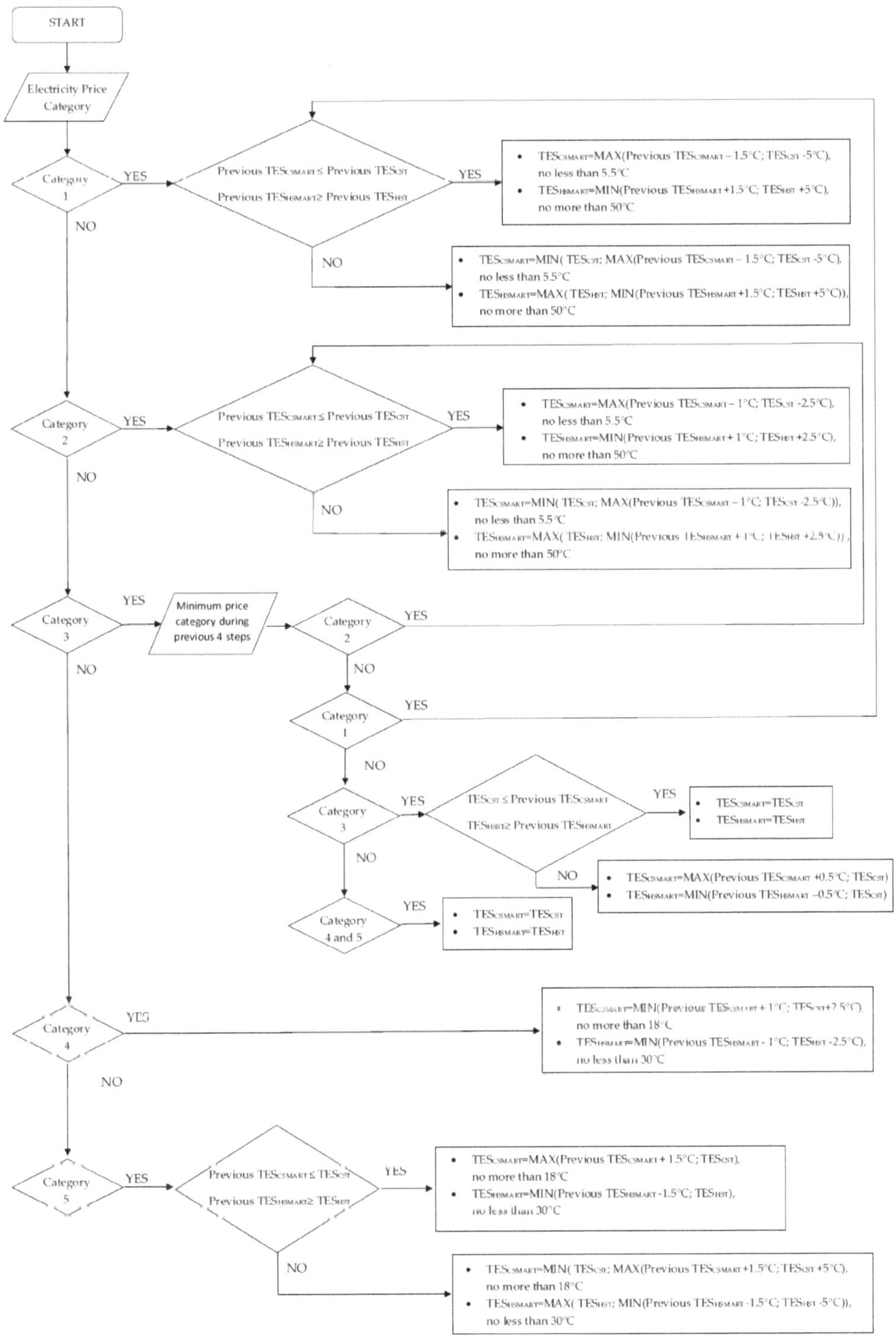

Figure 3. Smart control strategy (price only) algorithm.

In addition to adjustments to the TES tanks' temperatures, the algorithm also changes the DHW tank's temperatures. The compensation values for each price category are presented in Table 3.

Table 3. Smart control temperature compensation for DHW.

Price Category	1	2	3	4	5
DHW_{SMART}	+4 °C	+1 °C	-	−5 °C	−7 °C

2.3.3. Smart Control Strategy (Price and PV)

Energy consumption from simulations of hydronic systems with the price only strategy was added to the remaining circuits measured in the CSIRO study for each timestep. It created the electric demand profile of the entire household. Then, PV generation was deducted from this load, creating an electricity import and export profile. If the balance was negative (electricity export), tank temperatures were again adjusted. It was conducted only for timesteps with categories 1 and 2 according to the equations below:

$$TES_{CPV} = MIN(TES_{CSMART}, PreviousTES_{CPV}) - \frac{|E_e| \cdot COP_{avC}}{C_p \cdot V_{TESC}} \quad (3)$$

$$TES_{HPV} = MAX(TES_{HSMART}, PreviousTES_{HPV}) + \frac{|E_e| \cdot COP_{avH}}{C_p \cdot V_{TESH}} \quad (4)$$

$$DHW_{PV} = DHW_{SMART} + \frac{|E_e| \cdot COP_{avDHW}}{C_p \cdot V_{DHW}} \quad (5)$$

where E_e is electricity export from the household, COP_{av} is the annual average COP for the task from the smart strategy (price only) simulations, C_p is the specific heat of water (0.00116 kWh/(kg·K)), and V is a volume of the tank. This forces the heat pump to work during periods of overgeneration using surplus energy and charging tanks. Because this algorithm affects the timesteps with price categories 1 and 2, the excessive load is only consumed when the grid already suffers from overgeneration. Other categories are not subjected to the algorithm. Therefore, the household would still contribute to the grid supply when it is beneficial. The adjusted smart control strategy (price and PV) sets were imported to the TRNSYS model, and simulations for the last scenario were conducted. Table 4 presents a short summary of all cases.

Table 4. Summary of analysed scenarios.

System	Description
Conventional	AC electricity load from CSIRO study and DHW electricity load simulated in TRNSYS.
ASHP Standard	The electricity load of space conditioning and DHW heating simulated in the TRNSYS model of an ASHP with standard control strategy based on typical heat pump operation.
ASHP Smart (Price Only)	The electricity load of space conditioning and DHW heating simulated in the TRNSYS model of an ASHP with smart control strategy reacting to wholesale electricity prices.
ASHP Smart (Price + PV)	The electricity load of space conditioning and DHW heating simulated in the TRNSYS model of an ASHP with smart control strategy reacting to wholesale electricity prices and PV generation.
GSHP Standard	The electricity load of space conditioning and DHW heating simulated in the TRNSYS model of a GSHP with standard control strategy based on typical heat pump operation.
GSHP Smart (Price Only)	The electricity load of space conditioning and DHW heating simulated in the TRNSYS model of a GSHP with smart control strategy reacting to wholesale electricity prices.
GSHP Smart (Price + PV)	The electricity load of space conditioning and DHW heating simulated in the TRNSYS model of a GSHP with smart control strategy reacting to wholesale electricity prices and PV generation.

3. Results

3.1. Total Energy Consumption and Average COP

The first analysed outcomes of the study are total energy consumption and average annual COP for each task and scenario. The simulation outcomes are presented in Table 5.

Table 5. Annual total energy consumption and COP for each task and scenario.

System	DHW	Cooling	Heating	Total
Conventional	931 kWh 1	358 kWh 4.67	478 kWh 4.18	1766 kWh
ASHP Standard	356 kWh 3.41	374 kWh 4.86	387 kWh 5.28	1117 kWh
ASHP Smart (Price Only)	389 kWh 3.23	400 kWh 4.68	404 kWh 5.12	1192 kWh
ASHP Smart (Price + PV)	386 kWh 3.25	403 kWh 4.66	418 kWh 5.02	1206 kWh
GSHP Standard	400 kWh 3.03	291 kWh 6.14	363 kWh 5.56	1055 kWh
GSHP Smart (Price Only)	439 kWh 2.86	326 kWh 5.73	376 kWh 5.50	1141 kWh
GSHP Smart (Price + PV)	440 kWh 2.85	330 kWh 5.69	397 kWh 5.28	1167 kWh

All hydronic heating scenarios turned out to be much more efficient than conventional systems, reaching at least a 31.7% annual energy consumption reduction, mostly due to a higher COP for DHW preparation of heat pump systems. The GSHP systems were the most efficient in cooling and heating among all of the analysed options, using 18.6% and 23.9% less energy than conventional air conditioning for these tasks while working with a standard control strategy. The ASHP system with standard control used slightly more energy for cooling than AC, however, reaching a higher COP. This means that the ASHP transferred more thermal energy than the conventional system. Whereas for heating, the ASHP was more energy-conservative and reached a significantly higher COP than AC. It was also the most efficient way of preparing the DHW.

Both smart control strategies increased the energy consumption of hydronic systems. The lower COP is a result of forcing the heat pump systems to work in less favourable conditions, raising the water temperature for heating tasks and a decrease in cooling. Other contributors to increased energy consumption are higher tank heat loss and the fact that additional thermal energy in TES tanks could not be utilised and was wasted.

Looking only at total annual energy consumption, it can be said that hydronic systems are a good replacement for conventional solutions; however, smart control makes these systems less efficient. Nevertheless, the smart control strategy was not designed to decrease energy usage but to control the time when it occurs.

3.2. Space Conditioning and DHW Load Distribution

The energy consumption of the analysed systems was measured for each price category, and the results are presented in Figure 4.

There are no significant differences in the percentage load distribution for each price category of conventional and hydronic systems with standard control. However, hydronic systems with standard control systems used noticeably less energy in each price category than conventional AC and DHW heating. This means that hydronic systems can increase overall household efficiency. However, they do not bring any major help in stabilising the load. It is true that the system uses less energy during peak hours, but it also uses less energy during off-peak periods, making RES overgeneration even more problematic.

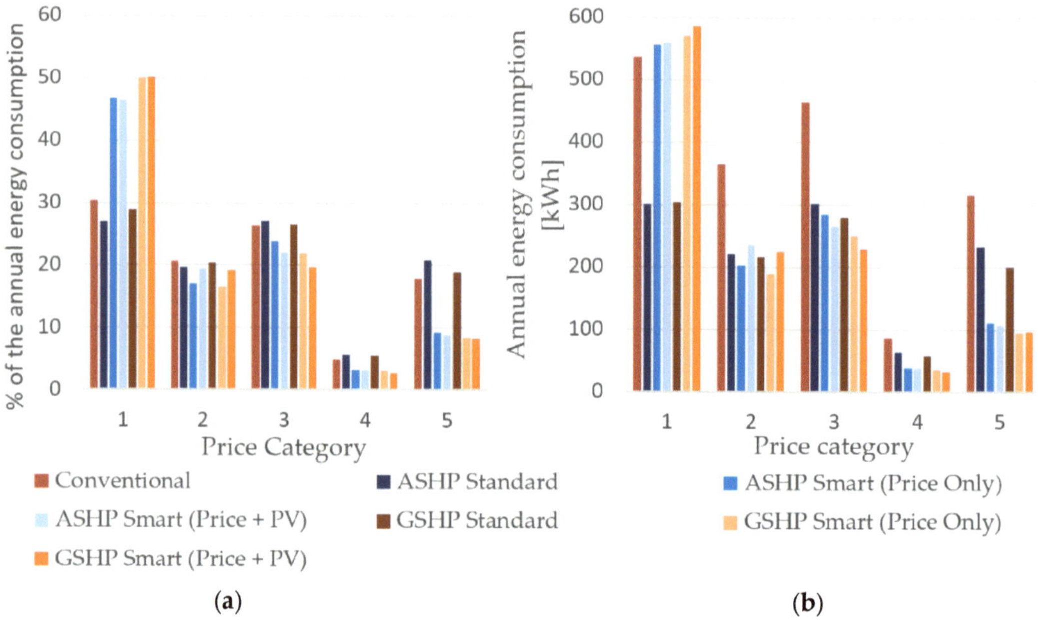

Figure 4. Space conditioning and DHW heating energy consumption in each category: (**a**) percentage; (**b**) total.

The smart control strategy caused hydronic systems to use around half of the electricity during price category 1 (ASHP 46–47%, GSHP 50%). The load distribution for the other categories dropped compared to conventional and hydronic standard systems. The largest decrease was noted in category 5. The electricity consumption in category 5 was halved compared to the standard control strategy and reduced by a third compared to the conventional system. The load in category 4 was also reduced. However, the change was not as radical. Categories 2 and 3 showed slightly less kWh usage using the smart strategy. Most of the load was shifted to category 1, where smart hydronic systems used even more energy than the conventional solution.

This means that even though a smart control strategy causes the hydronic system to use more energy, the consumption distribution is more favourable. The algorithm allowed the system to shift most of the load from categories 4 and 5 to category 1. Therefore, the household would contribute less to peak events and help utilise clean RES energy during periods of high generation.

3.3. Household Self-Consumption and Self-Sufficiency Levels

Another part of the analysis was to investigate the impact of hydronic systems on the utilisation of household-generated PV energy. For this purpose, self-consumption (SC) and self-sufficiency (SS) levels were calculated using Equations (6) and (7), where G_T is total generation, C_T is total consumption, Exp_T is total export, and Imp_T is total import. C_T is the sum of AC, DHW, and the remaining loads. The results are presented in Table 6.

$$\mathrm{SC} = \frac{G_T - Exp_T}{G_T} \quad (6)$$

$$\mathrm{SS} = \frac{C_T - Imp_T}{C_T} \quad (7)$$

Table 6. Household PV self-consumption and self-sufficiency levels.

System	SC	SS
Conventional	79.7%	23.5%
ASHP Standard	75.0%	26.1%
ASHP Smart (Price Only)	76.5%	26.0%
ASHP Smart (Price + PV)	81.2%	27.5%
GSHP Standard	75.4%	26.7%
GSHP Smart (Price Only)	76.1%	26.3%
GSHP Smart (Price + PV)	81.1%	27.8%

Due to the fact that hydronic systems use less electricity than conventional systems, SC levels of PV generation for the standard and smart (price only) strategies decreased. However, the smart (price + PV) algorithm allowed for an increase in self-consumption to slightly over 81%, which is over one percentage point more than the result with the conventional system. So this solution not only helped reduce electricity consumption but also increased the utilisation of household-generated energy. This is confirmed by the highest self-sufficiency levels for both heat pumps with such a control system. Despite that, the system based on the GSHP reached a slightly lower level of SC than the ASHP; the overall balance in household self-sufficiency for this heat pump was the greatest among all of the analysed cases. This was caused by lower energy consumption and a higher COP of the system during overproduction periods. The highest PV generation usually occurs at the same time as the highest outdoor temperature. Therefore, the ASHP needs to work harder to transfer the heat to the sink with a higher temperature in a cooling-dominated climate. A ground heat exchanger provides a much more stable sink temperature.

3.4. Electricity Import and Export Distribution

The entire household's electricity import and export loads were calculated and divided into price categories. These loads are the sum of AC, DHW, and the remaining loads minus PV generation. Table 7 presents these results.

Table 7. Annual household electricity import (red, positive values) and export (green, negative values) for each price category.

System	Category 1	Category 2	Category 3	Category 4	Category 5
Conventional	1080	661	545	275	559
	−110	−68	−53	−11	−10
ASHP Standard	862	530	450	228	482
	−128	−81	−68	−16	−16
ASHP Smart (Price Only)	1101	510	425	233	364
	−110	−79	−67	−15	−20
ASHP Smart (Price + PV)	1073	511	409	234	360
	−79	−47	−70	−15	−20
GSHP Standard	865	525	433	221	448
	−127	−81	−67	−15	−15
GSHP Smart (Price Only)	1118	497	405	221	318
	−113	−79	−68	−15	−20
GSHP Smart (Price + PV)	1100	498	390	216	350
	−80	−46	−70	−15	−20

Typical household electricity import for all hydronic heating cases and price categories 2–5 was lower than for the conventional system. The largest reduction in consumption was noted for category 5 for all four smart control strategies. The ASHP-based system helped to save 34.9–35.5%, while the GSHP-based system reduced the load by 37.4–37.6% during the highest peak events. Electricity import for the standard control strategy in category 1 dropped compared to the conventional solution; however, the smart control strategy allowed for reaching similar or slightly higher values. This confirms the earlier conclusions that smart hydronic systems are able to shift the load from peak to off-peak periods. Households with such systems would strain the grid much less than those with conventional air conditioning and water heating.

In addition to the electricity import, the smart (price + PV) control allowed for the decrease in electricity exports for categories 1 and 2, which was desirable in overgeneration limitation. Unfortunately, not all of the absorbed surplus energy was efficiently used. The ASHP smart (price + PV) system reduced exports by 60 kWh, but it saved only 45 kWh on electricity imports compared with ASHP smart (price only). The numbers for GSHP smart (price + PV) were even less optimistic, as the system reduced exports by 64 kWh and imports by 37 kWh. The most likely cause was the lower COP of the heat pumps that worked with extended temperature sets, as well as the fact that stored thermal energy was not utilised in time.

3.5. Space Conditioning and DHW Heating Fulfilment

As discussed earlier, one of the benefits of hydronic systems is that they can store energy in a TES tank instead of habitable space and use the reserves when the electricity load of heating and cooling devices needs to be limited. While earlier subsections showed improvements in energy consumption and distribution, Table 8 presents how smart strategies impacted the level of space conditioning fulfilment. The heating and cooling load has been considered to be fulfilled if the fan coil unit used to establish heating and cooling curves (Section 2.3.1) can cover the current space conditioning load with the current water temperature in the system.

Table 8. Hydronic heating and cooling load fulfilment.

System	% of Time	% of Cooling Load	% of Heating Load
ASHP Standard	99.88%	99.70%	99.98%
ASHP Smart (Price Only)	99.11%	98.57%	99.32%
ASHP Smart (Price + PV)	99.14%	98.56%	99.34%
GSHP Standard	99.91%	99.71%	99.98%
GSHP Smart (Price Only)	99.22%	98.48%	99.47%
GSHP Smart (Price + PV)	99.22%	98.50%	99.48%

Results for standard control strategies showed very minimal under-conditioning in both time and loads. Undoubtedly, the TRNSYS simulations' accuracy could be a major part of these numbers. Therefore, they can be treated as the basis for comparison with smart strategy cases.

Smart strategies lowered the level of satisfaction, but the decrease was not substantial. All systems still provided the required heating and cooling for over 99% of the time in the year. They also provided around 98.5% of the cooling load and 99.3–99.5% of the heating load. This means that the major amount of shifted electrical load barely affected indoor comfort. This achievement would not be possible for conventional air conditioning.

Overheating and over-cooling are not a problem due to the fact that space conditioning circuits are controlled independently. Therefore, even if the water temperature is higher than required for heating or lower than needed for cooling, the output can be adjusted to match the demand by various methods such as an on/off controller, variable speed pump, or variable airflow of a fan coil unit.

Also, DHW preparation was checked against the drop in the satisfactory level. The results, which are presented in Table 9, show that the temperature of the DHW outlet dropped below the recommended level of 50 °C in only marginal periods.

Table 9. Percent of time with DHW temperature below 50 °C.

System	% of Time
ASHP Standard	0.00%
ASHP Smart (Price Only)	0.24%
ASHP Smart (Price + PV)	0.26%
GSHP Standard	0.00%
GSHP Smart (Price Only)	0.30%
GSHP Smart (Price + PV)	0.31%

4. Study Discussion and Limitations

The presented research was focused on the estimation of the impact of replacement of a conventional cooling and heating system with a hydronic heat pump system with smart management on the electrical load generated by a typical South Australian household in the context of the transition towards net zero emissions.

In contrast to most of the research studies regarding the utilisation of hydronic heat pumps to stabilise the grid, it was decided to use measured electricity and hot water usage data. This allowed for the consideration of factors such as behavioural patterns and usage preferences, not only simulated cooling and heating loads. The peak load phenomenon is largely caused by the residential sector, especially AC. Therefore, it is crucial to include the household residents' factors in the analysis. Also, measured data allows for the accurate use of other inputs, such as electricity price and weather conditions, which are aligned and represent the same events. However, the usage of measured AC electricity consumption required the calculation of cooling and heating loads. This process required some assumptions and simplifications. This process could cause some distortion of the simulations' input data and results. Moreover, it has to be remembered that the typical household is only a theoretical concept. Most likely, none of the households participating in the CSIRO study had electricity use similar to the typical one. Therefore, extreme events can be distorted due to the results from multiple houses being averaged. Nevertheless, the considered solution aims to help not only households but mostly the national grid and its stability to allow further transition towards net zero emissions, which justifies this approach.

The peak cooling and heating loads, as well as the total energy consumption for space conditioning, may seem relatively low. As mentioned above, the distortion in peak loads could be one of the contributors. Moreover, typically in Australia, only the most used rooms are equipped with air conditioning. The low inertia of the AC system also means that they tend to be turned on and off as needed. However, hydronic systems work with much larger inertia, especially systems such as underfloor heating or cooling. Therefore, they usually maintain the temperature of the space at a relatively steady level for a longer period of time. Also, a hydronic heating plant room is a relatively large investment cost compared to an AC system. Therefore, it is usually more profitable to install a distribution system in the entire habitable space of the house. This could increase heating and cooling loads, also affecting energy consumption. These factors have not been taken into account in this project.

The smart control strategies used in this study were designed for this project. The parameters were adjusted multiple times until the results of the simulations were considered satisfactory. Too conservative an approach had too little influence on the base load profile to be considered meaningful. In turn, too aggressive values highly impacted load fulfilment levels, which could cause a decrease in the satisfaction level of users of such a system. One of the main advantages of hydronic heat pump systems over conventional AC is the ability to store energy and provide only the required cooling or heating load. Therefore, the balance between the grid needs and users' comfort needs to be carefully considered. It is possible that the smart strategies could be even better tuned. However, this was not the main scope of the study. Also, the use of proprietary control strategies can be considered a limitation of the results of this study. However, similar systems have already been developed, and they are in use. An example of such a system is the smart grid ready (SG Ready) protocol introduced in Germany. However, it has become a standard heat pump feature in devices manufactured by world-leading companies. The protocol allows for external control of heat pumps by two zero-voltage contacts [41]. The contact can trigger four heat pump states: blocked operation, normal operation, encouraged operation, and forced operation. Therefore, the principle of operation is similar to the system analysed in this study. The heat pump market in Australia is still small. If smart heat pump systems are implemented in Australia, it could be the already developed SG Ready protocol or a new program suited to the local needs. The program could include model predictive control (MPC), which was not accounted for in this study. The MPC could help in the accurate preparation of TES and DHW tanks before the anticipated peak event and prevent unnecessary energy losses if the peak is not predicted and the surplus RES energy can be utilised elsewhere.

5. Conclusions

The analysis of the effects of replacing a conventional system consisting of an air conditioning unit and electric water heater with a hydronic heat pump system with smart control in a typical South Australian household presented in this paper provides promising results in shifting residential electrical load. All of the analysed scenarios bring some benefits to the load generated by the household. Hydronic systems with standard control drastically decreased energy consumption, mostly due to providing much more efficient DHW preparation. Smart control with the price only algorithm helped to significantly reduce electricity consumption during the periods of the highest electricity price, which appears when there is insufficient generation. Most of this load is shifted to the periods with the lowest electricity price, indicating too much supply. A system based on a ground source heat pump controlled by this control strategy used half of the total annual electricity consumption during price category 1. A further version of the smart control system, which added a module reacting to excessive PV generation, allowed for self-consumption and self-sufficiency levels. It also limited the export of the surplus load during periods of overgeneration. The difference in performance between systems based on an ASHP and a GSHP was slight. The GSHP achieved marginally less consumption and was able to shift more load. Thanks to the storage ability of water in TES tanks, the actions focused on adjusting the heat pump output barely impacted the space conditioning load. The load was satisfied for over 99.1% of the time. Only about 1.5% of the cooling load and 0.5–0.7% of the heating load were not delivered. Compared to the amount of shifted electricity, these numbers are rather insignificant. Nevertheless, they could be improved by tuning the cooling and heating curves, adjusting the algorithm values, or upsizing the FCU to allow for more capacity during the periods when the system relies on stored energy.

Therefore, it can be concluded that a hydronic system with a heat pump and smart control in a typical South Australian household can be a great tool in managing residential load. The problem with grid stability intensifies. Each year, Australia increases the capacity of domestic PV systems and the number of small AC units. Along with other grid stability problems, the Australian electricity network faces a huge challenge in the coming years. Efforts related to the Australian plan of reaching net zero emissions in 2050 could be squandered if the power network fails. The analysis presented in this paper was conducted using data measured in 2016. Currently, the price fluctuations are larger. This could make the aforementioned solutions even more beneficial.

Author Contributions: Conceptualization, A.R. and R.N.; Methodology, A.R. and R.N.; Software, A.R. and R.N.; Validation, A.R. and R.N.; Formal analysis, A.R., R.N. and M.J.; Investigation, A.R. and R.N.; Resources, R.N.; Data curation, A.R. and R.N.; Writing—original draft, A.R.; Writing—review & editing, R.N. and M.J.; Visualization, A.R. and R.N.; Supervision, R.N. and M.J.; Project administration, R.N. and M.J.; Funding acquisition, R.N. All authors have read and agreed to the published version of the manuscript.

Funding: This research received no external funding.

Data Availability Statement: The original contributions presented in the study are included in the article, further inquiries can be directed to the corresponding authors.

Conflicts of Interest: The authors declare no conflict of interest.

References

1. Australian Government. *Australia's Long-Term Emissions Reduction Plan. A Whole-of-Economy Plan to Achieve Net Zero Emissions by 2050*; Australian Government Department of Industry, Science, Energy and Resources: Canberra, Australia, 2021.
2. Australian Government—Department of Climate Change, Energy, the Environment and Water. Residential Buildings. Available online: https://www.dcceew.gov.au/energy/energy-efficiency/buildings/residential-buildings (accessed on 20 April 2024).
3. Boretti, A. Energy storage needs for an Australian national electricity market grid without combustion fuels. *Energy Storage* **2019**, *2*, e92. [CrossRef]
4. Bayborodina, E.; Negnevitsky, M.; Franklin, E.; Washusen, A. Grid-Scale Battery Energy Storage Operation in Australian Electricity Spot and Contingency Reserve Markets. *Energies* **2021**, *14*, 8069. [CrossRef]
5. ENERGEIA. *Trajectory for Low Energy Buildings: Infrastructure and Customer Impacts*; Prepared for Department of Environment and Energy: Sydney, Australia, 2019.
6. Arraño-Vargas, F.; Shen, Z.; Jiang, S.; Fletcher, J.; Konstantinou, G. Challenges and Mitigation Measures in Power Systems with High Share of Renewables—The Australian Experience. *Energies* **2022**, *15*, 429. [CrossRef]
7. Fan, H.; MacGill, I.F.; Sproul, A.B. Statistical analysis of drivers of residential peak electricity demand. *Energy Build.* **2017**, *141*, 205–217. [CrossRef]
8. Australian PV Institute. Available online: https://pv-map.apvi.org.au/historical (accessed on 6 April 2024).
9. Brodribb, P.; McCann, M.; Franjić, J.; Dewerson, G. *Heat Pumps—Emerging Trends in the Australian Market*; Expert Group for Australian Government: Brighton, VIC, Australia, 2023.
10. Climate Council of Australia. Submission to the Senate Economics References Committee Inquiry into Residential Electrification. Available online: https://www.climatecouncil.org.au/wp-content/uploads/2023/10/Residential-Electrification-Submission.docx.pdf (accessed on 7 April 2024).
11. Narayanan, R.; Sethuvenkatraman, S.; Pippia, R. Energy and Comfort Evaluation of Fresh Air-Based Hybrid Cooling System in Hot and Humid Climates. *Energies* **2022**, *15*, 7537. [CrossRef]
12. Hashmi, S.A.; Ali, C.F.; Zafar, S. Internet of things and cloud computing-based energy management system for demand side management in smart grid. *Int. J. Energy Res.* **2020**, *45*, 1007–1022. [CrossRef]
13. Mahapatra, C.; Moharana, A.K.; Leung, V.C.M. Energy Management in Smart Cities Based on Internet of Things: Peak Demand Reduction and Energy Savings. *Sensors* **2017**, *17*, 2812. [CrossRef] [PubMed]
14. Hossein Motlagh, N.; Mohammadrezaei, M.; Hunt, J.; Zakeri, B. Internet of Things (IoT) and the Energy Sector. *Energies* **2020**, *13*, 494. [CrossRef]
15. Ono, T.; Hagishima, A.; Tanimoto, J. Evaluation of potential for peak demand reduction of residential buildings by household appliances with demand response. *Electron. Commun. Jpn.* **2022**, *105*, e12379. [CrossRef]
16. Energy Made Easy. Available online: https://www.energymadeeasy.gov.au/hot-topics/understanding-controlled-loads (accessed on 6 April 2024).
17. Energex-PeakSmart Air Conditioning. Available online: https://www.energex.com.au/manage-your-energy/cashback-rewards-program/peaksmart-air-conditioning (accessed on 6 April 2024).
18. Ember Pulse. Available online: https://emberpulse.com/ (accessed on 6 April 2024).
19. Smith, R.; Meng, K.; Dong, Z.; Simpson, R. Demand response: A strategy to address residential air-conditioning peak load in Australia. *J. Mod. Power Syst. Clean Energy* **2013**, *1*, 223–230. [CrossRef]
20. Tongsopit, S.; Junlakarn, S.; Wibulpolprasert, W.; Chaianong, A.; Kokchang, P.; Hoang, N.V. The economics of solar PV self consumption in Thailand. *Renew. Energy* **2019**, *138*, 395–408. [CrossRef]
21. Yu, H.J.J. System contributions of residential battery systems: New perspectives on PV self-consumption. *Energy Econ.* **2021**, *96*, 105151. [CrossRef]
22. Fina, B.; Roberts, M.B.; Auer, H.; Bruce, A.; MacGill, I. Exogenous influences on deployment and profitability of photovoltaics for self-consumption in multi-apartment buildings in Australia and Austria. *Appl. Energy* **2021**, *283*, 116309. [CrossRef]
23. Kamboj, S.; Narayanan, R. Ground Heat Exchangers for Cooling and Heating Applications in Buildings. In *Reference Module in Earth Systems and Environmental Sciences*; Elsevier: Amsterdam, The Netherlands, 2023.
24. Fischer, D.; Madani, H. On heat pumps in smart grids: A review. *Renew. Sustain. Energy Rev.* **2017**, *70*, 342–357. [CrossRef]
25. Péan, T. *Heat Pump Controls to Exploit the Energy Flexibility of Building Thermal Loads*; Universitat Politècnica de Catalunya: Barcelona, Spain, 2021.
26. Dallmer-Zerbe, K.; Fischer, D.; Biener, W.; Wille-Haussmann, B.; Wittwer, C. Droop controlled operation of heat pumps on clustered distribution grids with high PV penetration. In Proceedings of the 2016 IEEE International Energy Conference (ENERGYCON), Leuven, Belgium, 4–8 April 2016; pp. 1–6.

27. Kok, K.; Roossien, B.; MacDougall, P.; van Pruissen, O.; Venekamp, G.; Kamphuis, R.; Laarakkers, J.; Warmer, C. Dynamic pricing by scalable energy management systems - Field experiences and simulation results using PowerMatcher. In Proceedings of the 2012 IEEE Power and Energy Society General Meeting, San Diego, CA, USA, 22–26 July 2012; pp. 1–8.
28. Vorushylo, I.; Keatley, P.; Shah, N.; Green, R.; Hewitt, N. How heat pumps and thermal energy storage can be used to manage wind power: A study of Ireland. *Energy* **2018**, *157*, 539–549. [CrossRef]
29. Li, Y.; Mojiri, A.; Rosengarten, G.; Stanley, C. Residential demand-side management using integrated solar-powered heat pump and thermal storage. *Energy Build.* **2021**, *250*, 111234. [CrossRef]
30. Li, Y.; Rosengarten, G.; Mojiri, A. Performance sensitivity of hot and cold thermal storage with onsite photovoltaics and heat pumps. *Solar Energy* **2023**, *263*, 111946. [CrossRef]
31. CSIRO. Typical House Energy Use. Available online: https://ahd.csiro.au/other-data/typical-house-energy-use/ (accessed on 7 April 2024).
32. Rapucha, A.; Narayanan, R.; Jha, M. Impact of Smart Hydronic System with Heat Pump on Electricity Load of a Typical Queensland Household. In *Future Energy*; Green Energy and Technology; Springer: Cham, Switzerland, 2023; pp. 163–172.
33. Mitsubishi Electric MSZ-AP DataBook. Available online: https://www.mitsubishi-les.info/database/servicemanual/files/DataBook_2018_M_Pseries.pdf (accessed on 17 September 2022).
34. Meteostat. Adelaide Weather Data. Available online: https://meteostat.net/en/station/94672?t=2016-01-01/2016-12-31 (accessed on 7 April 2024).
35. Commonwealth of Australia. *Water Heating Data Collection and Analysis, Residential End Use Monitoring Program (REMP)*; Commonwealth of Australia: Canberra, Australia, 2012.
36. Ecoforest Geothermia. Technical Data Sheets-Ecoforest Heat Pumps. Available online: https://www.ecoforest.com/wp-content/uploads/2022/01/DS_Ecoforest_heat_pumps_EN_v2021_02_LQ.pdf (accessed on 7 April 2024).
37. Mitsubishi Electric Air to Water Heat Pump Systems-Data Book Vol.4. Available online: https://www.mitsubishi-les.info/database/servicemanual/files/201803_ATW_DATABOOK.pdf (accessed on 7 April 2024).
38. Fischer, D.; Lindberg, K.B.; Madani, H.; Wittwer, C. Impact of PV and variable prices on optimal system sizing for heat pumps and thermal storage. *Energy Build.* **2016**, *128*, 723–733. [CrossRef]
39. AEMO. Aggregated Price and Demand Data. Available online: https://aemo.com.au/energy-systems/electricity/national-electricity-market-nem/data-nem/aggregated-data (accessed on 8 April 2024).
40. Le, K.X.; Huang, M.J.; Wilson, C.; Shah, N.N.; Hewitt, N.J. Tariff-based load shifting for domestic cascade heat pump with enhanced system energy efficiency and reduced wind power curtailment. *Appl. Energy* **2020**, *257*, 113976. [CrossRef]
41. Fischer, D.; Wolf, T.; Triebel, M.A. Flexibility of heat pump pools: The use of SG-Ready from an aggregator's perspective. In Proceedings of the 12th IEA Heat Pump Conference 2017, Rotterdam, The Netherlands, 15–18 May 2017.

energies

MDPI

Article

The Role of Blockchain-Secured Digital Twins in Promoting Smart Energy Performance-Based Contracts for Buildings

Mohamed Nour El-Din [1,*], João Poças Martins [1,*], Nuno M. M. Ramos [2] and Pedro F. Pereira [2]

1 CONSTRUCT—GEQUALTEC, Faculty of Engineering (FEUP), University of Porto, 4200-465 Porto, Portugal
2 CONSTRUCT—LFC, Faculty of Engineering (FEUP), University of Porto, 4200-465 Porto, Portugal; nmmr@fe.up.pt (N.M.M.R.); fpfp@fe.up.pt (P.F.P.)
* Correspondence: mohamed.abushamma@gmail.com (M.N.E.-D.); jppm@fe.up.pt (J.P.M.)

Abstract: Energy performance-based contracts (EPCs) offer a promising solution for enhancing the energy performance of buildings, which is an overarching step towards achieving Net Zero Carbon Buildings, addressing climate change and improving occupants' comfort. Despite their potential, their execution is constrained by difficulties that hinder their diffusion in the architecture, engineering, construction, and operation industry. Notably, the Measurement and Verification process is considered a significant impediment due to data sharing, storage, and security challenges. Nevertheless, there have been minimal efforts to analyze research conducted in this field systematically. A systematic analysis of 113 identified journal articles was conducted to fill this gap. A paucity of research tackling the utilization of digital technologies to enhance the implementation of EPCs was found. Consequently, this article proposes a framework integrating Digital Twin and Blockchain technologies to provide an enhanced EPC execution environment. Digital Twin technology leverages the system by monitoring and evaluating energy performance in real-time, predicting future performance, and facilitating informed decisions. Blockchain technology ensures the integrity, transparency, and accountability of information. Moreover, a private Blockchain infrastructure was originally introduced in the framework to eliminate high transaction costs related to on-chain storage and potential concerns regarding the confidentiality of information in open distributed ledgers.

Keywords: performance contracts; digital twins; blockchain; smart contracts; zero carbon buildings; energy savings; energy performance

Citation: Nour El-Din, M.; Poças Martins, J.; Ramos, N.M.M.; Pereira, P.F. The Role of Blockchain-Secured Digital Twins in Promoting Smart Energy Performance-Based Contracts for Buildings. *Energies* **2024**, *17*, 3392. https://doi.org/10.3390/en17143392

Academic Editors: Zsuzsa Szalay and Bruno Peuportier

Received: 30 April 2024
Revised: 6 July 2024
Accepted: 8 July 2024
Published: 10 July 2024

1. Introduction

Efforts are being made to mitigate environmental impacts and promote sustainability in the architecture, engineering, construction, and operation (AECO) industry. Buildings represent around 40% of the European Union's (EU) total energy consumption and generate approximately 36% of Europe's greenhouse gases (GHGs), making the AECO industry one of the most polluting sectors [1]. Net Zero Carbon Buildings (NZCBs) have gained global recognition as a pioneering sustainable development approach to achieve the net zero goal for built environments by 2050 [2]. NZCBs are energy-efficient buildings utilizing on-site or off-site renewable energy sources and verified offsets to achieve equilibrium between energy demand and renewable energy supply, or neutralize carbon emissions linked to annual energy usage and provision [2,3]. Operational carbon is considered one of the primary metrics provided by the EN15978:2011 standard for the zero-carbon assessment, also called the whole-life carbon assessment (WLCA) [4]. Lowering operational carbon emissions can be accomplished by implementing solutions typically offered for Zero Energy Buildings (ZEBs) and Nearly Zero Energy Buildings (nZEBs), such as integrated designs, minimized plug loads, and energy-efficient retrofits [5]. In the same context, in the last decade, the European Union (EU) has developed policies to accelerate the cost-effective retrofitting of existing buildings, with the vision of a decarbonized building stock by 2050 [6]. Even though it is possible to construct new energy-efficient structures, most energy consumption is still attributed to current buildings, emphasizing the critical need to enhance their energy efficiency [7]. This situation makes investors, owners, and users face an immense challenge. It is necessary to invest in saving measures to improve the energy performance of buildings, which entails a considerable short-term financial commitment with relatively long payback periods. Furthermore, increasing and accelerating the extent of building renovation is crucial in post-COVID-19 economic recovery [8]. Energy Performance-based Contracting (EPC) is among the potential strategies to achieve this goal and improve the energy efficiency of buildings [9].

EPC, as defined by the Energy Efficiency Directive 2012/27/EU, is a type of "creative financing" for capital improvement that allows the funding of energy upgrades from cost reductions [10]. EPC has the potential to accelerate the pace of energy renovation for current buildings and encourage the application of energy-efficient measures in upcoming constructions [11]. These agreements are established between the client and a private entity that serves as a service provider, often known as an "Energy Service Company" (ESCO) [12]. This contract model offers a mutually beneficial outcome where owners can reduce energy expenses while providers profit from continuous incentives and a new service offering. EPCs are set to become more widespread as awareness of their benefits and cost savings increases.

The utilization of EPCs has not been extensively embraced in the built environment. Academics have identified challenges with accountability, the absence of standardized performance assessments, novel and unfamiliar financial concepts, and the added burden of upfront communication between parties [13]. Nevertheless, the continuous digitalization of the AECO industry and advancements in technologies such as Digital Twins (DTs) and Blockchain present a novel prospect for implementing performance-based building better [10]. DTs can facilitate performance-based contracting by establishing performance expectations via simulation, continuously monitoring and updating actual performance, and furnishing recommendations for maintenance and operation via analytics. In general, DTs can assist in accurately and equitably forecasting and evaluating performance, thereby surmounting a known obstacle to EPC implementation in the built environment [13].

Moreover, Blockchain can provide an immutable and transparent digital record of transactions. Although CO_2 emissions are generated due to the computational power required, the energy savings facilitated by EPC significantly outweigh these emissions. These savings are achieved by leveraging advanced predictive models and data management techniques that ensure substantial energy efficiency improvements [14]. By integrating Blockchain technology in EPC, the efficiency and reliability of energy savings verification are enhanced, and greater transparency and trust in the energy savings process are ensured, thus contributing to the overall goal of achieving NZCBs. The net effect is a substantial reduction in CO_2 emissions, affirming that the benefits of energy savings far exceed the carbon footprint of the Blockchain system itself [15].

Certain Blockchains (i.e., Ethereum) also allow for the implementation of smart contracts, which utilize scripts to establish tamper-proof transaction logic. Smart contracts are automated computer programs operating within a Blockchain protocol, enabled by the general-purpose computation capabilities of Blockchains. They encompass contractual arrangements, contract execution, and governance of preconditions for contractual obligations. Ethereum introduced the first Turing-complete scripting language with smart contract support, making it a prominent and highly utilized Blockchain platform for smart contracts [16]. Thus, the development of performance-based smart contracts has the potential to be deployed within the Blockchain framework to establish protocols for automating real-world processes. A fundamental challenge for performance-based buildings is accountability, an issue that Blockchain can address by ensuring protection mechanisms that help avoid the risks and costs of opportunistic behavior in construction supply chain collaboration [17]. However, few attempts have been made to study or implement performance-based smart contracts.

To address this issue, the article aims to propose a framework for integrating Digital Twin and Blockchain technologies and demonstrate how the interdependency of these technologies can facilitate the diffusion of EPCs in the AECO industry.

2. Background

2.1. Energy Performance-Based Contracts (EPCs)

EPCs emerged after the first oil crisis in North America in the 1970s. They have evolved as a cutting-edge finance strategy to lower energy use by recouping the expenses of providing energy-saving technologies [18,19]. In 2008, Europe faced a severe financial crisis that affected national economies and caused market uncertainties, especially in Mediterranean countries. Despite limited financial resources, the EU's energy policy has become more rigorous, with the vision of a decarbonized building stock by 2050 [20]. As a means of conserving energy to meet the EU energy policy objectives and to improve energy efficiency in buildings, EPC is considered a potential approach and has been used by many EU countries [9].

An EPC is a contract agreement between an ESCO and an energy user. The contract sets a specific energy-saving target, and the ESCO provides energy-efficient technologies, financing, installation, and maintenance through an appropriate business model. If executed successfully, the ESCO can recover its investment and earn a reasonable profit, which benefits both parties involved [21]. The EPC concept is illustrated in Figure 1. The EPC's business model outlines the obligations of both building owners and ESCOs when executing EPC projects and provides a means of distributing risk in such projects [22].

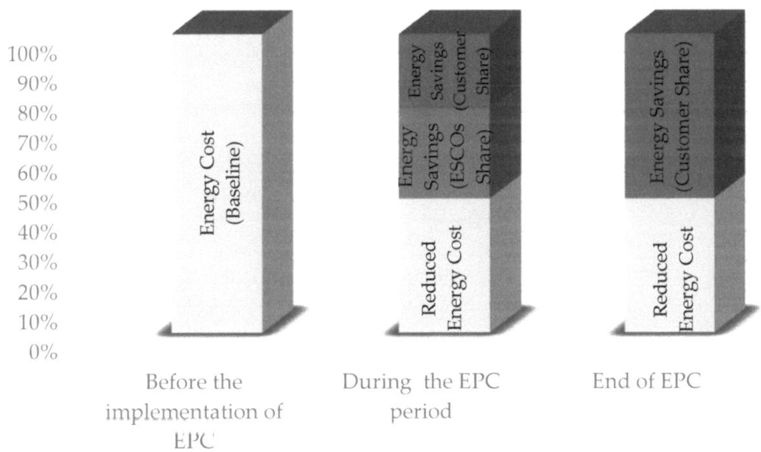

Figure 1. Energy Performance-based Contracting.

EPC projects are perceived as investments with high levels of risk [23], thus affecting their diffusion in the AECO industry. Despite working on providing accurate initial predictions, various risks and barriers hinder the successful implementation of EPCs. Inadequate financing options resulting from conservative lending practices, insufficient familiarity with performance-based project financing, and alterations in economic and market circumstances are considered financial challenges facing EPC projects [22]. In addition, a lack of performance savings standardized Measurement and Verification procedures, a lack of reliable data to optimize performance, alack of knowledge of the technology and its benefits, potential changes in governmental policies, climate change, and unanticipated or inappropriate building usage may influence the energy-conservation benefits of EPCs [24,25]. Ultimately, there is an overall lack of awareness about EPCs among AECO industry stakeholders.

2.2. Energy Service Companies (ESCOs)

As aforementioned, EPCs are agreements established between clients and service providers. The clients can be either public or private organizations, while the service providers are usually private companies known as Energy Service Companies [12]. Experts and scholars have identified the services provided by ESCOs as promising opportunities to fulfill consumers' energy requirements more sustainably [26]. The definition of ESCOs differs from country to country. However, an ESCO is typically responsible for implementing energy efficiency measures and ensuring their effectiveness. They are also responsible for monitoring the contract and may not receive a payment if they fail to meet the energy savings agreed upon in the contract [20]. Thus, their remuneration is associated with project performance.

ESCOs may differ in their operational methods in EPC projects, but the primary distinction lies in whether they offer funding for the project they are implementing or not [22]. ESCOs can obtain the necessary investment from their funds or through financing options offered by a third-party financial institution. The success of ESCOs is influenced by several crucial factors, such as the size and flexibility of the banking system involved in energy-performance contracting, the structure of the energy-efficiency market, the local institutional environment, the technical, financial, and business expertise of ESCO personnel, as well as potential clients and funders, and most notably, access to financing [27].

2.3. Measurement and Verification in EPC

The energy savings achieved through EPC projects are commonly disputed because they serve as the foundation for contract payments and adherence [28]. Disputes are one of the major risks that hinder the diffusion of EPCs in the AECO industry due to a lack of trust and intention to cooperate between stakeholders in EPC projects. In an ideal situation, disputes regarding energy savings in EPC projects are resolved through the Measurement and Verification (M&V) process [29]. M&V is intended to confirm the enhancements provided through energy conservation measures (ECMs) by evaluating the actual performance of buildings once the installation and construction work on the system is finished and a consistent level of operation has been achieved [30,31]. The evaluation and communication of the effectiveness of the installed ECMs according to an M&V plan is typically the responsibility of ESCOs [32]. A quality M&V is the means by which actual savings are quantified and could be considered an insurance policy [28].

Conventionally, M&V calculations have been carried out utilizing techniques chosen on a per-case basis, considering the ECM's characteristics, the projected savings, and the available site data [33]. However, a number of M&V protocols have been created to enhance uniformity and minimize ambiguity in gauging the energy savings derived from retrofitting existing buildings [30]. Two approaches that are commonly acknowledged are the International Performance Measurement and Verification Protocol (IPMVP) provided by the Efficiency Valuation Organization (EVO®) [34] and ASHRAE Guideline 14 [35]. M&V protocols prioritize quantitative requirements and may not cover all issues in a building's performance gap. On-site investigations may only uncover some technical issues and may not reflect all key causes [36].

Recent technological advancements, such as "smart" meters and energy management and information systems (EMISs), have enabled more rapid and cost-effective M&V processes. Utilizing more sophisticated data analytics techniques on more detailed datasets with shorter time intervals, which could be automated and conducted regularly, would offer viable solutions [37]. These technologies assist in conserving energy and provide functionalities that exceed conventional M&V approaches. Evolving M&V techniques towards responsive, dynamic, and precise approaches are commonly known as Measurement and Verification 2.0 (M&V 2.0) [38]. M&V 2.0 leverages metered data to improve real-time performance evaluation, tenant participation, and resource management through the use of analysis tools and algorithms. Additionally, hardware and software advancements have enhanced the precision of M&V functions, such as baseline modeling, detecting anomalous events, and establishing energy consumption benchmarks [30]. However, M&V requirements need to be more strictly prescribed in EPCs.

2.4. Role of Digital Technologies for Energy Performance in AECO

The incorporation of innovative technologies like Digital Twins, Internet of Things (IoT), and Artificial Intelligence (AI) is deemed as a very auspicious solution to tackle the issues confronted by the AECO industry [39], which include inadequate compliance with regulations, poor performance, ineffective communication, fragmentation of information flow, and a lack of trust among different stakeholders [40].

The utilization of DT technology enables an improvement in evaluating buildings' real-life performance by duplicating their actions and behavior in various situations, ensuring that all involved stakeholders remain informed and up-to-date. The process involves amalgamating data collected from multiple sources, including sensors (IoT), Building Information Modelling (BIM) models, and simulations, enabling the examination of the building's energy efficiency and indoor environmental quality. It also facilitates the evaluation of the impact of different design and operational tactics [41] that affect decision-making for smart asset management [13]. Thus, DTs have the potential to evaluate the energy performance of buildings, paving the way for solutions to the risks associated with EPC.

In all DT applications, IoT is regarded as the fundamental technology. Indeed, a recent study has predicted that over 90 percent of all IoT platforms will have Digital Twinning capability by 2028 [42]. IoT uses sensors to collect data from real-world objects, which can be used to create a digital duplicate of a physical object. The digital replica can be scrutinized, optimized, and manipulated. With the assistance of IoT, which continuously updates data, DT applications can produce a virtual, real-time model of a physical object. In the same context, AI can support DTs as an advanced analytical tool that can automatically scrutinize the collected data, furnish valuable insights, generate predictions about the potential outcomes, and suggest how to avoid potential problems [43]. Thus, IoT and AI technologies could provide granular data using automated data analytics, which is necessary to evaluate EPCs better.

Furthermore, Blockchain, a popular type of Distributed Ledger Technology (DLT), provides trustworthiness, security, quality, and data openness. Decentralized Applications (dApps) are web applications that utilize Blockchain technology to store and manage their interactions [44]. These dApps depend on distributed ledgers and decentralized databases, which will eliminate the reliance on a single trusted source and establish a secure framework for sharing lifecycle information. This is especially crucial in a complicated ecosystem where stakeholders engage with DTs. While ensuring the required integrity, confidentiality, and availability, this approach can address the data exchange challenges [45]. With Blockchain employed, transactions' legitimacy would be guaranteed, and cryptography and consensus mechanisms can be employed to facilitate the validation and traceability of high-value transactions. Thus, incorporating Blockchain technology in the AECO industry and its fusion with DTs and BIM for managing lifecycle information holds enormous potential to address concerns regarding trust, transparency, and communication [46], tackling trust issues between EPC stakeholders.

To that extent, digital technologies have an important role in advancing the AECO industry. Nonetheless, academics have urged investigating how novel business and financing models of performance contracts can be combined with emerging automation technologies such as DTs and IoT [47], but little research has explored this in detail. Table 1 summarizes the added value of using DT and Blockchain technologies for promoting EPC for buildings, highlighting how they address issues not fully resolved by other technologies.

Table 1. The added value of DT and Blockchain technologies for EPC in buildings.

Aspect	Added Value	
	Digital Twin	**Blockchain**
Enhanced Performance Evaluation	Creates high-fidelity virtual models of buildings, allowing real-time monitoring, simulation, and optimization of energy performance, which lead to more accurate performance evaluation and proactive maintenance, which are critical for successful EPC implementation [13].	Ensures data integrity and transparency, facilitating secure and tamper-proof recording of energy performance data, enhancing stakeholder trust, and streamlining the verification process [14].
Improved Data Management	Integrates various data sources into a single platform for comprehensive analysis and better decision-making for energy optimization [48].	Blockchain secures data storage and sharing, addressing data manipulation and unauthorized access concerns that are particularly beneficial for managing large volumes of energy data generated by smart buildings [40].
Automation and Smart Contracts	The combination of DTs and Blockchain enables the automation of EPC processes through smart contracts, which automatically execute and enforce contract terms based on predefined conditions and real-time data [49]. This reduces administrative overhead, minimizes disputes, and ensures timely and accurate performance-based payments, thereby increasing the efficiency and reliability of EPCs [14].	

3. EPC in AECO

3.1. Search Methodology

The selection for this review was restricted to published or in-press journal articles and review articles. The review covered articles from three electronic databases: Scopus, Web of Science, and ScienceDirect. It was conducted using the reference management program Mendeley. The retrieved articles were from databases that guarantee the quality and reliability of indexed scientific journals (e.g., Science Citation Index (SCI), Science Citation Index Expanded (SCI-E), or Engineering Index (EI)). The "article title/abstract/keyword" field was used for the search. The terminology used in the literature search was influenced by a preceding search using generic terms. In this search, journals containing the keywords combination (("performance contract*" OR "performance-based contract*") AND ("energy savings" OR "energy performance" OR "energy efficiency" OR "energy performance gap" OR ESCO OR "energy service compan*")) in the title, abstract and keywords were selected. The selection process was based on the following inclusion criteria:

- Publication year: 2013 to 2023;
- Document type: articles and review articles;
- Source type: journals;
- Language: English;
- Others: subject areas limited to engineering, energy, and environmental sciences.

Articles were recorded and tracked for each limitation applied, with records of the initial count of articles and the number excluded by each limitation. Selected studies from the three databases were exported to Mendeley for filtering to eliminate duplicated records. Full-text documents were collected for articles that met the inclusion criteria based on their title and abstract. If the relevance was unclear to the research objectives, the article was still considered relevant, and the full text was collected. A backward-snowballing process was also used to identify older articles that could provide the corresponding information. A flowchart representing the search methodology is shown in Figure 2.

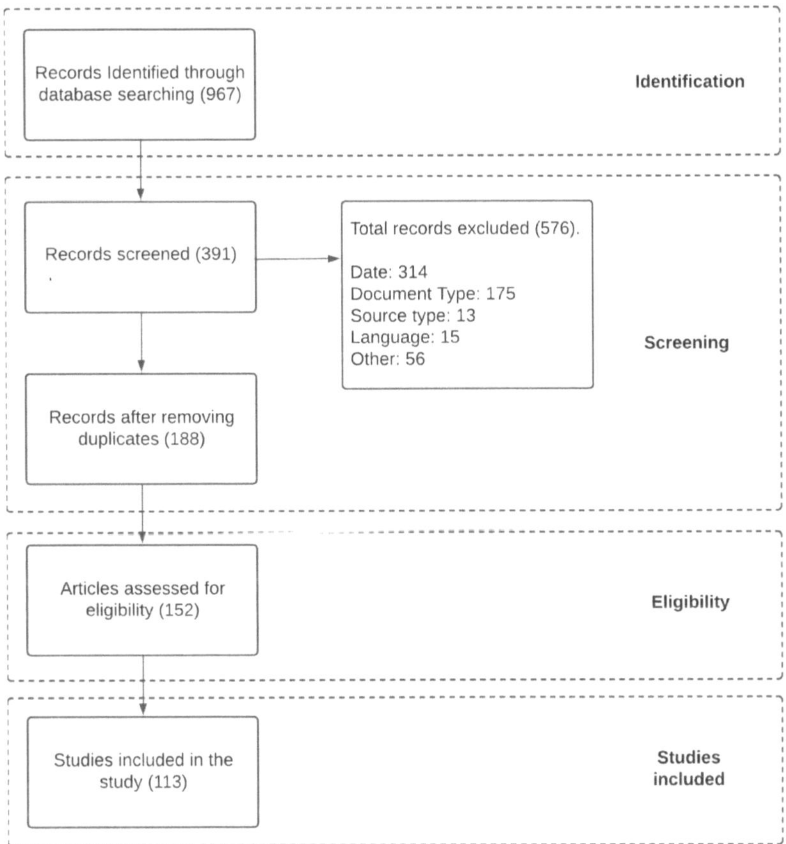

Figure 2. Systematic review methodological flowchart.

The database search yielded 967 publications, of which 391 were considered for filtering based on the inclusion criteria. After removing duplicates, 188 publications were retrieved, and after excluding non-related topics, 152 articles were left for further analysis. A more in-depth examination of the title, abstract, and keywords was conducted to ensure that only articles related to EPC in the AECO industry were included, resulting in 113 relevant articles for the study.

3.2. *Search Results/Analysis*

In order to better analyze the selected articles, it was essential to categorize articles that have applications for buildings from conceptual articles. As shown in Figure 3, two main groups were identified. The buildings group includes 81 articles tackling the building sector (e.g., commercial, residential, and generic types of buildings) [12,20,24,50–126], while the remaining 32 articles tackled conceptual studies that focus on non-building related EPC applications (e.g., review articles) [22,28,127–156]. There are few review articles in this field. Zhang and Yuan (2019) [137] and Shang et al. (2017) [142] are two notable studies addressing challenges related to EPC; however, specific difficulties of EPC implementation in achieving Net Zero Carbon Buildings, particularly the data sharing, storage, and security issues in the M&V process, were not comprehensively addressed, which highlights the importance of developing research work that provides more in-depth and up-to-date insights on these specific challenges. Thus, in the context of the challenges mentioned above, the categorized data were further studied to provide several insights about the current state of EPC diffusion in the AECO industry, highlighting current limitations and future opportunities.

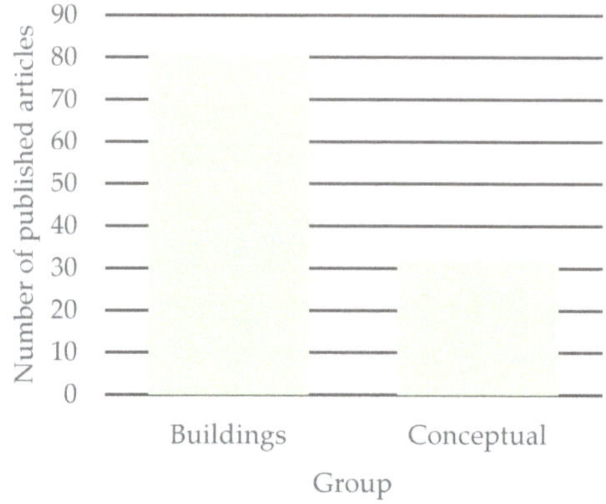

Figure 3. Grouping of articles under study.

The data presented in Figure 4, regarding the number of EPC-related publications in the AECO industry, demonstrate interesting trends over the past decade. The number of publications in both the buildings and conceptual categories has an upward trend, but it is not clearly a steady increase since 2013, with a peak of 13 publications in the buildings category in 2018 and 9 publications in the conceptual category in 2016. Interestingly, while the number of publications in the buildings category remained relatively consistent in recent years, the conceptual category saw a decline in publications from four in 2019 to one in 2021. Nonetheless, both categories saw a slight increase in 2022, followed by a slight decrease again in 2023. These findings indicate approximately an average of 10 publications per year, which highlights the need for further research to optimize the effectiveness and implementation of EPCs.

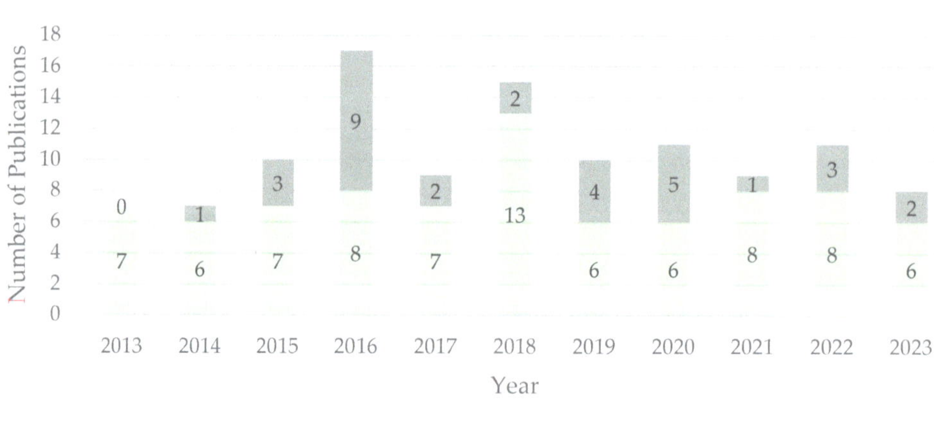

Figure 4. Number of energy performance-based contract-related publications in the AECO industry by year, from 2013 to 2023.

From a deeper aspect of the studies included in the building group, the distribution of studies related to EPCs was classified according to the building type, which reveals some interesting insights into the current focus of EPC research, as shown in Figure 5. The data show that most studies have focused on commercial buildings, with 59 papers (63%) dedicated to this building type. In contrast, only 18 studies (19%) have focused on residential buildings, indicating a significant research gap in this area. The remaining 17 studies (18%) are categorized as generic, which may include studies that do not differentiate between building types. The significant disparity in the number of studies between commercial and residential buildings suggests that implementation and optimization of EPCs in commercial buildings has received greater attention than in residential buildings. This distribution highlights the need for more research focused on EPCs in residential buildings, as these buildings represent a significant portion of the overall building stock and can benefit from implementing EPCs in terms of energy savings and environmental impact.

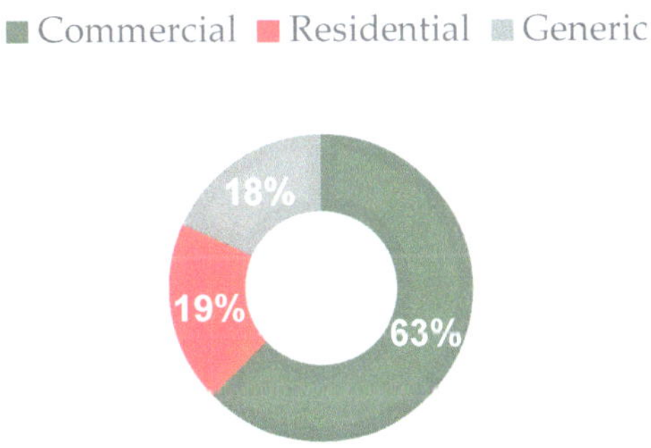

Figure 5. Distribution of studies according to building type.

Moreover, mapping EPC publications related to buildings in the AECO industry by country of application in the case studies provides valuable insights into the research trends and priorities in this field. The data in Table 2 reveal that China and the USA lead the number of publications, with 17 and 13, respectively, constituting about 37.5 percent of the publications. Italy, France, and the UK also have a significant number of publications, with six, five, and four, respectively. It is worth noting that some countries, such as Switzerland, Germany, and Portugal, have only one publication each, indicating that EPC research in these countries may still be in the early stages. This mapping provides a valuable starting point for further EPC research analysis and comparison in different countries and regions.

Table 2. Mapping EPC publications related to buildings in the AECO industry by country.

Country	No. of Publications	Country	No. of Publications
China	17	Poland	1
USA	13	Iran	1
Italy	6	Switzerland	1
France	5	Germany	1
UK	4	Portugal	1
Canada	3	Croatia	1
Malaysia	3	Greece	1
Netherlands	3	Denmark	1
UAE	3	Ukraine	1
Taiwan	2	Slovakia	1
Norway	2	Russia	1
Spain	2	Latvia	1
Australia	2	Turkey	1
Hong Kong	2		

Furthermore, published articles within the building category were classified into four main research topic areas, as shown in Figure 6. The provided data offered valuable insights into the current focus of EPC research within the industry. The data indicate that EPC development has been the most prominent research topic, with almost 36 percent of published articles showing a strong interest in exploring and improving the processes, methodologies, and strategies associated with EPC development [20,157,158]. Following closely behind, EPC execution challenges, legal and contractual constraints, financial mechanisms and business models evaluation analysis, and market studies of EPC and ESCO diffusion have garnered attention, with around 32.5 percent of published articles [12,159,160]. This demonstrates the significance of addressing the practical and contractual aspects of EPC implementation and the financial considerations and market dynamics associated with such contracts. Published articles focusing on understanding the decision-making processes and managing, identifying, and classifying risk and uncertainty factors in EPC projects have also been substantial research topics, with 23 percent of published articles [161–164].

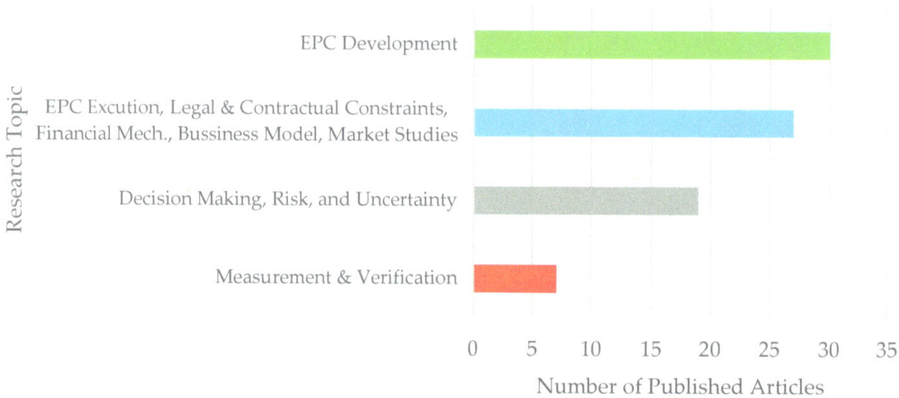

Figure 6. EPC main research topics in the AECO industry.

Lastly, the effective M&V of the energy savings achieved by EPC projects has been largely ignored. However, few studies have acknowledged that M&V is critical for implementing EPC projects. Ke et al. [165] explored the analysis of building energy consumption parameters and M&V of energy savings. The study focused on calibrating a building energy model that facilitates M&V energy savings. The calibrated model was utilized to analyze the impact of changes to energy consumption parameters on the overall energy consumption in the building, providing valuable insights into energy management strategies. Park et al. [166] presented a methodological approach for calibrating building energy-performance simulation models to ensure accurate M&V energy savings. In 2019, Newsham [33] utilized a case study to assess the effectiveness of M&V methods. Specifically, a simple regression-based M&V approach was applied to analyze whole building energy data. Alfaris et al. [167] explored the energy performance of retrofitted buildings undergoing an EPC during the COVID-19 pandemic. The study focused on the approach to be taken and its impact on monitoring the energy profile following the IPMVP. The data collected were then compared with the baseline model to assess the effectiveness of the EPC.

Piccinini et al. [31] and Agenis et al. [168] provided M&V applications for an EPC. The former proposed a novel Reduced Order Model (ROM) framework that facilitates the estimation of energy savings in building retrofits. The ROM was incorporated in the IPMVP to support the M&V of energy savings. The study demonstrated and validated the ROM's ability to forecast energy consumption in an operating educational building. The latter proposed an automated method for selecting the most relevant baseline model based on the IPMVP, which was generalized to handle cases where contracts involve multiple buildings of various types or consumption ranges. The method identified a common best model using new dimensionless indicators, which was useful in buildings with different energy profiles.

Despite their potential benefits, M&V studies have yet to fully leverage the advantages of integrating digital technologies such as DTs, IoTs, Blockchain, and AI into the M&V plan. By utilizing these technologies, it is possible to adjust the physical project's real-time behavior according to the virtual model's performance assessments. This could significantly improve the accuracy and efficiency of M&V, paving the way for the widespread adoption of EPCs in the AECO industry.

3.3. Identifying Research Limitations in EPC

This article clearly shows that the building sector is the focus of many studies. However, only a small percentage of these—8.5%—is dedicated to the vital aspect of Measurement and Verification [31,33,109,165–168]. Additionally, no studies related to the residential sector have been found to address M&V. Advanced M&V, or M&V 2.0, has only been studied in two research articles [33,109]. Moreover, only a fifth of studies related to the building sector are concerned with the residential sector. Interestingly, although energy performance certificates are frequently mentioned in M&V-related articles, they are only discussed in the abstract in 70%, and no studies have explored the relationship between M&V and EPC contract terms. Only one study has explored the potential of Digital Twin and Blockchain technologies or smart contracts concerning EPCs [13]. However, the study's use of Digital Twins did not fully utilize their potential to adjust the physical product's real-time behavior according to the virtual model's performance assessments.

These research gaps highlight the need for further investigation and exploration of M&V, EPCs, and their relationship with Digital Twin and Blockchain technologies to promote the diffusion of EPC projects in the AECO industry.

4. A Framework for Delivering a Smart EPC Using Digital Twin and Blockchain Technologies

4.1. Overview

In this section, the authors aim to develop a framework that applies to energy performance contracts for building projects. The fundamental idea of this framework is to utilize digitalization by integrating DT and Blockchain technologies to deliver a smart EPC. The framework facilitates a trustworthy M&V environment that encounters trust problems and disputes that arise between stakeholders resulting from poor and inaccurate performance management and evaluation in EPCs, which disincentivizes its diffusion in the AECO industry.

The proposed framework uses multi-layered architecture applicable to EPCs in building projects. The following sections will describe the logical structure of the framework. Section 4.2 provides an overview of the framework's architecture. Section 4.3 emphasizes the Digital Twin layer of an asset and its sublayers. Section 4.4 explains the Blockchain service layer and its sublayers. Section 4.5 describes what a Virtual Data Room provides for stakeholders.

4.2. Framework Architecture

In the same context of this research, the proposed framework illustrated in Figure 7 consists of three main layers: (1) the Digital Twin of an Asset; (2) the Blockchain Service layer that is characterized by two sublayers, the private Blockchain infrastructure and the Consortium Blockchain; and (3) the Virtual Data Room. The framework's logic starts with the Digital Twin layer, which provides the base of the M&V environment. The DT layer focuses on integrating static and dynamic building data. The static data of the building is provided through the as-built BIM model (i.e., IFC files). Dynamic data is fed by a real-time data stream captured by sensors attached to physical assets, which is then transmitted to the virtual space in the Digital Twin. This approach to conveying information facilitates swift recognition of underperformance issues, thereby enhancing the ability to take immediate actions and make informed decisions to help optimize a building's energy efficiency through bi-directional dynamic data communication and analytics.

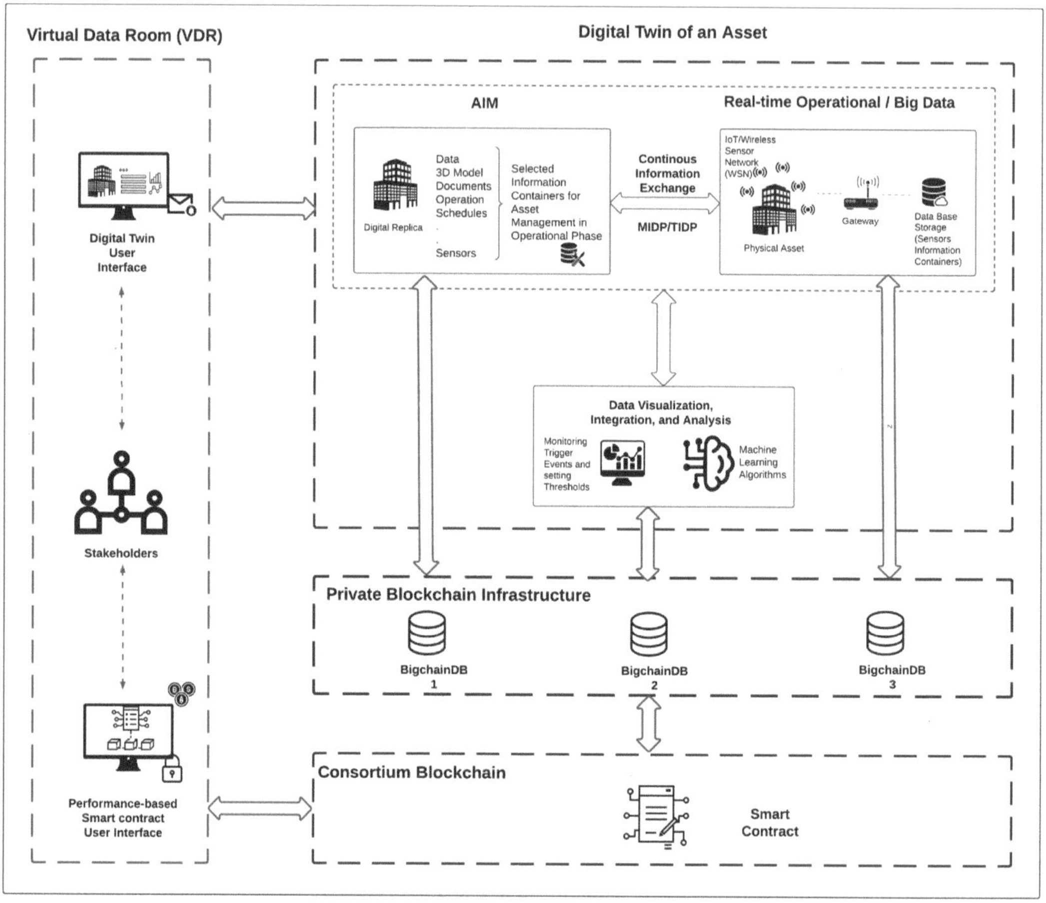

Figure 7. Blockchain-secured Digital Twin framework for smart EPCs.

Choosing an adequate infrastructure for Blockchain application depends on several factors including robustness, information privacy, development and maintenance costs, and speed [169]. The increasing use of AIM and DTs raises concerns about data security and privacy, particularly when the collected data include private information about asset performance and users [170]. These digital models may contain sensitive data, such as occupancy and consumption of water and electricity, which should remain confidential. Additionally, real-time building data, such as indoor air quality, comfort levels, number of occupants, or actual storage levels, can be linked to expected performance and have contractual implications [171]. In this study, since the data from buildings' energy performance can be sensitive for privacy and legal reasons, a private Blockchain is selected.

To the authors' knowledge, the Blockchain service layer is the first integration of a private Blockchain infrastructure using BigchainDB software v 2.2.2 [172] and a Consortium Blockchain (e.g., Ethereum) to a building Digital Twin for an energy performance-based smart contract in the AECO industry. The use of BigchainDB software as a private Blockchain infrastructure offers a combination of the perks of a typical Blockchain and a typical distributed database, such as decentralization, immutability, owner-controlled assets, low latency, high transaction rate, no transaction fee, the permission of access for stakeholders, indexing, and querying of structured data [173]. The authors believe that these advantages help mitigate major challenges—hindering the diffusion of using Digital Twin Blockchain-based energy performance-based contracts—posed by data storage due to high transaction costs resulting from the high intensity of sensor data (i.e., in public Blockchains, every transaction of adding sensor data to the network is subjected to a transaction fee that is determined through the computation of the required computing resources (gas amount and the multiplication of this value by the gas price) and possible concerns regarding the confidentiality of information (e.g., energy data) in open distributed ledgers. In addition, the performing complex performance evaluation and optimization is impossible to execute on Consortium (public) Blockchains.

Finally, the Virtual Data Room provides a user interface that allows each stakeholder permission to access their information and to interact with performance data, triggered actions, and contract functions in a user-friendly environment. Overall, the authors believe the proposed framework to be effective for implementation to manage the performance of energy applications in buildings, enabling real-time adjustments, flexibility, and independent decision-making for interventions and operations.

4.3. Digital Twin of an Asset

The Digital Twin layer serves as a dynamic and interconnected platform that enables real-time monitoring, analysis, simulation, and optimization, facilitating enhanced operational efficiency, predictions, informed decision-making, and effective resource utilization of buildings. This layer comprises three sublayers: (1) The physical/built asset that involves the interaction of physical subsystems, such as sensors and actuators, which operate within the system to monitor and control various aspects of the building's functions through capturing real-time conditions such as temperature, humidity, light intensity, and occupancy. (2) The Asset Information Model (AIM) that contains only required information containers transferred from the as-built BIM model, which are based on the requirements of the building's energy performance management besides essential sensor information containers to provide a real-time update on energy-related parameters. The process of selectively filtering the information containers from the Project Information Model (PIM) to the AIM prevents the inclusion of unnecessary information that could burden various stakeholders. (3) The data visualization, integration, and analysis sublayer offers comprehensive solutions for processing raw data into actionable information using data analytics techniques rooted in statistical theory, Artificial Intelligence, and Machine Learning algorithms.

Overall, in the DT layer, a baseline model is initially developed, and real-time updated models are subsequently generated by collecting current operational indicators. These models are employed to quantify energy savings and assess the fulfillment of predefined energy performance contract conditions. Moreover, it enables the integration of a dynamic display feature within the Virtual Data Room, which will be further elaborated upon. Once a substantial dataset has been accumulated, the data are integrated into the system's algorithm to enable automated control and feedback adjustment. Simultaneously, the system can forecast future scenarios and offer recommendations.

In light of the challenges encountered due to the absence of standardized guidelines governing the practical implementation process of DT, the approach used in this proposed framework is based on a previously established standardized DT framework developed by the authors [174]. The standardized DT framework integrates the BIM ISO 19650 standards [175–178] in the DT processes to facilitate interoperability between the used digital technologies (i.e., Digital Twins and Blockchain).

4.4. Blockchain Service Layer

The key challenge of utilizing Digital Twin technology to enhance the energy performance evaluation and dispute elimination in EPCs is to make the process tamper-proof. Thus, developing secure data sharing and a smart contract is essential for effectively managing a complex system consisting of interconnected Digital Twins and various stakeholders. For this purpose, the Blockchain layer serves as the foundation layer for the Digital Twin that ensures the integrity, transparency, privacy, and accountability of information through the framework. The architecture demonstrates how the Blockchain securely and reliably handles all transactions within the Digital Twin, making the information from the Digital Twin trustworthy. As a result, it can be utilized for smart contract execution and/or payment with confidence.

A core step in evaluating energy performance successfully is to provide accurate and safely exchanged data. IoT provides an accurate and automated data source, which helps eliminate human errors. As mentioned, the collected data are stored in a central repository named "Information containers", which is presented as a part of the BIM processes defined by the ISO 19650 series. On the other hand, the Blockchain ensures the safety of the data exchanged and stored for smart contract execution. However, storing performance data within the smart contract on public Blockchains always poses major challenges, such as high transaction costs and concerns regarding data privacy [13]. In the same context, if data are stored locally off-chain (centralized), the usefulness of the Blockchain is diminished, or the Blockchain network does not store all the data from IoT or DTs, but only stores and shares data required for evaluation. Thus, a trade-off exists between increased trust at the expense of higher on-chain data storage costs or opting for off-chain data storage with reduced trust.

The core idea of this framework is to add a private Blockchain sublayer, accessed through permission, that works as an intermediate layer between the Digital Twin layer and the smart contract layer. The European Data Protection Supervisor (EDPS) emphasizes the need for managing personal data—such as altering, deleting, and selectively disclosing it—to protect individuals' privacy [179]. Ideally, to facilitate data deletion, Blockchain participants would need to establish a mutually agreed-upon process for collectively executing lawful requests to erase personal data from decentralized ledgers [180]. From a technological standpoint, research on eliminating Blockchain's immutability while maintaining security is still in its early stages [181]. More explicitly, the clash between immutability and privacy/data protection rights makes absolute immutability a significant obstacle to the adoption of Blockchain technology when personal data are involved [182]. From this viewpoint, recent progress in incorporating mutability, governed by strict, pre-approved rules, is attractive to both regulators and businesses [183]. Off-chaining techniques are currently viewed as essential in Blockchain-based application development due to their significant advantages, such as lowering Blockchain data storage needs, thereby reducing scalability issues and ensuring compliance with the General Data Protection Regulation (GDPR) [181,184,185]. Moreover, academic research utilizing off-chaining techniques used as a private Blockchain infrastructure for storing actual information [186,187], have been suggested for aligning Blockchains with the GDPR privacy requirements [181].

In this study, the selection of a private Blockchain architecture provides further control over the ledger itself, including decommissioning, as it explicitly enables their right to be forgotten (also known as erasure) [188] when the ledger itself ceases to exist. Although data encryption on a public Blockchain provides a layer of security, which may be sufficient in many use cases, in this study, we explore solutions that allow sensitive information to be deleted once the purpose of the ledger is achieved. The private Blockchain infrastructure utilizes BigchainDB, offering a privileged opportunity for control and privacy over the network. The strength of using BigchainDB is that its properties combine the advantages of Blockchain (e.g., decentralization, Byzantine fault tolerance, immutability, and owner-controlled assets) and typical distributed databases (e.g., low latency, indexing and querying of structured data, and high transaction rates) [173]. Furthermore, in this framework, information containers are stored off-chain to various BigchainDB databases (e.g., information models, documents, and sensor data) to provide confidential information access rights to stakeholders, which maintains data privacy and offers flexibility and scalability, accommodating diverse data formats and volumes. In contrast, based on the EPC evaluation period, only performance indicators essential for contract evaluation and automatic execution are stored on-chain. This selective on-chain storage optimizes Blockchain resources, enhancing transaction throughput and minimizing storage costs. Moreover, guaranteeing traceable storage and data sharing from the sensor to the DT within the

Blockchain network ensures that all data transactions within the DT are reliable and trustworthy and guarantees critical updates necessary for prompt decision-making that can be selectively shared within the Blockchain network. In addition, this database can gather data throughout the entire lifecycle of an asset, and by leveraging the inherent benefits of a real Digital Twin, this approach harnesses bi-directional data exchange by establishing a link between algorithmic decisions stored on the Blockchain and their consequential effects on both the models and the corresponding physical asset in the physical realm.

Moreover, since Blockchains cannot connect to real-world data and events on their own, Decentralized Oracle Networks (DONs) will be used to combine on-chain code (smart contract) and off-chain infrastructure (BigchainDB) [189]. Blockchain oracles are "entities that connect Blockchains to external systems, thereby enabling smart contracts to execute based upon inputs and outputs from the real world" [190]. In other words, oracles serve as intermediaries between Blockchain systems and the external world [191]. Oracles serve as valuable tools for reducing the necessity of costly transactions on a Blockchain, such as storing and utilizing data within smart contracts [192]. Furthermore, oracles are foundational in providing environmental data sourced from sensor readings, satellite imagery, and sophisticated ML calculations to smart contracts. These contracts, in turn, enable the distribution of rewards to individuals involved in reforestation efforts or practicing sustainable consumption [190].

To this end, the private Blockchain infrastructure developed will facilitate the use of data on the smart contract published on the Consortium (public) Blockchain by sharing only the required information through semantic path access for energy performance compliance.

4.5. Virtual Data Room

A Virtual Data Room (VDR) is a user interface that facilitates access with permission to information for each stakeholder. It enables stakeholders to interact with performance data, visualize simulations, and initiate actions through the Digital Twin user interface. In addition to an EPC smart contract user interface to track and analyze all data transactions, it utilizes contract functions within a user-friendly environment. This interface enhances collaboration and streamlines communication among stakeholders, enabling them to effectively navigate and leverage the relevant information for their respective roles and responsibilities. The VDR optimizes the overall user experience, fostering efficient decision-making and promoting effective coordination among stakeholders throughout the contract period.

5. Conclusions

This research aimed to promote the use of EPCs in the AECO industry by utilizing advancements in digital technologies. This was achieved by conducting a systematic analysis of 113 published journal articles. The results showed limitations related to M&V and EPCs' interplay with Digital Twin and Blockchain technologies in the building sector. M&V received minimal attention in studies, with only 8.5% dedicated to this aspect, and none of them addressed M&V in the residential sector. Advanced M&V, or M&V 2.0, was explored in only two research articles. Moreover, although EPCs are frequently mentioned, their relationship with M&V and contract terms remains unexplored. Only a single study investigated the potential of DT and Blockchain technologies for EPCs, but it underutilized the capabilities of DTs and suffered from high real-time transaction costs of data in the Blockchain network.

These research gaps highlight the necessity for further investigation into M&V, EPCs, and their integration with Digital Twin and Blockchain technologies to facilitate the implementation of EPC projects in the AECO industry. In response, the architecture of a framework that combines Digital Twin and Blockchain technologies to create an improved environment for executing EPCs was proposed. The proposed framework consists of three main layers: the Digital Twin of an asset, the Blockchain service layer (including a private Blockchain infrastructure and a Consortium Blockchain), and the Virtual Data Room. The framework shows the potential to enhance Measurement and Verification (M&V) in energy performance-based smart contracts in the AECO industry.

The proposed framework combines the Digital Twin layer with the Blockchain service layer, integrating static and dynamic building data to identify underperformance and facilitate informed decision-making. The Blockchain service layer includes a private Blockchain infrastructure and a Consortium Blockchain, addressing challenges related to data storage, transaction costs, and information confidentiality. The framework incorporates a private Blockchain infrastructure (BigchainDB) as an initial addition, aiming to eliminate the significant transaction costs associated with on-chain storage and address potential concerns about the confidentiality of information in open distributed ledgers. The Virtual Data Room provides stakeholders with a user-friendly interface to access their authorized information and interact with performance data. The framework aims to effectively manage energy applications in buildings, enabling real-time adjustments, flexibility, and autonomous decision-making for interventions and operations.

In future work, the authors recommend formulating a detailed framework for information flow between current framework layers and providing a proof-of-concept that delivers insights into the potential of using the proposed framework for a better EPC digitalized environment.

Author Contributions: Conceptualization, M.N.E.-D., J.P.M. and N.M.M.R.; methodology, M.N.E.-D., J.P.M. and N.M.M.R.; software, M.N.E.-D.; validation, P.F.P., J.P.M. and N.M.M.R.; formal analysis, M.N.E.-D., P.F.P., J.P.M. and N.M.M.R.; investigation, M.N.E.-D. and P.F.P.; data curation, M.N.E.-D.; writing—original draft preparation, M.N.E.-D.; writing—review and editing, M.N.E.-D. and P.F.P.; supervision, J.P.M. and N.M.M.R.; project administration, J.P.M. and N.M.M.R.; funding acquisition, M.N.E.-D., J.P.M. and N.M.M.R. All authors have read and agreed to the published version of the manuscript.

Funding: This work was financially supported by: programmatic funding—UI/BD/151302/2021 of the CONSTRUCT, Instituto de I&D em Estruturas e Construções, funded by national funds through the FCT and the last author would like to acknowledge the support of FCT—Fundação para a Ciência e a Tecnologia through the individual Scientific Employment Stimulus 2021.02686.

Data Availability Statement: The original contributions presented in the study are included in the article, further inquiries can be directed to the corresponding authors.

Conflicts of Interest: The authors declare no conflicts of interest.

References

1. European Commission Energy Performance of Buildings Directive. Available online: https://energy.ec.europa.eu/topics/energy-efficiency/energy-efficient-buildings/energy-performance-buildings-directive_en (accessed on 12 April 2022).
2. Ohene, E.; Chan, A.P.C.; Darko, A.; Nani, G. Navigating toward Net Zero by 2050: Drivers, Barriers, and Strategies for Net Zero Carbon Buildings in an Emerging Market. *Build. Environ.* **2023**, *242*, 110472. [CrossRef]
3. Becqué, R.; Weyl, D.; Stewart, E.; Mackres, E.; Jin, L.; Shen, X. *Accelerating Building Decarbonization: Eight Attainable Policy Pathways to Net Zero Carbon Buildings For All*; World Resources Institute: Washington, DC, USA, 2019.
4. *EN 15978:2011*; Sustainability of Construction Works—Assessment of Environmental Performance of Buildings—Calculation Method. European Standard: Brussels, Belgium, 2011.
5. Tirelli, D.; Besana, D. Moving toward Net Zero Carbon Buildings to Face Global Warming: A Narrative Review. *Buildings* **2023**, *13*, 684. [CrossRef]
6. *Directive 2018/844*; Directive (EU) 2018/844 of the European Parliament and of the Council of 30 May 2018 Amending Directive 2010/31/EU on the Energy Performance of Buildings and Directive 2012/27/EU on Energy Efficiency. European Union: Brussels, Belgium, 2018.
7. Tan, B. Design of Balanced Energy Savings Performance Contracts. *Int. J. Prod. Res.* **2020**, *58*, 1401–1424. [CrossRef]
8. Zangheri, P.; Armani, R.; Kakoulaki, G.; Bavetta, M.; Martirano, G.; Pignatelli, F.; Baranzelli, C. *Building Energy Renovation for Decarbonisation and Covid-19 Recovery A Snapshot at Regional Level*; Publications Office of the European Union: Luxembourg, 2020.
9. Augustins, E.; Jaunzems, D.; Rochas, C.; Kamenders, A. Managing Energy Efficiency of Buildings: Analysis of ESCO Experience in Latvia. *Energy Procedia* **2018**, *147*, 614–623. [CrossRef]
10. E3P EPC—Energy Performance Contracting. Available online: https://e3p.jrc.ec.europa.eu/articles/energy-performance-contracting (accessed on 18 March 2023).

11. Lugarić, T.R.; Dodig, D.; Bogovac, J. Effectiveness of Blending Alternative Procurement Models and Eu Funding Mechanisms Based on Energy Efficiency Case Study Simulation. *Energies* **2019**, *12*, 1612. [CrossRef]
12. Piterou, A.; Coles, A.M. A Review of Business Models for Decentralised Renewable Energy Projects. *Bus. Strateg. Environ.* **2021**, *30*, 1468–1480. [CrossRef]
13. Hunhevicz, J.J.; Motie, M.; Hall, D.M. Digital Building Twins and Blockchain for Performance-Based (Smart) Contracts. *Autom. Constr.* **2022**, *133*, 103981. [CrossRef]
14. Gürcan, O.; Agenis-Nevers, M.; Batany, Y.M.; Elmtiri, M.; Le Fevre, F.; Tucci-Piergiovanni, S. An Industrial Prototype of Trusted Energy Performance Contracts Using Blockchain Technologies. In Proceedings of the 2018 IEEE 20th International Conference on High Performance Computing and Communications; IEEE 16th International Conference on Smart City; IEEE 4th International Conference on Data Science and Systems (HPCC/SmartCity/DSS), Exeter, UK, 28–30 June 2018; IEEE: New York, NY, USA, 2019; Volume I, pp. 1336–1343. [CrossRef]
15. Sharma, P.K.; Kumar, N.; Park, J.H. Blockchain Technology toward Green IoT: Opportunities and Challenges. *IEEE Netw.* **2020**, *34*, 263–269. [CrossRef]
16. Xu, Q.; Aung, K.M.M.; Zhu, Y.; Yong, K.L. *Studies in Computational Intelligence 715 New Advances in the Internet of Things*; Yager, R.R., Pascual Espada, J., Eds.; Springer International Publishing: Cham, Switzerland, 2018; ISBN 9783319581897.
17. Qian, X.; Papadonikolaki, E. Shifting Trust in Construction Supply Chains through Blockchain Technology. *Eng. Constr. Archit. Manag.* **2021**, *28*, 584–602. [CrossRef]
18. Martiniello, L.; Morea, D.; Paolone, F.; Tiscini, R. Energy Performance Contracting and Public-Private Partnership: How to Share Risks and Balance Benefits. *Energies* **2020**, *13*, 3625. [CrossRef]
19. Liu, H.; Hu, M.; Zhang, X. Energy Costs Hosting Model: The Most Suitable Business Model in the Developing Stage of Energy Performance Contracting. *J. Clean. Prod.* **2016**, *172*, 2553–2566. [CrossRef]
20. Natividade, J.; Cruz, C.O.; Silva, C.M. Improving the Efficiency of Energy Consumption in Buildings: Simulation of Alternative EnPC Models. *Sustainability* **2022**, *14*, 4228. [CrossRef]
21. Qin, Q.; Liang, F.; Li, L.; Wei, Y.M. Selection of Energy Performance Contracting Business Models: A Behavioral Decision-Making Approach. *Renew. Sustain. Energy Rev.* **2017**, *72*, 422–433. [CrossRef]
22. Pätäri, S.; Sinkkonen, K. Energy Service Companies and Energy Performance Contracting: Is There a Need to Renew the Business Model? Insights from a Delphi Study. *J. Clean. Prod.* **2014**, *66*, 264–271. [CrossRef]
23. Garbuzova-Schlifter, M.; Madlener, R. AHP-Based Risk Analysis of Energy Performance Contracting Projects in Russia. *Energy Policy* **2016**, *97*, 559–581. [CrossRef]
24. Lee, P.; Lam, P.T.I.; Lee, W.L. Performance Risks of Lighting Retrofit in Energy Performance Contracting Projects. *Energy Sustain. Dev.* **2018**, *45*, 219–229. [CrossRef]
25. Zhang, M.; Wang, M.; Jin, W.; Xia-Bauer, C. Managing Energy Efficiency of Buildings in China: A Survey of Energy Performance Contracting (EPC) in Building Sector. *Energy Policy* **2018**, *114*, 13–21. [CrossRef]
26. Hannon, M.J.; Bolton, R. UK Local Authority Engagement with the Energy Service Company (ESCo) Model: Key Characteristics, Benefits, Limitations and Considerations. *Energy Policy* **2015**, *78*, 198–212. [CrossRef]
27. Taylor, R.P.; Govindarajalu, C.; Levin, J.; Meyer, A.S.; Ward, W.A. *Financing Energy Efficiency: Lessons from Brazil, China, India, and Beyond*; The International Bank for Reconstruction and Development/TheWorld Bank: Washington, DC, USA, 2008; ISBN 9780821373040.
28. Shonder, J.A.; Avina, J.M. New Directions in Measurement and Verification for Performance Contracts. *Energy Eng. J. Assoc. Energy Eng.* **2016**, *113*, 7–17. [CrossRef]
29. Burman, E.; Mumovic, D. *Measurement and Verification Models for Cost-Effective Energy-Efficient Retrofitting*; Elsevier Ltd.: Amsterdam, The Netherlands, 2017; ISBN 9780081011287.
30. Alrobaie, A.; Krarti, M. A Review of Data-Driven Approaches for Measurement and Verification Analysis of Building Energy Retrofits. *Energies* **2022**, *15*, 7824. [CrossRef]
31. Piccinini, A.; Hajdukiewicz, M.; Keane, M.M. A Novel Reduced Order Model Technology Framework to Support the Estimation of the Energy Savings in Building Retrofits. *Energy Build.* **2021**, *244*, 110896. [CrossRef]
32. Department of Energy & Climate Change. *Guide to Energy Performance Contracting Best Practices*; Department of Energy & Climate Change: London, UK, 2015.
33. Newsham, G.R. Measurement and Verification of Energy Conservation Measures Using Whole-Building Electricity Data from Four Identical Office Towers. *Appl. Energy* **2019**, *255*, 113882. [CrossRef]
34. EVO. *International Performance Measurement and Verification Protocol (IPMVP)*; EVO: Washington, DC, USA, 2016.
35. ASHRAE. *Guideline 14: Measurement of Energy, Demand, and Water Savings*; ASHRAE: Atlanta, GA, USA, 2014.
36. Jain, N.; Burman, E.; Stamp, S.; Mumovic, D.; Davies, M. Cross-Sectoral Assessment of the Performance Gap Using Calibrated Building Energy Performance Simulation. *Energy Build.* **2020**, *224*, 110271. [CrossRef]
37. Granderson, J.; Touzani, S.; Claudine, C.; Sohn, M.; Fernandes, S. *Assessment of Automated Measurement and Verification (M&V) Methods*; Lawrence Berkeley National Laboratory: Berkeley, CA, USA, 2015.
38. Gallagher, C.V.; Leahy, K.; O'Donovan, P.; Bruton, K.; O'Sullivan, D.T.J. IntelliMaV: A Cloud Computing Measurement and Verification 2.0 Application for Automated, near Real-Time Energy Savings Quantification and Performance Deviation Detection. *Energy Build.* **2019**, *185*, 26–38. [CrossRef]

39. Lee, D.; Lee, S.H.; Masoud, N.; Krishnan, M.S.; Li, V.C. Integrated Digital Twin and Blockchain Framework to Support Accountable Information Sharing in Construction Projects. *Autom. Constr.* **2021**, *127*, 103688. [CrossRef]
40. Teisserenc, B.; Sepasgozar, S. Adoption of Blockchain Technology through Digital Twins in the Construction Industry 4.0: A PESTELS Approach. *Buildings* **2021**, *11*, 670. [CrossRef]
41. Deng, M.; Menassa, C.C.; Kamat, V.R. From BIM to Digital Twins: A Systematic Review of the Evolution of Intelligent Building Representations in the AEC-FM Industry. *J. Inf. Technol. Constr.* **2021**, *26*, 58–83. [CrossRef]
42. Research and Markets Digital Twins Market by Technology, Twinning Type, Cyber-to-Physical Solutions, Use Cases and Applications in Industry Verticals 2022–2027. Available online: https://www.researchandmarkets.com/reports/4805605/li-ion-battery-global-market-trajectory-and (accessed on 22 March 2023).
43. Lv, Z.; Xie, S. Artificial Intelligence in the Digital Twins: State of the Art, Challenges, and Future Research Topics. *Digit. Twin* **2021**, *1*, 12. [CrossRef]
44. Darabseh, M.; Poças Martins, J. Transforming Construction Entities from Traditional Management to Autonomous Management Using Blockchain. In *Trends on Construction in the Digital Era—Proceedings of ISIC 2022*; Correia, A.G., Azenha, M., Cruz, P.J.S., Novais, P., Pereira, P., Eds.; Springer Nature: Cham, Switzerland, 2023; pp. 111–121.
45. Putz, B.; Dietz, M.; Empl, P.; Pernul, G. EtherTwin: Blockchain-Based Secure Digital Twin Information Management. *Inf. Process. Manag.* **2021**, *58*, 102425. [CrossRef]
46. Abioye, S.O.; Oyedele, L.O.; Akanbi, L.; Ajayi, A.; Davila Delgado, J.M.; Bilal, M.; Akinade, O.O.; Ahmed, A. Artificial Intelligence in the Construction Industry: A Review of Present Status, Opportunities and Future Challenges. *J. Build. Eng.* **2021**, *44*, 103299. [CrossRef]
47. Mourtzis, D.; Boli, N.; Alexopoulos, K.; Różycki, D. A Framework of Energy Services: From Traditional Contracts to Product-Service System (PSS). *Procedia CIRP* **2018**, *69*, 746–751. [CrossRef]
48. Xiong, T.; Cheng, Q.; Yang, C.; Yang, X.; Lin, S. Application of Digital Twin Technology in Intelligent Building Energy Efficiency Management System. In Proceedings of the 2021 International Conference on E-Commerce and E-Management (ICECEM), Dalian, China, 24–26 September 2021; pp. 393–396. [CrossRef]
49. Celik, Y.; Petri, I.; Rezgui, Y. Leveraging BIM and Blockchain for Digital Twins. In Proceedings of the 2021 IEEE International Conference on Engineering, Technology and Innovation (ICE/ITMC), Cardiff, UK, 21–23 June 2021; pp. 1–10. [CrossRef]
50. Rad, M.A.H.; Jalaei, F.; Golpour, A.; Varzande, S.S.H.; Guest, G. BIM-Based Approach to Conduct Life Cycle Cost Analysis of Resilient Buildings at the Conceptual Stage. *Autom. Constr.* **2021**, *123*, 103480. [CrossRef]
51. Xue, F.; Lu, W.; Chen, Z.; Webster, C.J. From LiDAR Point Cloud towards Digital Twin City: Clustering City Objects Based on Gestalt Principles. *ISPRS J. Photogramm. Remote Sens.* **2020**, *167*, 418–431. [CrossRef]
52. Mehmood, F.; Edwards, D.; Lai, J.; Parn, E.A.; Riaz, Z. Engineering-out Hazards: Digitising the Management Working Safety in Confined Spaces. *Facilities* **2019**, *37*, 196–215. [CrossRef]
53. Zhang, Y.; Jia, Q. A Simulation-Based Policy Improvement Method for Joint-Operation of Building Microgrids With Distributed Solar Power and Battery. *IEEE Trans. Smart Grid* **2018**, *9*, 6242–6252. [CrossRef]
54. Violante, W.; Cañizares, C.A.; Trovato, M.A.; Forte, G. An Energy Management System for Isolated Microgrids With Thermal Energy Resources. *IEEE Trans. Smart Grid* **2020**, *11*, 2880–2891. [CrossRef]
55. Corbett, J.; Wardle, K.; Chen, C. Toward a Sustainable Modern Electricity Grid: The Effects of Smart Metering and Program Investments on Demand-Side Management Performance in the US Electricity Sector 2009-2012. *IEEE Trans. Eng. Manag.* **2018**, *65*, 252–263. [CrossRef]
56. Song, Y.; Mao, F.; Liu, Q. Human Comfort in Indoor Environment: A Review on Assessment Criteria, Data Collection and Data Analysis Methods. *IEEE Access* **2019**, *7*, 119774–119786. [CrossRef]
57. Mondal, A.; Misra, S.; Obaidat, M.S. Distributed Home Energy Management System With Storage in Smart Grid Using Game Theory. *IEEE Syst. J.* **2017**, *11*, 1857–1866. [CrossRef]
58. Canterino, F.; Cagliano, R.; Longoni, A.; Bartezzaghi, E. The Interplay between Smart Manufacturing Technologies and Work Organization. *Int. J. Oper. Prod. Manag.* **2019**, *39*, 913–934. [CrossRef]
59. Pärn, E.A.; Ahmed, A.; Al-saeed, Y.W. An 80-Year Projection of NZEB Strategies in Extreme Climatic Conditions of Iraq. *Int. J. Build. Pathol. Adapt.* **2020**, *38*, 472–492. [CrossRef]
60. Brager, G.S.; de Dear, R.J. Thermal Adaptation in the Built Environment: A Literature Review. *Energy Build.* **1998**, *27*, 83–96. [CrossRef]
61. Rejikumar, G.; Raja Sreedharan, V.; Arunprasad, P.; Jinil, P.; Sreeraj, K.M. Industry 4.0: Key Findings and Analysis from the Literature Arena. *Benchmarking Int. J.* **2019**, *26*, 2514–2542. [CrossRef]
62. Zhai, S.; Zhou, H.; Wang, Z.; He, G. Analysis of Dynamic Appliance Flexibility Considering User Behavior via Non-Intrusive Load Monitoring and Deep User Modeling. *CSEE J. Power Energy Syst.* **2020**, *6*, 41–51. [CrossRef]
63. Zhong, B.; Gan, C.; Luo, H.; Xing, X. Ontology-Based Framework for Building Environmental Monitoring and Compliance Checking under BIM Environment. *Build. Environ.* **2018**, *141*, 127–142. [CrossRef]
64. Liu, D.; Liu, B.; Wang, C.; Jin, W.; Zha, Q.; Shi, G.; Wang, D.; Sang, X.; Ni, C. Ionic Liquid-Assisted Exfoliation of Two-Dimensional Metal-Organic Frameworks for Luminescent Sensing. *ACS Sustain. Chem. Eng.* **2020**, *8*, 2167–2175. [CrossRef]
65. Yu, G.; Mao, Z.; Hu, M.; Li, Z.; Sugumaran, V. BIM+ Topology Diagram-Driven Multiutility Tunnel Emergency Response Method. *J. Comput. Civ. Eng.* **2019**, 33, 04019038. [CrossRef]

66. Coraddu, A.; Oneto, L.; Baldi, F.; Cipollini, F.; Atlar, M.; Savio, S. Data-Driven Ship Digital Twin for Estimating the Speed Loss Caused by the Marine Fouling. *Ocean Eng.* **2019**, *186*, 106063. [CrossRef]
67. Pan, Y.H.; Qu, T.; Wu, N.Q.; Khalgui, M.; Huang, G.Q. Digital Twin Based Real-Time Production Logistics Synchronization System in a Multi-Level Computing Architecture. *J. Manuf. Syst.* **2021**, *58*, 246–260. [CrossRef]
68. Howell, S.; Rezgui, Y.; Beach, T. Integrating Building and Urban Semantics to Empower Smart Water Solutions. *Autom. Constr.* **2017**, *81*, 434–448. [CrossRef]
69. Gain, U. Applying Frameworks for Cognitive Services in IIoT. *J. Syst. Sci. Syst. Eng.* **2021**, *30*, 59–84. [CrossRef]
70. Verner, I.M.; Cuperman, D.; Reitman, M. Robot Online Learning to Lift Weights: A Way to Expose Students to Robotics and Intelligent Technologies. *Int. J. Online Eng.* **2017**, *13*, 174–182. [CrossRef]
71. *ASHRAE-55 Thermal Environmental Conditions for Human Occupancy*; American Society of Heating, Refrigerating and Air Conditioning Engineers, Inc.: Atlanta, GA, USA, 2017.
72. Moretti, N.; Xie, X.; Merino, J.; Brazauskas, J.; Parlikad, A.K. An Openbim Approach to Iot Integration with Incomplete As-Built Data. *Appl. Sci.* **2020**, *10*, 8287. [CrossRef]
73. Okakpu, A.; GhaffarianHoseini, A.; Tookey, J.; Haar, J.; Ghaffarian Hoseini, A. An Optimisation Process to Motivate Effective Adoption of BIM for Refurbishment of Complex Buildings in New Zealand. *Front. Archit. Res.* **2019**, *8*, 646–661. [CrossRef]
74. Wang, P.; Wu, P.; Wang, J.; Chi, H.-L.; Wang, X. A Critical Review of the Use of Virtual Reality in Construction Engineering Education and Training. *Int. J. Environ. Res. Public Health* **2018**, *15*, 1204. [CrossRef] [PubMed]
75. Bank, B.; Chabassier, J. Model-Based Digital Pianos: From Physics to Sound Synthesis. *IEEE Signal Process. Mag.* **2019**, *36*, 103–114. [CrossRef]
76. Malik, A.A.; Brem, A. Digital Twins for Collaborative Robots: A Case Study in Human-Robot Interaction. *Robot. Comput. Integr. Manuf.* **2021**, *68*, 102092. [CrossRef]
77. Tong, X.; Liu, Q.; Pi, S.; Xiao, Y. Real-Time Machining Data Application and Service Based on IMT Digital Twin. *J. Intell. Manuf.* **2020**, *31*, 1113–1132. [CrossRef]
78. Ciano, M.P.; Pozzi, R.; Rossi, T.; Strozzi, F. Digital Twin-Enabled Smart Industrial Systems: A Bibliometric Review. *Int. J. Comput. Integr. Manuf.* **2020**, *34*, 690–708. [CrossRef]
79. Angrish, A.; Starly, B.; Lee, Y.-S.; Cohen, P.H. A Flexible Data Schema and System Architecture for the Virtualization of Manufacturing Machines (VMM). *J. Manuf. Syst.* **2017**, *45*, 236–247. [CrossRef]
80. Xu, Y.; Sun, Y.; Liu, X.; Zheng, Y. A Digital-Twin-Assisted Fault Diagnosis Using Deep Transfer Learning. *IEEE Access* **2019**, *7*, 19990–19999. [CrossRef]
81. Nakamura, T. Digital Twin Computing Initiative. *NTT Tech. Rev.* **2020**, *18*, 13–18. [CrossRef]
82. De Dear, R.J.; Brager, G.S. Thermal Comfort in Naturally Ventilated Buildings: Revisions to ASHRAE Standard 55. *Energy Build.* **2002**, *34*, 549–561. [CrossRef]
83. Dolgui, A.; Ivanov, D.; Sokolov, B. Reconfigurable Supply Chain: The X-Network. *Int. J. Prod. Res.* **2020**, *58*, 4138–4163. [CrossRef]
84. Radončić, N.; Sattlegger, E.; Lacourse-Dontigny, X.; Mitsch, T. Designing a State-of-the-Art Monitoring System in Challenging Operating Conditions. *Geomech. Tunnelbau* **2021**, *14*, 54–62. [CrossRef]
85. Faqih, F.; Zayed, T. Defect-Based Building Condition Assessment. *Build. Environ.* **2021**, *191*, 107575. [CrossRef]
86. Petri, I.; Kubicki, S.; Rezgui, Y.; Guerriero, A.; Li, H. Optimizing Energy Efficiency in Operating Built Environment Assets through Building Information Modeling: A Case Study. *Energies* **2017**, *10*, 1167. [CrossRef]
87. Dambrot, S.M. Symbiotic Autonomous Systems, Digital Twins and Artificial Intelligence: Emergence and Evolution. Mondo Digitale. 2019. Available online: https://mondodigitale.aicanet.net/2019-1/articoli/03_MD80_Symbiotic_Autonomous_Digital_Twins_and_Artificial_Intelligence.pdf (accessed on 29 April 2024).
88. Gao, H.; Koch, C.; Wu, Y. Building Information Modelling Based Building Energy Modelling: A Review. *Appl. Energy* **2019**, *238*, 320–343. [CrossRef]
89. Zhang, Y.; Hu, H.; Xu, F. Social Network Visual Simulation for Process Reengineering of Construction Change Management under Building Information Modelling Technology. *J. Intell. Fuzzy Syst.* **2020**, *39*, 1471–1480. [CrossRef]
90. Novembri, G.; Rossini, F.L. Swarm Modelling Framework to Improve Design Support Systems Capabilities. *J. Inf. Technol. Constr.* **2020**, *25*, 398–415. [CrossRef]
91. Zhou, Y.; Zhang, C.; Han, X.; Lin, Y. Monitoring Combustion Instabilities of Stratified Swirl Flames by Feature Extractions of Time-Averaged Flame Images Using Deep Learning Method. *Aerosp. Sci. Technol.* **2021**, *109*, 106443. [CrossRef]
92. Lin, Y.-C.; Cheung, W.-F. Developing WSN/BIM Based Environmental Monitoring Management System for Parking Garages in Smart Cities. *J. Manag. Eng.* **2020**, *36*, 04020012. [CrossRef]
93. Alhamami, A.; Petri, I.; Rezgui, Y.; Kubicki, S. Promoting Energy Efficiency in the Built Environment through Adapted BIM Training and Education. *Energies* **2020**, *13*, 2308. [CrossRef]
94. Sierla, S.; Kyrki, V.; Aarnio, P.; Vyatkin, V. Automatic Assembly Planning Based on Digital Product Descriptions. *Comput. Ind.* **2018**, *97*, 34–46. [CrossRef]
95. Hwangbo, S.; Sin, G. Design of Control Framework Based on Deep Reinforcement Learning and Monte-Carlo Sampling in Downstream Separation. *Comput. Chem. Eng.* **2020**, *140*, 106910. [CrossRef]
96. Zhang, C.; Zhou, G.; Hu, J.; Li, J. Deep Learning-Enabled Intelligent Process Planning for Digital Twin Manufacturing Cell. *Knowl.-Based Syst.* **2020**, *191*, 105247. [CrossRef]

97. Guo, H.; Chen, M.; Mohamed, K.; Qu, T.; Wang, S.; Li, J. A Digital Twin-Based Flexible Cellular Manufacturing for Optimization of Air Conditioner Line. *J. Manuf. Syst.* **2021**, *58*, 65–78. [CrossRef]
98. Nativi, S.; Mazzetti, P.; Craglia, M. Digital Ecosystems for Developing Digital Twins of the Earth: The Destination Earth Case. *Remote Sens.* **2021**, *13*, 2119. [CrossRef]
99. *US Energy Information Administration (EIA) March 2022 Monthly Energy Review*; US Energy Information Administration (EIA): Washington, DC, USA, 2022.
100. Qiu, S.; Mias, C.; Guo, W.; Geng, X. HS2 Railway Embankment Monitoring: Effect of Soil Condition on Underground Signals. *SN Appl. Sci.* **2019**, *1*, 537. [CrossRef]
101. Orozco-Messana, J.; Iborra-Lucas, M.; Calabuig-Moreno, R. Neighbourhood Modelling for Urban Sustainability Assessment. *Sustainability* **2021**, *13*, 4654. [CrossRef]
102. Agnusdei, G.P.; Elia, V.; Gnoni, M.G. Is Digital Twin Technology Supporting Safety Management? A Bibliometric and Systematic Review. *Appl. Sci.* **2021**, *11*, 2767. [CrossRef]
103. Bortoluzzi, B.; Efremov, I.; Medina, C.; Sobieraj, D.; McArthur, J.J. Automating the Creation of Building Information Models for Existing Buildings. *Autom. Constr.* **2019**, *105*, 102838. [CrossRef]
104. Shalabi, F.; Turkan, Y. IFC BIM-Based Facility Management Approach to Optimize Data Collection for Corrective Maintenance. *J. Perform. Constr. Facil.* **2017**, *31*, 04016081. [CrossRef]
105. Tao, F.; Zhang, M.; Nee, A.Y.C. Background and Concept of Digital Twin. In *Digital Twin Driven Smart Manufacturing*; Academic Press: Cambridge, MA, USA, 2019; pp. 3–28.
106. Wang, Q.; Lee, B.D.; Augenbroe, G.; Paredis, C.J.J. An Application of Normative Decision Theory to the Valuation of Energy Efficiency Investments under Uncertainty. *Autom. Constr.* **2017**, *73*, 78–87. [CrossRef]
107. Hasan, I.; Gardezi, S.S.S.; Manzoor, B.; Arshid, M.U. Sustainable Consumption Patterns Adopting BIM-Enabled Energy Optimization-A Case Study of Developing Urban Centre. *Polish J. Environ. Stud.* **2022**, *31*, 3095–3103. [CrossRef]
108. Oliver, E. The Role of Real Time Data in Monitoring and Verification. *Energy Eng. J. Assoc. Energy Eng.* **2018**, *115*, 26–36. [CrossRef]
109. Afroz, Z.; Burak Gunay, H.; O'Brien, W.; Newsham, G.; Wilton, I. An Inquiry into the Capabilities of Baseline Building Energy Modelling Approaches to Estimate Energy Savings. *Energy Build.* **2021**, *244*, 111054. [CrossRef]
110. Kim, A.; Haberl, J.; Anderson, S. Comparison between Current Industry Methods and an Energy Simulation Model for Quantifying Energy Service Projects. *J. Archit. Eng.* **2016**, *22*, 04015016. [CrossRef]
111. Winther, T.; Gurigard, K. Energy Performance Contracting (EPC): A Suitable Mechanism for Achieving Energy Savings in Housing Cooperatives? Results from a Norwegian Pilot Project. *Energy Effic.* **2017**, *10*, 577–596. [CrossRef]
112. Dong, B.; Lam, K.P.; Huang, Y.C.; Dobbs, G.M. *A Comparative Study of the IFC and GbXML Informational Infrastructures for Data Exchange in Computational Design Support Environments*; IBPSA International Building Performance Simulation Association: Rapid City, SD, USA, 2007; pp. 1530–1537.
113. Dall'O', G.; Ferrari, S.; Bruni, E.; Bramonti, L. Effective Implementation of ISO 50001: A Case Study on Energy Management for Heating Load Reduction for a Social Building Stock in Northern Italy. *Energy Build.* **2020**, *219*, 110029. [CrossRef]
114. Röck, M.; Hollberg, A.; Habert, G.; Passer, A. LCA and BIM: Visualization of Environmental Potentials in Building Construction at Early Design Stages. *Build. Environ.* **2018**, *140*, 153–161. [CrossRef]
115. Ounis, S.; Aste, N.; Butera, F.M.; Pero, C.D.; Leonforte, F.; Adhikari, R.S. Optimal Balance between Heating, Cooling and Environmental Impacts: A Method for Appropriate Assessment of Building Envelope's U-Value. *Energies* **2022**, *15*, 3570. [CrossRef]
116. Van den Brom, P.; Hansen, A.R.; Gram-Hanssen, K.; Meijer, A.; Visscher, H. Variances in Residential Heating Consumption—Importance of Building Characteristics and Occupants Analysed by Movers and Stayers. *Appl. Energy* **2019**, *250*, 713–728. [CrossRef]
117. Helsinki Helsinki—Energy and Climate Atlas. Available online: https://kartta.hel.fi/3d/atlas/#/ (accessed on 22 February 2024).
118. Stadtmodell 3D-Stadtmodell. Geomatik + Vermessung Stadt Zürich. Available online: https://www.stadt-zuerich.ch/content/ted/de/index/geoz/plan-und-datenbezug/3d-stadtmodell.html# (accessed on 22 February 2024).
119. Zhang, Y.; Kasahara, S.; Shen, Y.; Jiang, X. Smart Contract-Based Access Control for the Internet of Things. *arXiv* **2018**, arXiv:1802.04410. [CrossRef]
120. Aranda, J.; Tsitsanis, T.; Georgopoulos, G.; Longares, J.M. Innovative Data-Driven Energy Services and Business Models in the Domestic Building Sector. *Sustainability* **2023**, *15*, 3742. [CrossRef]
121. Naderi, H.; Heydari, M.H.; Parchami Jalal, M. Risk Analysis in Implementing Building Energy Performance Projects: Hybrid DANP-VIKOR Model Analysis—A Case Study in Iran. *Buildings* **2023**, *13*, 2066. [CrossRef]
122. Mohamad Munir, Z.H.; Ahmad Ludin, N.; Junedi, M.M.; Ahmad Affandi, N.A.; Ibrahim, M.A.; Mat Teridi, M.A. A Rational Plan of Energy Performance Contracting in an Educational Building: A Case Study. *Sustainability* **2023**, *15*, 1430. [CrossRef]
123. Gončarovs, K.; Jegiazarjana, K. Beyond Well-Being: The Assessment of the Energy Renovation in Latvia by the Residents. *Environ. Clim. Technol.* **2023**, *27*, 813–823. [CrossRef]
124. Medved, P. EPCHC-energy performance contracting (EPC) model for historic city centres. *Acta Innov.* **2023**, *47*, 28–40. [CrossRef]
125. Meraghni, S.; Terrissa, L.S.; Yue, M.; Ma, J.; Jemei, S.; Zerhouni, N. A Data-Driven Digital-Twin Prognostics Method for Proton Exchange Membrane Fuel Cell Remaining Useful Life Prediction. *Int. J. Hydrogen Energy* **2021**, *46*, 2555–2564. [CrossRef]

126. Micolier, A.; Taillandier, F.; Taillandier, P.; Bos, F. Li-BIM, an Agent-Based Approach to Simulate Occupant-Building Interaction from the Building-Information Modelling. *Eng. Appl. Artif. Intell.* **2019**, *82*, 44–59. [CrossRef]
127. Liu, G.; Zheng, S.; Xu, P.; Zhuang, T. An ANP-SWOT Approach for ESCOs Industry Strategies in Chinese Building Sectors. *Renew. Sustain. Energy Rev.* **2018**, *93*, 90–99. [CrossRef]
128. Acuner, E.; Cin, R.; Onaygil, S. Energy Service Market Evaluation by Bayesian Belief Network and SWOT Analysis: Case of Turkey. *Energy Effic.* **2021**, *14*, 62. [CrossRef]
129. Zhang, W.; Yuan, H. A Bibliometric Analysis of Energy Performance Contracting Research from 2008 to 2018. *Sustainability* **2019**, *11*, 3548. [CrossRef]
130. Shang, T.; Yang, L.; Liu, P.; Shang, K.; Zhang, Y. Financing Mode of Energy Performance Contracting Projects with Carbon Emissions Reduction Potential and Carbon Emissions Ratings. *Energy Policy* **2020**, *144*, 111632. [CrossRef]
131. Shang, T.; Liu, P.; Guo, J. How to Allocate Energy-Saving Benefit for Guaranteed Savings EPC Projects? A Case of China. *Energy* **2020**, *191*, 116499. [CrossRef]
132. Roshchanka, V.; Evans, M. Scaling up the Energy Service Company Business: Market Status and Company Feedback in the Russian Federation. *J. Clean. Prod.* **2016**, *112*, 3905–3914. [CrossRef]
133. Shang, T.; Sun, X.; Liu, P.; Gao, J. Cracking the Achilles' Heel of Energy Performance Contracting Projects: The Credit Risk Identification Method for Clients. *Int. J. Green Energy* **2020**, *17*, 196–207. [CrossRef]
134. Shang, T.; Zhang, K.; Liu, P.; Chen, Z. A Review of Energy Performance Contracting Business Models: Status and Recommendation. *Sustain. Cities Soc.* **2017**, *34*, 203–210. [CrossRef]
135. Lee, P.; Lam, P.T.I.; Lee, W.L.; Chan, E.H.W. Analysis of an Air-Cooled Chiller Replacement Project Using a Probabilistic Approach for Energy Performance Contracts. *Appl. Energy* **2016**, *171*, 415–428. [CrossRef]
136. Zhou, Y.; Evans, M.; Yu, S.; Sun, X.; Wang, J. Linkages between Policy and Business Innovation in the Development of China's Energy Performance Contracting Market. *Energy Policy* **2020**, *140*, 111208. [CrossRef]
137. Coleman, P. Escalation Rates in Energy Savings Performance Contracts. *Energy Eng. J. Assoc. Energy Eng.* **2015**, *112*, 66–77. [CrossRef]
138. Ruan, H.; Gao, X.; Mao, C. Empirical Study on Annual Energy-Saving Performance of Energy Performance Contracting in China. *Sustainability* **2018**, *10*, 1666. [CrossRef]
139. Xing, G.; Qian, D.; Guo, J. Research on the Participant Behavior Selections of the Energy Performance Contracting Project Based on the Robustness of the Shared Savings Contract. *Sustainability* **2016**, *8*, 730. [CrossRef]
140. Ning, Y.; Cherian, J.; Sial, M.S.; Álvarez-Otero, S.; Comite, U.; Zia-Ud-Din, M. Green Bond as a New Determinant of Sustainable Green Financing, Energy Efficiency Investment, and Economic Growth: A Global Perspective. *Environ. Sci. Pollut. Res.* **2023**, *30*, 61324–61339. [CrossRef] [PubMed]
141. Yuan, X.; Ma, R.; Zuo, J.; Mu, R. Towards a Sustainable Society: The Status and Future of Energy Performance Contracting in China. *J. Clean. Prod.* **2016**, *112*, 1608–1618. [CrossRef]
142. Leffel, B. Climate Consultants and Complementarity: Local Procurement, Green Industry and Decarbonization in Australia, Singapore, and the United States. *Energy Res. Soc. Sci.* **2022**, *88*, 102635. [CrossRef]
143. Nolden, C.; Sorrell, S. The UK Market for Energy Service Contracts in 2014–2015. *Energy Effic.* **2016**, *9*, 1405–1420. [CrossRef]
144. Lu, Z.; Shao, S. Impacts of Government Subsidies on Pricing and Performance Level Choice in Energy Performance Contracting: A Two-Step Optimal Decision Model. *Appl. Energy* **2016**, *184*, 1176–1183. [CrossRef]
145. Deng, X.; Zheng, S.; Xu, P.; Zhang, X. Study on Dissipative Structure of China's Building Energy Service Industry System Based on Brusselator Model. *J. Clean. Prod.* **2017**, *150*, 112–122. [CrossRef]
146. Franco, D.V.; Segers, J.-P.; Herlaar, R.; Hannema, A.R. Trends in Sustainable Energy Innovation—Transition Teams for Sustainable Innovation. *J. Innov. Manag.* **2022**, *10*, 22–46. [CrossRef]
147. Zhang, W.; Yuan, H. Promoting Energy Performance Contracting for Achieving Urban Sustainability: What Is the Research Trend? *Energies* **2019**, *12*, 1443. [CrossRef]
148. Peng, Y.; Wei, Y.; Bai, X. Scaling Urban Sustainability Experiments: Contextualization as an Innovation. *J. Clean. Prod.* **2019**, *227*, 302–312. [CrossRef]
149. Wang, Z.; Xu, G.; Lin, R.; Wang, H.; Ren, J. Energy Performance Contracting, Risk Factors, and Policy Implications: Identification and Analysis of Risks Based on the Best-Worst Network Method. *Energy* **2019**, *170*, 1–13. [CrossRef]
150. Aasen, M.; Westskog, H.; Korneliussen, K. Energy Performance Contracts in the Municipal Sector in Norway: Overcoming Barriers to Energy Savings? *Energy Effic.* **2016**, *9*, 171–185. [CrossRef]
151. Wacinkiewicz, D.; Słotwiński, S. The Statutory Model of Energy Performance Contracting as a Means of Improving Energy Efficiency in Public Sector Units as Seen in the Example of Polish Legal Policies. *Energies* **2023**, *16*, 5060. [CrossRef]
152. Akkoç, H.N.; Onaygil, S.; Acuner, E.; Cin, R. Implementations of Energy Performance Contracts in the Energy Service Market of Turkey. *Energy Sustain. Dev.* **2023**, *76*, 101303. [CrossRef]
153. Pätäri, S.; Annala, S.; Jantunen, A.; Viljainen, S.; Sinkkonen, A. Enabling and Hindering Factors of Diffusion of Energy Service Companies in Finland—Results of a Delphi Study. *Energy Effic.* **2016**, *9*, 1447–1460. [CrossRef]
154. Guo, K.; Zhang, L. Guarantee Optimization in Energy Performance Contracting with Real Option Analysis. *J. Clean. Prod.* **2020**, *258*, 120908. [CrossRef]

155. Shang, T.; Zhang, K.; Liu, P.; Chen, Z.; Li, X.; Wu, X. What to Allocate and How to Allocate?-Benefit Allocation in Shared Savings Energy Performance Contracting Projects. *Energy* **2015**, *91*, 60–71. [CrossRef]
156. Lee, P.; Lam, P.T.I.; Lee, W.L. Risks in Energy Performance Contracting (EPC) Projects. *Energy Build.* **2015**, *92*, 116–127. [CrossRef]
157. Hufen, H.; De Bruijn, H. Getting the Incentives Right. Energy Performance Contracts as a Tool for Property Management by Local Government. *J. Clean. Prod.* **2016**, *112*, 2717–2729. [CrossRef]
158. Brouns, J.; Nassiopoulos, A.; Bourquin, F.; Limam, K. Dynamic Building Performance Assessment Using Calibrated Simulation. *Energy Build.* **2016**, *122*, 160–174. [CrossRef]
159. Yang, J.B.; Chou, H.Y. Key Challenges in Executing Energy-Savings Performance Contracts in Public Buildings: Taiwan Experience. *J. Chin. Inst. Eng. Trans. Chinese Inst. Eng. A* **2017**, *40*, 482–491. [CrossRef]
160. Stuart, E.; Larsen, P.H.; Goldman, C.A.; Gilligan, D. A Method to Estimate the Size and Remaining Market Potential of the U.S. ESCO (Energy Service Company) Industry. *Energy* **2014**, *77*, 362–371. [CrossRef]
161. Töppel, J.; Tränkler, T. Modeling Energy Efficiency Insurances and Energy Performance Contracts for a Quantitative Comparison of Risk Mitigation Potential. *Energy Econ.* **2019**, *80*, 842–859. [CrossRef]
162. Fu, S.; Zhou, H.; Xiao, Y.-Z. Optimum Selection of Energy Service Company Based on Intuitionistic Fuzzy Entropy and VIKOR Framework. *IEEE Access* **2020**, *8*, 186572–186584. [CrossRef]
163. Prabatha, T.; Hewage, K.; Sadiq, R. An Energy Performance Contract Optimization Approach to Meet the Competing Stakeholder Expectations under Uncertainty: A Canadian Case Study. *Sustainability* **2022**, *14*, 4334. [CrossRef]
164. Carpino, C.; Bruno, R.; Carpino, V.; Arcuri, N. Uncertainty and Sensitivity Analysis to Moderate the Risks of Energy Performance Contracts in Building Renovation: A Case Study on an Italian Social Housing District. *J. Clean. Prod.* **2022**, *379*, 134637. [CrossRef]
165. Ke, M.T.; Yeh, C.H.; Jian, J.T. Analysis of Building Energy Consumption Parameters and Energy Savings Measurement and Verification by Applying EQUEST Software. *Energy Build.* **2013**, *61*, 100–107. [CrossRef]
166. Park, S.; Norrefeldt, V.; Stratbuecker, S.; Grün, G.; Jang, Y.S. Methodological Approach for Calibration of Building Energy Performance Simulation Models Applied to a Common "Measurement and Verification" Process. *Bauphysik* **2013**, *35*, 235–241. [CrossRef]
167. AlFaris, F.; Juaidi, A.; Abdallah, R.; Peña-Fernández, A.; Manzano-Agugliaro, F. Energy Performance Analytics and Behavior Prediction during Unforeseen Circumstances of Retrofitted Buildings in the Arid Climate. *Energy Rep.* **2021**, *7*, 6182–6195. [CrossRef]
168. Agenis-Nevers, M.; Wang, Y.; Dugachard, M.; Salvazet, R.; Becker, G.; Chenu, D. Measurement and Verification for Multiple Buildings: An Innovative Baseline Model Selection Framework Applied to Real Energy Performance Contracts. *Energy Build.* **2021**, *249*, 111183. [CrossRef]
169. Darabseh, M.; Martins, J.P. Blockchain Orchestration and Transformation for Construction. *Smart Cities* **2023**, *6*, 652–675. [CrossRef]
170. Turk, Ž.; Klinc, R. Potentials of Blockchain Technology for Construction Management. *Procedia Eng.* **2017**, *196*, 638–645. [CrossRef]
171. Darabseh, M.; Martins, J. Framework for the Use of Blockchain to Support the Development of Asset Information Models. In Proceedings of the 15th World Congress on Engineering Asset Management, Pioneiros, Brazil, 15–18 August 2021.
172. BigchainDB GmbH. BigchainDB 2020. v 2.2.2. Available online: https://www.bigchaindb.com/ (accessed on 7 July 2024).
173. BigchainDB GmbH. *Whitepaper: BigchainDB 2.0 The Blockchain Database*; BigChainDB: Berlin, Germany, 2018.
174. Nour El-Din, M.; Pereira, P.F.; Poças Martins, J.; Ramos, N.M.M. Digital Twins for Construction Assets Using BIM Standard Specifications. *Buildings* **2022**, *12*, 2155. [CrossRef]
175. *EN ISO 19650-1*; Organization and Digitization of Information about Buildings and Civil Engineering Works, Including Building Information Modelling (BIM)—Information Management Using Building Information Modelling—Part 1: Concepts and Principles. European Committee for Standardization: Geneva, Switzerland, 2018.
176. *EN ISO 19650-2*; Organization and Digitization of Information about Buildings and Civil Engineering Works, Including Building Information Modelling (BIM)—Information Management Using Building Information Modelling—Part 2: Delivery Phase of the Assets. European Committee for Standardization: Geneva, Switzerland, 2018.
177. *EN ISO 19650-3*; Organization and Digitization of Information about Buildings and Civil Engineering Works, Including Building Information Modelling (BIM)—Information Management Using Building Information Modelling—Part 3: Operational Phase of the Assets. European Committee for Standardization: Geneva, Switzerland, 2020.
178. *EN ISO 19650-5*; Organization and Digitization of Information about Buildings and Civil Engineering Works, Including Building Information Modelling (BIM)—Information Management Using Building Information Modelling—Part 5: Security-Minded Approach to Information Manage. European Committee for Standardization: Geneva, Switzerland, 2020.
179. European Data Protection Supervisor (EDPS). *Opinion 5/2018, Preliminary Opinion on Privacy by Design*; EDPS: Bruxelles, Belgium, 2018.
180. Wirth, C.; Kolain, M. Privacy by BlockChain Design: A Blockchain-Enabled GDPR-Compliant Approach for Handling Personal Data. In Proceedings of the 1st ERCIM Blockchain Workshop 2018, Reports of the European Society for Socially Embedded Technologies, Amsterdam, The Netherlands, 8–9 May 2018.
181. Politou, E.; Casino, F.; Alepis, E.; Patsakis, C. Blockchain Mutability: Challenges and Proposed Solutions. *IEEE Trans. Emerg. Top. Comput.* **2021**, *9*, 1972–1986. [CrossRef]
182. Schwerin, S. Blockchain and Privacy Protection in the Case of the European General Data Protection Regulation (GDPR): A Delphi Study. *J. Br. Blockchain Assoc.* **2018**, *1*, 1–77. [CrossRef]

183. Sater, S. Blockchain and the European Union's General Data Protection Regulation: A Chance to Harmonize International Data Flows. *SSRN Electron. J.* **2018**. [CrossRef]
184. Finck, M. Blockchains and Data Protection in the European Union. *Eur. Data Prot. Law Rev.* **2018**, *4*, 17–35. [CrossRef]
185. Eberhardt, J.; Tai, S. On or off the Blockchain? Insights on off-Chaining Computation and Data. In Proceedings of the Service-Oriented and Cloud Computing, Oslo, Norway, 27–29 September 2017; Volume 10465, pp. 3–15.
186. Meiklejohn, S. Top Ten Obstacles along Distributed Ledgers Path to Adoption. *IEEE Secur. Priv.* **2018**, *16*, 13–19. [CrossRef]
187. García-Barriocanal, E.; Sánchez-Alonso, S.; Sicilia, M.A. Deploying Metadata on Blockchain Technologies. In Proceedings of the Metadata and Semantic Research. MTSR 2017. Communications in Computer and Information Science, Tallinn, Estonia, 28 November–1 December 2017; Springer: Cham, Switzerland, 2017; Volume 755, pp. 38–49.
188. Regulation 2016/679. Regulation (EU) 2016/679 of the European Parliament and of the Council of 27 April 2016 on the Protection of Natural Persons with Regard to the Processing of Personal Data and on the Free Movement of Such Data, and Repealing Directive 95/46/EC (General Data Protection Regulation). 2017. Available online: http://data.europa.eu/eli/reg/2016/679/oj (accessed on 17 April 2024).
189. Lo, S.K.; Xu, X.; Staples, M.; Yao, L. Reliability Analysis for Blockchain Oracles. *Comput. Electr. Eng.* **2020**, *83*, 106582. [CrossRef]
190. Chainlink What Is a Blockchain Oracle? Available online: https://chain.link/education/blockchain-oracles#decentralized-oracles (accessed on 17 April 2024).
191. Al-Breiki, H.; Rehman, M.H.U.; Salah, K.; Svetinovic, D. Trustworthy Blockchain Oracles: Review, Comparison, and Open Research Challenges. *IEEE Access* **2020**, *8*, 85675–85685. [CrossRef]
192. Kochovski, P.; Gec, S.; Stankovski, V.; Bajec, M.; Drobintsev, P.D. Trust Management in a Blockchain Based Fog Computing Platform with Trustless Smart Oracles. *Futur. Gener. Comput. Syst.* **2019**, *101*, 747–759. [CrossRef]

energies

MDPI

Article

Harnessing Artificial Neural Networks for Financial Analysis of Investments in a Shower Heat Exchanger

Sabina Kordana-Obuch *, Mariusz Starzec and Beata Piotrowska

Department of Infrastructure and Water Management, Rzeszow University of Technology, al. Powstańców Warszawy 6, 35-959 Rzeszow, Poland; mstarzec1990@prz.edu.pl (M.S.); b.piotrowska@prz.edu.pl (B.P.)
* Correspondence: sk@prz.edu.pl

Abstract: This study focused on assessing the financial efficiency of investing in a horizontal shower heat exchanger. The analysis was based on net present value (*NPV*). The research also examined the possibility of using artificial neural networks and SHapley Additive exPlanation (SHAP) analysis to assess the profitability of the investment and the significance of individual parameters affecting the *NPV* of the project related to installing the heat exchanger in buildings. Comprehensive research was conducted, considering a wide range of input parameters. As a result, 1,215,000 *NPV* values were obtained, ranging from EUR −1996.40 to EUR 36,933.83. Based on these values, artificial neural network models were generated, and the one exhibiting the highest accuracy in prediction was selected ($R^2 \approx 0.999$, $RMSE \approx 57$). SHAP analysis identified total daily shower length and initial energy price as key factors influencing the profitability of the shower heat exchanger. The least influential parameter was found to be the efficiency of the hot water heater. The research results can contribute to improving systems for assessing the profitability of investments in shower heat exchangers. The application of the developed model can also help in selecting appropriate technical parameters of the system to achieve maximum financial benefits.

Keywords: net present value (*NPV*); waste heat recovery; machine learning; multilayer perceptron; SHAP analysis; Python programming language

Citation: Kordana-Obuch, S.; Starzec, M.; Piotrowska, B. Harnessing Artificial Neural Networks for Financial Analysis of Investments in a Shower Heat Exchanger. *Energies* **2024**, *17*, 3584. https://doi.org/10.3390/en17143584

Academic Editors: Bruno Peuportier and Zsuzsa Szalay

Received: 26 June 2024
Revised: 17 July 2024
Accepted: 18 July 2024
Published: 21 July 2024

1. Introduction

One of the most significant challenges facing contemporary generations is the need to meet the growing demand for energy while simultaneously caring for the environment and minimizing the negative impact of energy systems on the planet [1,2]. Population growth and ongoing urbanization are just some of the factors exacerbating this issue. Additionally, climate change forces communities to alter consumer behaviors [3] and take swift action aimed at reducing greenhouse gas emissions. It is, therefore, evident that the current heating systems should undergo a thorough transformation towards more ecological, sustainable, and efficient solutions [4,5]. However, meeting these challenges will not be possible without developing technologies based on renewable energy sources (RESs). This applies to both well-known RESs, such as solar, wind, and geothermal energy [6], as well as so-called third-generation renewable energy sources, for example, warm wastewater [7]. It is also essential to promote the effective use of energy through continuous public education on the conscious use of resources. In this aspect, the financial efficiency of individual solutions is particularly important [8], as implementing effective energy-saving measures can bring financial benefits to both individual energy consumers and entire communities.

Buildings are of particular importance in terms of the potential for reducing energy consumption and associated costs. As reported by the International Energy Agency (IEA) [9], the use of buildings accounts for up to 30% of global final energy consumption and approximately 26% of energy-related emissions. This should not be surprising, as Ratajczak et al. [10] noted, in some cases, people spend up to 90% of their lives indoors. It should also be noted that over the past few years, the components of the energy usage balance in buildings have been changing. As a result of technological progress and increased environmental awareness in society, buildings are better insulated and equipped with more efficient central heating, air conditioning, and lighting systems. Consequently, the share of domestic hot water (DHW) heating in the overall energy consumption balance in buildings has increased. In Poland, this share averages over 17% [11]. Research from other countries [12] indicates that it can be significantly higher. Much also depends on the type of building because, in passive buildings, the energy demand for preparing domestic hot water may significantly exceed the energy demand for space heating [13]. It is important to note that a significant volume of hot water consumption is used in bathtubs and showers. Some studies suggest that this can be as much as 70% [14]. Therefore, particular attention should be paid to solutions that guarantee a reduction in energy demand for heating water used for these purposes.

Such solutions undoubtedly include shower heat exchangers [15,16]. These devices allow for the recovery of part of the thermal energy present in greywater released from the shower. Warm greywater, flowing through the heat exchanger, transfers its carried thermal energy to the counter-flowing cold water. After preheating, it is then directed to the DHW heater and/or the shower mixing valve. However, the best energy benefits, and thus the highest financial savings, are achieved by preheating all the water used in the shower [17]. It should be noted that the use of greywater heat exchangers can enhance the comfort of using the shower installation by increasing the efficiency of the domestic hot water heater. This can result from both reducing the volumetric flow of heated water and raising the temperature of the water supplied to this device. Considering that ensuring comfortable conditions for using the shower installation with minimal energy consumption is an aspect that cannot be ignored [18], the use of shower heat exchangers seems not only beneficial but also necessary, especially since greywater is available all year round [19], even in the winter season, when energy demand is highest.

The financial efficiency of using a shower heat exchanger in a residential building depends on several factors determining the amount of energy consumption for DHW heating for showering. These primarily include the hot water flow rate and the duration of shower installation use [20]. The input and output water temperatures of the DHW heater, as well as the type and efficiency of this device, are also significant. Besides energy demand for heating water, potential financial savings resulting from the application of a shower heat exchanger are influenced by its effectiveness and unit energy prices. The unit's effectiveness mainly depends on its type, as well as the functioning conditions of the shower installation and the connection method of preheated water pipes [21]. On the other hand, energy prices depend on the type of fuel used and the applicable tariffs for purchasing and distributing a given energy carrier. Over a longer period of heat recovery system use, geopolitical and economic situations should also be considered. Therefore, the results of the financial analyses conducted by various researchers differ significantly, and they draw contradictory conclusions. For example, Selimli and Eljetlawi [22] noted that the investment in such a device could pay off within two years. Other analyses [20] indicated, however, that depending on the conditions of the installation's use, this period can be significantly extended. Considering the above, it seems necessary to conduct a thorough analysis identifying the key factors influencing the profitability of investing in a shower heat exchanger. Their relevance should be verified in the context of specific operating conditions of the shower installation and the risk associated with changes in market or operational conditions that may cause deviations in actual financial flows from forecasted values.

Considering all possible scenarios of the shower installation's operation and determining the values of financial efficiency indicators for the full set of input data is, however, time-consuming and laborious. Consequently, financial analyses are often limited to strictly defined values of input parameters corresponding to typical operating conditions of the shower installation. Potential buyers and future users of such devices are deprived of knowledge about the solution's profitability in case these conditions change. A partial solution to this problem is provided by sensitivity analysis, including scenario analysis, which can give a fuller picture of the investment's profitability under different conditions [23]. However, even it does not guarantee a comprehensive consideration of all potential conditions and events that may affect the investment's profitability. It is not always possible to predict future changes, establish precise values of parameters subject to change over time, or consider interactions between scenarios. To gain a fuller picture of the investment's profitability in a shower heat exchanger and better understand the potential risks associated with it, more advanced techniques and tools are necessary. Such tools undoubtedly include artificial neural networks (ANNs). Thanks to the ANN models' ability to process large amounts of data, detect patterns, and forecast outcomes, they can help investors make more accurate investment decisions. Their use can also help automate some calculations and eliminate potential human errors. Considering the above, the objectives of this article include the following:

- Evaluation of the financial efficiency of using a horizontal shower heat exchanger;
- Assessment of the possibility of using ANNs to evaluate the profitability of investing in a shower heat exchanger;
- Assessment of the significance of individual parameters influencing the financial analysis results.

This paper continues and expands on the research described by Starzec et al. [21]. The mentioned research [21], however, focused on the use of ANN models to forecast the effectiveness of the heat exchanger, while the research described in this paper relates to the investment profitability expressed by the net present value (*NPV*).

2. Materials and Methods

2.1. Research Steps

In the first stage of the analyses, a prototype of a horizontal shower heat exchanger installed at the greywater outlet underwent experimental testing. The analysis was carried out for the most energy-justified design variant of the heat recovery system, assuming that preheated water flows to both the DHW heater and the mixing valve. Both the research stand and the technical parameters of the selected shower heat exchanger were presented in detail by the authors in an earlier publication [21] concerning the assessment of the suitability of artificial neural networks for assessing the effectiveness (ε) of a shower heat exchanger. Based on the research results obtained and described by Starzec et al. [21], the temperatures of preheated water were also determined, which formed the basis of the research described in this study (Appendix A, Figure A1). Based on these, research was carried out on the financial efficiency of using a shower heat exchanger for a total of 1,215,000 combinations of input parameters.

Next, ANN models were developed, adopting ten input parameters and one output parameter in the form of the *NPV* value. Then, using the selected ANN model and the SHapley Additive exPlanation (SHAP) analysis, a hierarchy of significance of the adopted input parameters affecting the financial efficiency of using the horizontal heat exchanger was determined. For this purpose, Python programming language was used. As a result, it was possible to identify the most important parameters that have the greatest impact on financial benefits derived from the usage of this device. Figure 1 illustrates the logical flowchart of the research process.

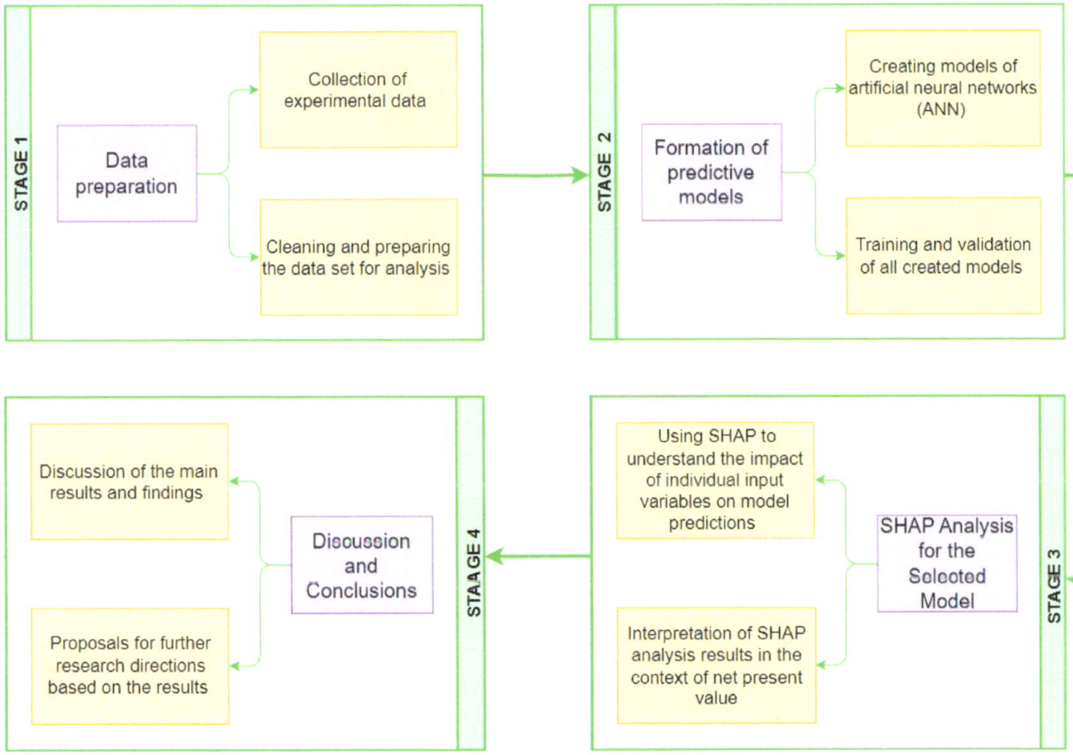

Figure 1. Research logic flowchart.

2.2. Net Present Value

The financial analysis of the investment in a shower heat exchanger was based on the *NPV* values. Net present value is a key tool in investment analysis and project evaluation, which is why it is widely used in the analysis of projects involving the use of RESs [24,25]. This method allows one to assess the profitability of a project by taking into account future cash flows and discounting them to the current value, which is described by Equation (1).

$$NPV = \sum_{t=1}^{n} \frac{CF_t}{(1+r)^t} - INV_0, \tag{1}$$

where *NPV* is the net present value of a project, EUR; CF_t is the cash flows in year t, EUR; INV_0 is the initial investment outlay, EUR; r is the discount rate; and n is the system lifespan, years.

A positive *NPV* value indicates that under the given conditions of system operation, the financial benefits from implementing the project will exceed the costs. If the costs are too high compared to the forecast revenues and the *NPV* turns out to be negative, the investment will be unprofitable.

Regarding investing in a shower heat exchanger, the value of financial flows (CF_t) results from financial savings related to reducing energy consumption in the building. The annual decrease in energy consumption for heating DHW was estimated based on Equation (2).

$$\Delta EC = \frac{365 \cdot l_s \cdot \left[q_{0h} \cdot \rho_0 \cdot c_{p0} \cdot (T_{hw} - T_{cw}) - q_{1h} \cdot \rho_1 \cdot c_{p1} \cdot (T_{hw} - T_{pw}) \right]}{\eta \cdot 3.6 \cdot 10^9}, \tag{2}$$

where ΔEC is the annual energy savings used to heat DHW, kWh (MWh); l_s is the total daily shower length, min (s); q_{0h} is the flow rate of DHW to the shower mixing valve in the variant without a shower heat exchanger, L/min (m^3/s); q_{1h} is the flow rate of DHW to the shower mixing valve in the variant with a shower heat exchanger, L/min (m^3/s); T_{hw} is the temperature of DHW, °C (K); T_{cw} is the temperature of cold water, °C (K); T_{pw} is the temperature of preheated water, °C (K); η is the efficiency of a DHW heater; ρ_0 and ρ_1 are the densities of water, kg/m^3; and c_{p0} and c_{p1} are the specific heat capacities of water, J/(kg·K).

The value of the *NPV* indicator in the case of investing in a shower heat exchanger is influenced by a number of factors related to the conditions of using the shower installation, the costs of supplying the building with energy, and the assumed level of risk expressed in the value of the discount rate (r). The amount of initial investment outlay (INV_0) is also important. In order to create a database of *NPV* values, based on which artificial neural network models were generated in the next stage of research, 1,215,000 combinations of input parameters were identified and analyzed. Table 1 lists the values of the input parameters that were considered when creating the above combinations. These values have been selected to take into account different scenarios and conditions of using the shower installation and to reflect different levels of risk. Considering that among the parameters determining the effectiveness (ε) of the shower heat exchanger, which ranged from 19.20% to 35.44%, the most important is the mixed water flow rate (q) and the linear bottom slope of the shower heat exchanger (i) [21], the analysis included five and eight values of these parameters, respectively. The adoption of a larger number of bottom slope values (i) results from the fact that the impact of changing the value of this parameter on the effectiveness (ε) of the horizontal shower heat exchanger is less predictable and more diverse over the entire range of its values. On the other hand, due to the limited importance of cold water and greywater temperatures (T_{cw} and T_{dw}), it was decided to reduce the number of considered values of these parameters to three. As a result, the analysis of the financial efficiency of the investment in the shower heat exchanger was based on 360 preheated water temperatures (T_{pw}) corresponding to different operating conditions of the greywater heat recovery system (Appendix A/Figure A1). The selected values of the above parameters corresponded to the combinations that, in the previous stage of research, were characterized by the greatest fit to the ANN models [21]. The calculations assumed a constant DHW temperature (T_{hw}) of 55 °C. This value corresponds to the minimum DHW temperature at the outlet from the sanitary facilities in Polish conditions, which was specified in the Regulation of the Minister of Infrastructure and Development [26].

Table 1. Values of the input parameters.

Input Parameter	Unit	Values
Cold water temperature (T_{cw})	°C	8, 14, 20
Greywater temperature (T_{dw})	°C	30, 35, 40
Initial investment outlay (INV_0)	EUR	400, 1200, 2000
Initial energy price (C_e)	EUR/kWh	0.02, 0.10, 0.18, 0.26, 0.34
Annual change in energy price (i_e)	%	−2.5, 0, 2.5, 5, 7.5
Domestic hot water heater efficiency (η)	%	80, 90, 100
Linear bottom slope of the shower heat exchanger (i)	%	0, 0.33, 0.66, 1, 2, 2.5, 3.5, 4
Discount rate (r)	%	0, 2.5, 5, 7.5, 10
Mixed water flow rate (q)	L/min	3, 4.5, 6.5, 8.5, 10
Total daily shower length (l_s)	min	10, 50, 90
System lifespan (n)	years	15

Additionally, the analysis took into account five different values of the initial energy price (C_e), the annual percentage change in the energy price (i_e), and the discount rate (r). In the case of parameters such as DHW heater efficiency (η), total daily shower length (l_s), and initial investment outlay (INV_0), it was decided to limit the amount of input variable values to three. In the case of the first of the mentioned parameters (η), this is due to the small range of its values. In the remaining cases (l_s, INV_0), the fact that these parameters change the value of the *NPV* indicator linearly was taken into account. Representative values of the above parameters were selected to consider the different number of system users and their various preferences as to the length of the shower, the varied scope of required installation works, different types of instantaneous DHW heaters (gas, electric), as well as different market trends.

2.3. Artificial Neural Networks

Artificial neural networks can be a useful tool in assessing the effectiveness of greywater heat recovery units. Thanks to the ability to process large amounts of data and identify complex patterns, ANNs can be used to analyze and model thermal processes occurring in these systems. Artificial neural networks can also be used to assess the financial efficiency of using shower heat exchangers. ANN models can predict energy savings and costs associated with installing and operating these systems. Using data on water consumption, energy costs, shower heat exchanger specifications, and operating conditions, ANNs can accurately assess the potential financial benefits of heat recovery in various scenarios. Such analyses can help both individual users and companies make informed investment decisions while ensuring optimization of expenses and maximizing the return on investment in greywater heat recovery technologies.

Artificial neural networks are inspired by the biological structures of the brain, which consist of neurons connected by synapses. ANNs are complex mathematical and computer structures that consist of many layers of artificial neurons capable of learning and processing information. They are widely used in various fields, such as image recognition [27], financial forecasting [28], and in the field of technical sciences [29]. The process of creating an artificial neural network begins with the definition of the problem, i.e., determining the purpose and scope of application of the neural network, for example, whether the network is to recognize images, predict the values of a given parameter, classify data, etc. The next step is to collect data (e.g., laboratory tests) that will be used to learn the ANN models. Next comes data preparation, where data cleanliness and normalization are crucial. The data must be in the right format, and the values must be normalized for the network to learn effectively. The next step is to choose the network architecture, which includes deciding on the number of layers, the number of neurons in each layer, and the type of activation function. Once the architecture is defined, the network training process begins, which involves adjusting neuron weights based on training data using optimization algorithms such as backpropagation. After training, there is a validation and testing stage, where the model is tested on a set of data that was not used during training to assess its performance and generalizability. Finally, the model is put into real use, where it can process new data and provide predictions or classifications according to its intended purpose [30].

In this study, the input data set comprised a total of ten parameters determining the effectiveness of the horizontal shower heat exchanger and the profitability of its application in a residential building (Table 1), assuming a 15-year system lifespan (n). Based on the flow rate (q), the slope of the unit (i), and the temperatures of cold and greywater (T_{cw} and T_{dw}), the values of the preheated water temperature (T_{pw}) were determined through experimental studies. The experimental results of the heat exchanger conducted at the Laboratory of Measurement Techniques and Water and Wastewater Transport Control at Rzeszow University of Technology can be found in the publication by Starzec et al. [21]. The measured T_{pw} values were then used to calculate the *NPV* indicator, which was the output variable of the model. The calculated *NPV* values are presented in Section 3.1. The normalized input data formed the basis for training MultiLayer Perceptron (MLP) neural networks. To ensure the accuracy and reliability of the model, the data set was divided into three parts: training set, validation set, and test set. The training set accounted for 70% of the total data, and the validation and testing sets accounted for 15% each. This division allows for appropriate tuning of the model and its subsequent evaluation of previously unknown data, which allows for a more objective assessment of its effectiveness. The data were divided randomly, which enabled effective verification of prediction results [31]. This approach allows for an easy assessment of how well the model generalizes to unseen data. If it performs well on the training, validation, and test sets, it indicates that it can make accurate predictions based on new data. Therefore, the model that provided the best fit of computational data with the model's predictions, as well as the most favorable values of model evaluation metrics across all three data sets, was chosen.

The developed artificial neural network models were evaluated using the root mean square error (*RMSE*) and coefficient of determination (R^2). The *RMSE* measures the average size of the model's prediction errors. The lower the *RMSE* value, the better the model fit. The value of the *RMSE* index is determined according to Equation (3). The R^2 index measures how well observed outcomes are predicted by the model. Its value ranges from 0 to 1, where 1 means a perfect fit of the model to the data. Equation (4) defines the coefficient of determination (R^2) [32].

$$RMSE = \sqrt{\frac{1}{m}\sum_{i=1}^{m}(y_i - \hat{y}_i)^2}, \tag{3}$$

$$R^2 = 1 - \frac{\sum(y_i - \hat{y}_i)^2}{\sum(y_i - \overline{y})^2}, \tag{4}$$

where m is the number of data sets, y_i is the measured value, $\hat{y}_i$ is the predicted value, and $\overline{y}$, is the mean value of data set.

2.4. SHapley Additive exPlanations

SHAP analysis seems to be a valuable tool for assessing the financial efficiency of using shower heat exchangers. Thanks to this method, it is possible to understand in detail how particular features influence the model's predictions [33], which is crucial when assessing the costs and benefits associated with the implementation of this type of technology. SHAP analysis allows for the identification of key factors influencing energy and financial savings, which enables more accurate investment planning and optimization of installation processes. Thanks to this tool, investors and decision-makers can make more informed decisions, minimizing risks and maximizing economic benefits from the use of shower heat exchangers.

SHAP analysis is a tool for explaining the results of complex machine learning models. SHAP is based on the concept of Shapley values from game theory, which is used to separate the output of a model into the contributions of individual features. In the context of machine learning, these values represent the contribution of each input feature to the model's prediction. Thanks to this, SHAP analysis allows for the interpretation of the performance of models that are often perceived as "black boxes" due to their complexity.

The basis of SHAP analysis is the calculation of the Shapley value for each feature. These values measure how adding a given feature to all possible combinations of other features affects the model result. In practice, SHAP algorithms calculate these values by simulating all possible combinations of features and their impact on the result. In this way, SHAP values provide global and local explanations of the model's performance. Global explanations show which features are most important overall, and local explanations help understand how particular features affect specific predictions.

One of the key benefits of SHAP analysis is its universality and accuracy. This method is applicable to a wide range of models, including decision trees, artificial neural networks, and linear models. Moreover, SHAP values are intuitive: the sum of individual feature contributions plus the "base value" equals the model output, which makes the interpretations easy to understand even for people who are not machine learning experts.

SHAP analysis is widely used in various fields where machine learning is used. SHAP analysis is used in power engineering [34], construction [35], environmental engineering [36], socio-economic sciences [37], and many other areas where understanding how predictive models work is crucial to making informed decisions.

3. Results

3.1. Net Present Value

Based on the data presented in Table 1, 1,215,000 *NPV* values were determined, which ranged from EUR −1996.40 to EUR 36,933.83 (Figure 2). Such a large dispersion of results indicates significant uncertainty as to expected future financial flows. Changes in the forecast values of input parameters, both those related to the costs of supplying energy to the property and those characterizing water consumption for showering, may significantly affect the outcomes of the financial analysis. This confirms the validity of analyzing the profitability of a project in a wide range of input data for effective strategic planning and risk management. The statistical analysis of the results showed that for the entire studied data set, the median net present value was EUR −90.70. Therefore, it is clear that in the case of most of the combinations of input parameters considered, the project would turn out to be unprofitable. This results from both the relatively low effectiveness of the heat exchanger and the wide range of adopted values of input parameters, including extremely unfavorable conditions of use of the heat recovery system. For a more complete picture of the distribution of *NPV* values, quartile values were also determined. The first quartile was equal to EUR −965.60, which means that lower *NPV* values were obtained for 25% of the parameter combinations. On the other hand, the third quartile was equal to EUR 1433.33. This shows that in 25% of cases, higher net present values were obtained. The distribution of *NPV* values is, therefore, right-skewed. This suggests that despite the predominance of cases for which $NPV < 0$, there are a significant number of scenarios with very high profitability, which significantly increases the predicted average *NPV* value. This distribution suggests a high level of risk while indicating the possibility of achieving significant financial profits under appropriate conditions of the system's operation.

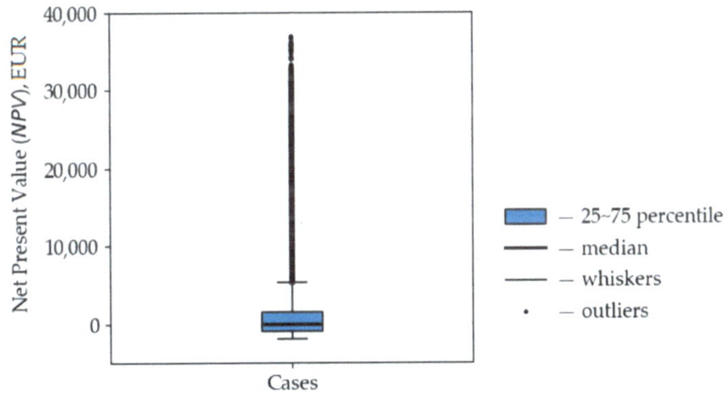

Figure 2. Distribution of the *NPV* values.

To thoroughly comprehend the impact of individual input variables on the profitability of the project and the dispersion of the obtained results, Figure 3 shows the distribution of the *NPV* values depending on the adopted values of these parameters. The analysis of individual box-and-whisker plots, prepared on the basis of 12.50–33.33% of all results (depending on the number of considered values of a given parameter), confirms that the median *NPV* value is close to zero in many cases. However, there are exceptions indicating a significant impact of selected parameters on the profitability of the project. Such exceptions undoubtedly include initial energy price (C_e), total daily shower length (l_s), and initial investment outlay (INV_0). The research analyzed initial energy prices ranging from 0.02 to 0.34 EUR/kWh. Adopting such a wide range of these parameters allows for the estimation of the profitability of the project with respect to various energy carriers. Analyzing Figure 3d, it can be seen that in the case of the lowest C_e values, the investment would be unprofitable in the vast majority of the considered scenarios because the third quartile turned out to be negative. For the next value considered (C_e = 0.10 EUR/kWh), the median *NPV* still remained negative. It is worth noting that this price is slightly higher than the current frozen price of natural gas in Poland. It can, therefore, be concluded that the implementation of a shower heat exchanger in a system where DHW is prepared using a gas water heater is unfavorable in financial terms. However, such low energy prices are unlikely over an extended period, which creates an opportunity to increase the financial efficiency of using this device. An increase in the initial energy price is equivalent to an increase in the value of the *NPV*. Already at C_e = 0.18 EUR/kWh, the median of the results becomes positive, and at the highest C_e value considered, almost the entire box is above the zero value. In the case of Poland, higher prices refer to electricity prices. It can, therefore, be concluded that shower heat exchangers will be a much more attractive option in the case of installations equipped with an electric DHW heater. In some cases, it will still be necessary to consider the risk of loss, but as the initial energy price increases, this risk will decrease while the potential return on investment simultaneously increases. An increase in the initial energy price additionally results in greater dispersion of results, especially the highest *NPV* values. This proves that the net present value is significantly sensitive to this parameter. Even small changes in the initial energy price can lead to significant changes in the projected *NPV* values. On the other hand, it confirms that there is the potential to achieve very high profits if the conditions of use of the installation and market conditions are appropriate.

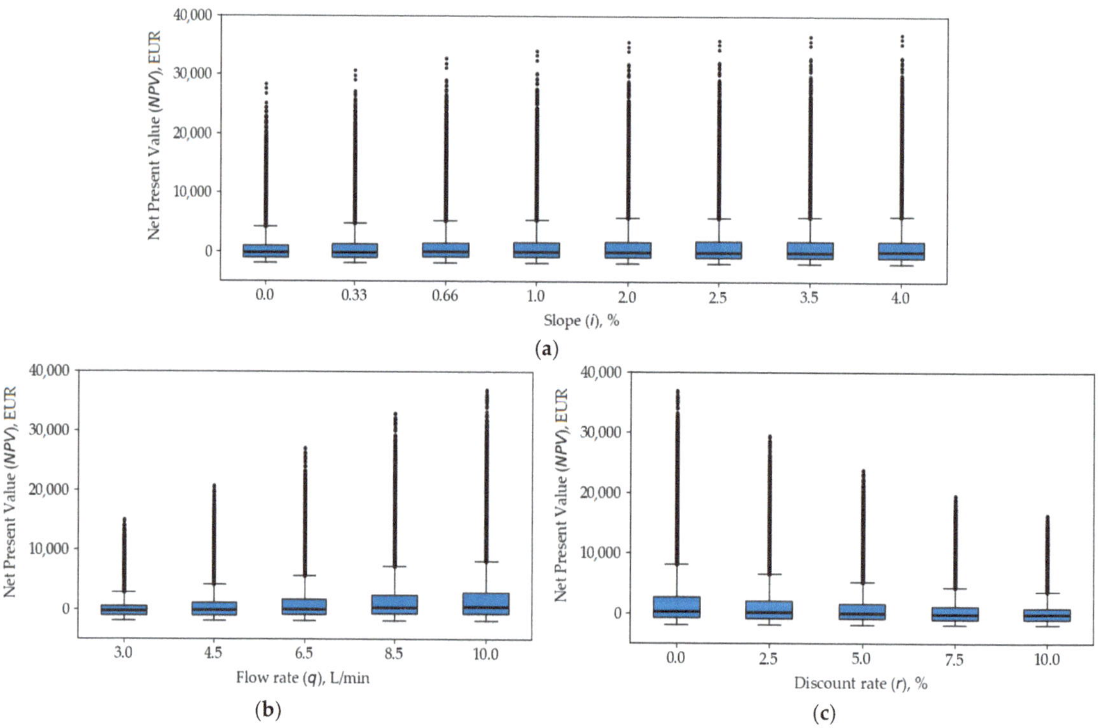

Figure 3. *Cont.*

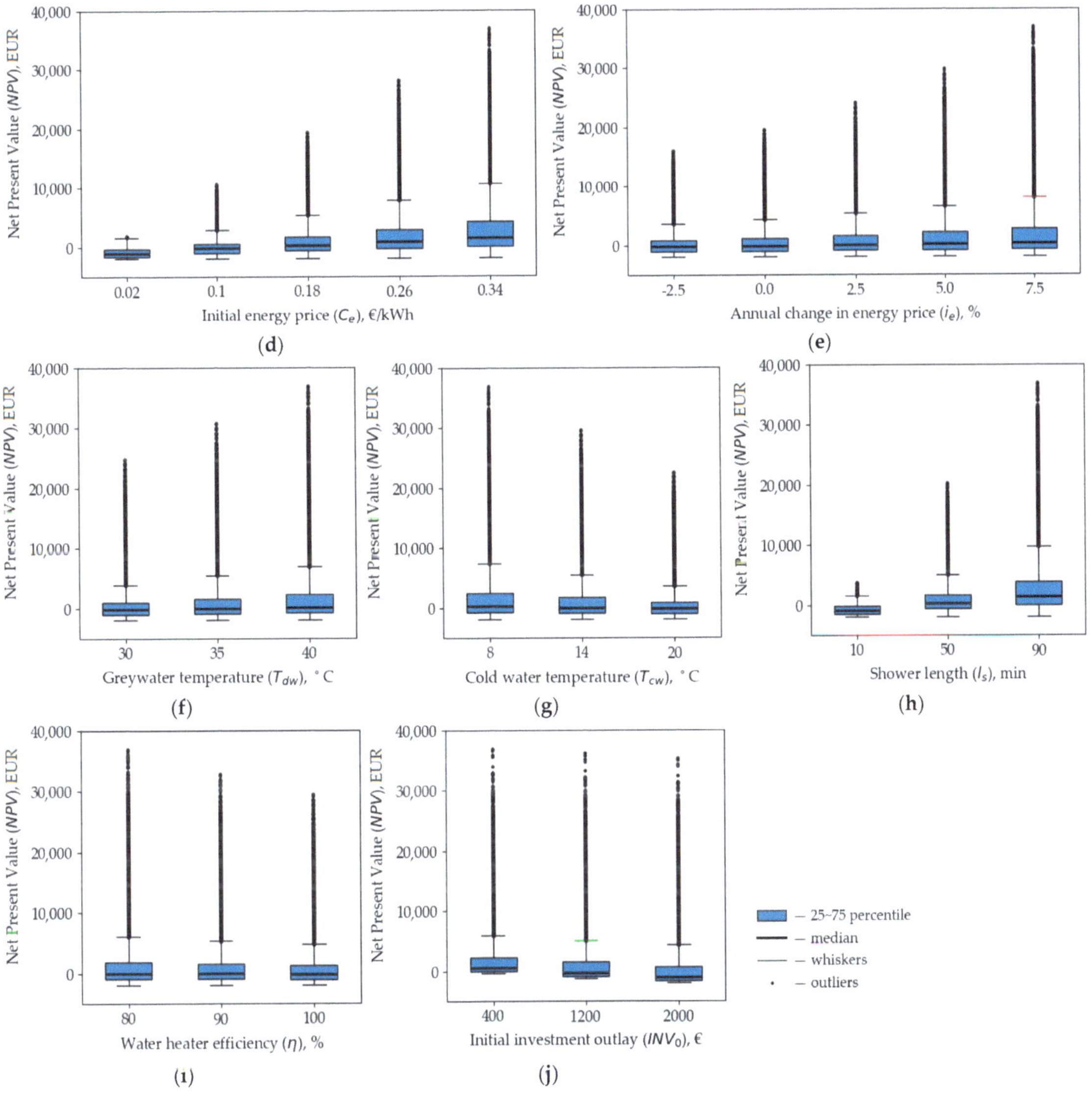

Figure 3. Distribution of *NPV* values against individual input parameters: (**a**) linear bottom slope; (**b**) mixed water flow rate; (**c**) discount rate; (**d**) initial energy price; (**e**) annual change in energy price; (**f**) greywater temperature; (**g**) cold water temperature; (**h**) total daily shower length; (**i**) domestic hot water heater efficiency; (**j**) initial investment outlay.

Total daily shower length (l_s) is also important (Figure 3h). The lowest value of the l_s corresponds to the situation when the shower is used by one person or two people who prefer short showers. In such conditions, water consumption is low, even in the case of shower heads with a high flow rate of mixed water. This causes the investment to have low financial efficiency. In over 75% of the considered scenarios, the project turned out to be unprofitable. It can, therefore, be concluded that in the case of single-person households, investing in a shower heat exchanger will be financially unprofitable. Increasing the time of water consumption from the shower head to 50 min per day resulted in the median of the results reaching a positive value. This means that in the case of several-person households, investing in a horizontal shower heat exchanger may prove profitable, especially when the DHW is heated with an electric water heater. Increasing the total daily shower length to 90 min means that approximately 75% of all scenarios guarantee profits from the implementation of the installation. The implementation of the analyzed shower heat exchanger will, therefore, be particularly beneficial for people who prefer long showers. It is also worth considering the possibility of installing a common heat exchanger on the greywater outflow from two neighboring apartments. Favorable financial effects could also be achieved if the unit was installed in buildings with higher water consumption than residential buildings, for example, sports facilities, health care facilities, campsites, dormitories, prisons, etc.

Clear differences in the determined *NPV* values and their distributions are also visible in the case of the initial investment outlay (INV_0). In the case of a relatively cheap heat exchanger, the initial investment outlay varies depending on the scope of required installation and renovation work. If the heat exchanger has been considered at the building design stage or is to be located in the immediate vicinity of the shower and the DHW heater, these costs will not be excessive. However, if the installation of the heat exchanger requires significant modifications to the internal water supply and sewage system as well as related construction and installation works, the initial investment outlay may increase significantly, and the costs of installing the unit may even exceed the cost of its purchase several times. In the case of the lowest investment costs considered (INV_0 = EUR 400), most scenarios turned out to be profitable (68.40%). Additionally, even in the event of extremely unfavorable circumstances, the potential financial losses will not be significant. An increase in the amount of initial investment outlay results in a clear reduction in the profitability of the investment. Even with expenditures of EUR 1200, more than half of the considered scenarios turned out to be financially unfavorable (55.44%). In the case of an increase in INV_0 to EUR 2000, approximately 67.45% of the parameter combinations were considered unfavorable. It follows that the implementation of greywater heat exchangers in Polish conditions will require subsidies, tax reliefs, or other forms of financial support for investments in solutions that increase the energy efficiency of buildings.

Upon reviewing the distributions illustrated in Figure 3, it becomes clear that the remaining input parameters have a much lower impact on the results of the investment profitability analysis. It is clearly visible that in the case of a decrease in the bottom linear slope of the heat exchanger (i), the flow rate of water and greywater through the unit (q), the annual change in the energy price (i_e), and the greywater temperature (T_{dw}), the directions of changes in these parameters are consistent with the direction of changes in the *NPV*. On the other hand, in the case of parameters such as the discount rate (r), cold water temperature (T_{cw}), and DHW heater efficiency (η), these directions are opposite. To precisely determine the influence of parameter changes on the analysis outcomes, it is necessary to use more advanced tools. For this reason, artificial neural networks were used in the next stage of research.

3.2. Artificial Neural Networks

To evaluate the feasibility of using machine learning methods to assess the financial efficiency of using a horizontal shower heat exchanger, artificial neural network models were generated. To evaluate these models, the *RMSE* and R^2 coefficients were used, which were determined in accordance with Equations (3) and (4). These indicators were computed for the training, validation, and test datasets.

During the research process, several neural network models were developed and tested. Among them, the model that was selected showed optimal values of the *RMSE* and R^2 indicators, which proved its highest efficiency. The values of the *RMSE* and R^2 indices for the selected model are summarized in Table 2.

Table 2. Values of R^2 and *RMSE* indicators for the selected artificial neural network model.

ANN Model Architecture	Training Set		Validation Set		Testing Set	
	R^2	*RMSE*	R^2	*RMSE*	R^2	*RMSE*
10-15-9-11-1	0.999	56.745	0.999	57.205	0.999	56.981

The selected artificial neural network model achieves a coefficient of determination (R^2) close to 1.0 and a low root mean square error (*RMSE*) for all three data sets. In practical terms, this implies that the model explains the variability of the data very well, and almost all the data are accurately fitted by the model (Figure 4). When both of these metrics are consistent across all data sets, we can conclude that the model is highly accurate in prediction. Furthermore, the lack of significant differences in the results for the training, validation, and testing sets suggests that the model is not overfitted. Overfitting usually leads to high R^2 on the training set but low on the testing set. The analysis results prove that the developed *ANN* model is effective in predicting data and can be used with a high level of confidence.

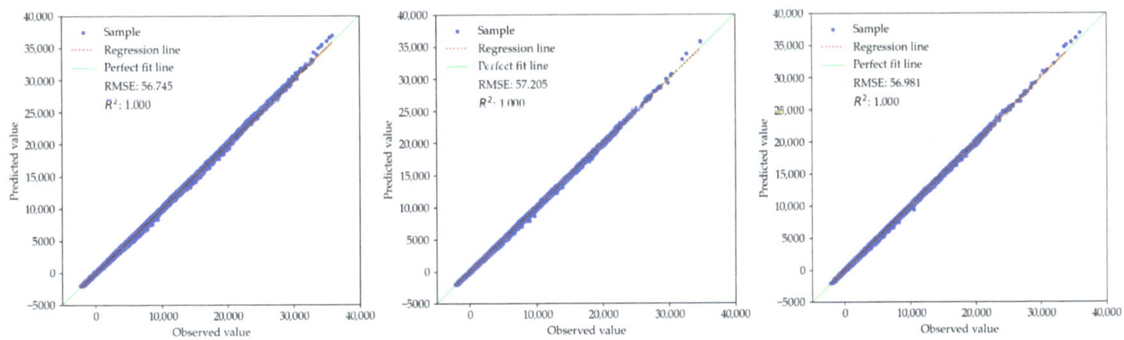

(**a**) (**b**) (**c**)

Figure 4. Comparing actual and forecasted *NPV* values: (**a**) training data set; (**b**) validation data set; (**c**) testing data set.

3.3. SHAP Analysis

In order to explain the impact of individual input parameters on the financial efficiency of the project, a SHAP analysis was conducted. The results are presented in Figure 5. Figure 5a presents global feature importance, and Figure 5b presents a local explanation summary. Global SHAP analysis aims to understand the overall behavior of the model, while local analysis focuses on understanding individual predictions. The Y-axis in Figure 5b shows the individual input parameters whose impact on the NPV value was analyzed, while the X-axis shows the SHAP values. The order of the input parameters along the vertical axis corresponds to the impact on the profitability of investing in a horizontal shower heat exchanger. The legend attached to the chart indicates the value of the input parameter in a given observation in relation to the entire range of its values adopted for the analysis. Each point corresponds to a single analyzed case, with the points on the right side indicating a positive impact on the *NPV* value, while the points on the left side represent a negative impact on the analysis results.

The SHAP analysis confirmed that in the case of the considered shower heat exchanger, total daily shower length (l_s) and initial energy price (C_e) are of key importance for the profitability of the project. This is evidenced by the average SHAP values, which for both of these parameters exceed EUR 800. Slightly higher values were obtained for the first parameter. It should be noted, however, that the annual change in energy price (i_e) will also have an impact on energy prices in the subsequent years of system operation. Although this parameter ranks only fifth in the hierarchy of importance of input parameters, the sum of average SHAP values for C_e and i_e will exceed EUR 1200. Therefore, it can be concluded that energy prices throughout the entire operational period will be of key importance for the achieved *NPV* values. Initial energy prices and their annual changes are related primarily to the type of basic energy carrier, as well as production and distribution costs, geopolitical conditions, and implemented government policies. For this reason, the course of changes in energy prices over the entire operational period is difficult to predict. Although attempts are made to create advanced models for forecasting future energy prices, these forecasts are always subject to a certain degree of uncertainty. The average user of a shower heat exchanger is not able to predict the dynamics of these changes on his own. For this reason, analyzing a wide range of potential energy prices is crucial for assessing the profitability of investing in a shower heat exchanger. The analysis should also focus on various total daily shower lengths (l_s), as this input parameter also has a considerable impact on the financial viability of the project. The habits of system users should be analyzed, and potential changes in their numbers in the future should be considered.

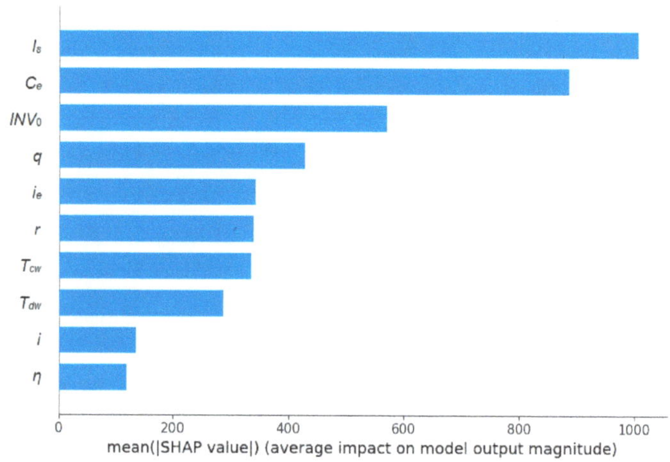

(a)

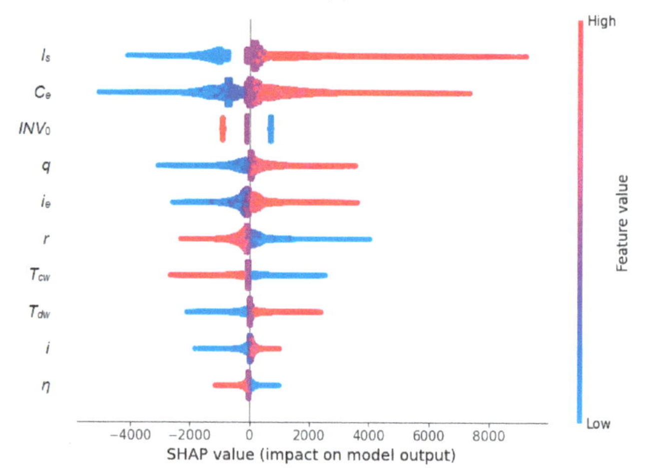

(b)

Figure 5. The influence of various input variables on the output variable: (**a**) mean (| SHAP value |); (**b**) SHAP summary plot (designations as in Table 1).

Also, analyzing individual cases, it can be seen that omitting the l_s and C_e parameters results in the highest SHAP values. For example, the maximum difference between the forecasted net present value considering ten input variables and the forecasted *NPV* without taking into account the total daily shower length (l_s) amounted to EUR 9226.15. In the case of the initial energy price (C_e), it was EUR 7349.38. In both cases, the directions of changes in the input parameters and the net present value are consistent, which means that an increase in the value of each of them increases the profitability of the project. An analogous pattern was also observed in the case of a mixed water flow rate (q), annual change in energy price (i_e), greywater temperature (T_{dw}), and linear bottom slope of the heat exchanger (i). In the case of the latter, however, the influence on the investment's economic viability is negligible. The average SHAP value does not reach EUR 150, and the maximum value for individual cases slightly exceeds EUR 1000. This is surprising because, in the previous stage of the analysis, the course of the impact of changes in this parameter on the effectiveness of the heat exchanger turned out to be the most diverse and least predictable of all the considered input parameters [21]. However, in the case of estimating financial savings, parameters that were not associated with the effectiveness of the shower heat exchanger turned out to be more significant. In addition to the linear bottom slope (i), cold water and greywater temperatures were also relatively low in the hierarchy. The only exception is the mixed water flow rate, for which SHAP values range from EUR −3064.84 to EUR 3556.30. This is probably due to the fact that the flow rate of water and greywater through the heat exchanger determines the amount of heat that can be recovered. A high flow rate (q) can increase the heat transfer rate of the heat exchanger, leading to greater energy savings and, consequently, increased financial savings and *NPV* value. For this reason, an increase in the value of q results in an increase in the net present value, even though the effectiveness of the heat exchanger is reduced in such a situation.

The parameters whose directions of change are opposite to the direction of changes in the *NPV* indicator include the initial investment outlay (INV_0). The average SHAP value, in this case, is close to EUR 600, which places this parameter in third place in the hierarchy of influence on the analysis results and confirms the need for co-financing the investment in a shower heat exchanger. Potential buyers should also look for savings in the costs of installing a heat recovery unit, e.g., by selecting a location that will limit the scope of required installation work. The parameters that had a negative impact on the model result also included the discount rate (r) and cold water temperature (T_{cw}), for which the average strength of the influence turned out to be similar. These parameters ranked sixth and seventh in the hierarchy, respectively. The last position went to the efficiency of the hot water heater (η). Also, in this case, an increase in the value of the input parameter results in a decrease in the value of the *NPV*. However, its influence on the model prediction result is negligible, which may be due to the fact that the efficiency of hot water heaters is relatively high in the entire adopted range of values of this parameter.

This study additionally analyzed local SHAP values for three characteristic cases (two extreme cases and the central one). The SHAP values for these individual observations are presented in Figure 6. For the parameter configuration shown in Figure 6a, the sum of the SHAP values is EUR 2820.76 with the calculated and predicted *NPV* values of EUR −1996.40 and EUR −2024.82, respectively. The initial investment outlay (INV_0), which reduces the *NPV* value in relation to the base value of EUR 795.94 by as much as EUR 906.81, has the greatest impact on the model's predictions. It is worth noting that INV_0 takes the highest considered value in this configuration. On the contrary, total daily shower length (l_s) and initial energy price (C_e), which also have a major influence on the prediction results, have the lowest values in the analyzed ranges. The remaining seven input parameters are responsible for reducing the base value by a total of EUR 577.81.

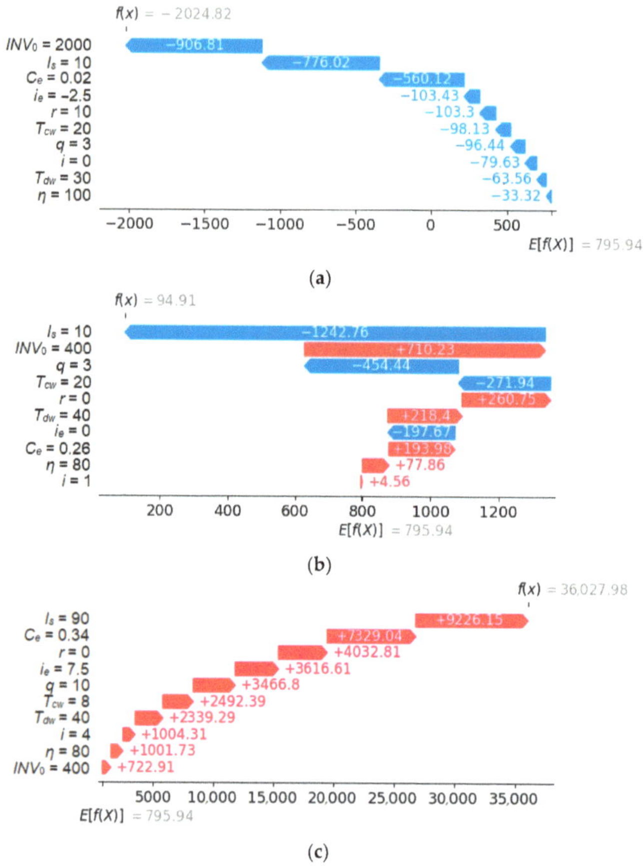

Figure 6. Force plot visualization: (**a**) NPV = EUR −1996.4; (**b**) NPV = EUR −27.6; (**c**) NPV = EUR 36,933.8 (designations as in Table 1).

Figure 6b refers to the median of the obtained SHAP values. In this case, the forecast *NPV* is EUR 94.91, and the parameter that has a key impact on this result is total daily shower length (l_s). Adopting the lowest l_s value reduces the base value by almost EUR 1250. The low flow rate (q), high cold water temperature (T_{cw}), and no changes in the energy price during system operation compared to the initial price also have a negative impact on the *NPV* prediction result. On the other hand, adopting a relatively low initial investment outlay (INV_0) results in an increase in the base value by over EUR 700. A positive, although less significant, impact on the *NPV* value is also demonstrated by the lack of changes in the value of money over time, high greywater temperature, high initial energy price, slightly exceeding the current price of electricity for households in Poland, and low efficiency of the DHW heater. The arrangement of the heat exchanger with a slight slope in the direction of greywater flow also has an impact, but it is negligible.

The last figure (Figure 6c) concerns the extremely favorable case where all input parameters contribute to an increase in the predicted net present value with respect to the base value. The hierarchy of significance of individual parameters in terms of the strength of their impact on the subject of analysis differs significantly from that presented for previous observations. Although the very long total daily shower length (l_s) and the high initial energy price (C_e) have the highest impact on the subject of the analysis, the discount rate (r) also ranks relatively high. Adopting a zero value of this parameter results in no reduction in the value of financial flows (savings) in the subsequent years of operation compared to the current year. A high annual change in energy price (i_e) and a high flow rate of media through the unit (q) guarantee an increase in the base value by over EUR 3000. Next in the ranking were cold water and greywater temperatures (T_{cw} and T_{dw}), linear bottom slope (i), and DHW heater efficiency (η). It may seem surprising that the initial investment outlay (INV_0) has the smallest impact on the model prediction in this case. This is probably because, due to significant water consumption for showering and high energy prices, the potential financial savings are so high that the low value of the initial investment outlay loses its importance.

Analyzing the selected observations, it was noticed that the strength of the impact of individual input variables on the *NPV* prediction result may differ significantly from the average values presented in Figure 5a. These differences may be due to the unique conditions prevailing in each case. In a situation where the projected financial savings are very high, the importance of initial investment outlay will not be so significant. However, if the building has low water consumption and energy prices are low, as is the case with the use of gas hot water heaters, even a low investment outlay may turn out to be a key factor determining the financial efficiency of the project. The average SHAP values shown in Figure 5a only provide a general picture. However, if you want to better adapt the recommendations to specific conditions, an individual analysis should be carried out.

4. Discussion

4.1. Investment Profitability

Considering the current geopolitical situation, potential energy crises, as well as ongoing climate change, saving energy in buildings should be a priority for both local government officials and ordinary citizens. Meanwhile, as noted by Su et al. [38], people prefer to place greater emphasis on issues related to building construction, while rising energy costs and the instability of energy supplies require urgent actions to reduce dependence on fossil fuels. The introduction of technologies that reduce energy consumption for water heating, such as shower heat exchangers, can significantly reduce energy consumption in households, contributing to improved energy security. Considering that financial and environmental issues are interconnected and mutually reinforcing [39], it can be assumed that investments in greywater heat recovery technologies can support both household budgets and sustainable development. The assessment of the financial efficiency of investments is particularly important, as noted by Ober et al. [40], because some user groups place less emphasis on pro-environmental actions. In such cases, the profitability of the investment project becomes a key aspect of its evaluation.

For the considered horizontal heat exchanger, the net present value ranges in a relatively wide range from EUR −1996.40 to EUR 36,933.83, indicating significant uncertainty regarding future cash flows. The analysis showed that more than half of the calculated *NPV* values were negative, which is an unfavorable result compared to other types of shower heat exchangers [17]. However, even for this heat exchanger, there is a significant number of scenarios with very high profitability. With the highest initial investment outlay (INV_0 – EUR 2000), the discounted payback period for the most profitable cases does not exceed two years. With lower INV_0 values, it may be shorter than a year. Therefore, it is crucial to have a thorough understanding of the parameters influencing the *NPV* value and to manage them in a way that maximizes potential benefits and minimizes risks when assessing the financial efficiency of such an initiative. A comprehensive evaluation of the influence of these parameters on the profitability of the project can help balance uncertainties regarding the outcomes. It can also ensure achieving stable and profitable results.

The results of the analyses evaluating the efficiency of investments in horizontal shower heat exchangers showed that not all parameters considered in calculating the *NPV* have a notable impact on the project's financial viability. The SHAP analysis proved that statistically, the total daily shower length and the initial energy price have the greatest impact on the *NPV* value. It is also worth noting that these parameters were significant in both global and local analysis, and their impact on the prediction result was consistently positive. This means that as their values increase, the profitability of the investment also increases. Increasing the total daily shower length leads to higher water consumption, which increases the energy demand for its heating, thereby potentially enhancing energy savings through the application of a shower heat exchanger. Similarly, a higher energy price enhances the profitability of investments in energy-efficient technologies, resulting in increased savings and higher *NPV* values. Additionally, although the annual change in energy price ranks fifth in the hierarchy of input parameter significance, the sum of average SHAP values for C_e and i_e exceeds EUR 1200. This indicates the crucial importance of energy prices and, therefore, the type of DHW heater used throughout the entire life cycle of the shower heat exchanger.

The average user of a shower heat exchanger does not influence energy prices and is unable to independently predict the dynamics of their changes. However, the implementation of technologies based on renewable energy sources can reduce dependence on conventional energy sources and minimize the risk associated with fluctuations in their prices. On the other hand, it is different for parameters characterizing the use of the shower installation, including l_s. As mentioned earlier, the longer the total daily shower length, the higher the water consumption in the installation, and consequently, the greater the financial savings. An increase in this parameter can result, for example, from an increase in the number of system users [13]. In contrast, in the opposite situation, the *NPV* value would decrease. A similar issue can arise when the flow rate of mixed water from the showerhead is reduced. While the average impact of q on predicting *NPV* values is significantly lower than l_s, in some cases, it can still be very significant, as confirmed by the analysis of individual observations. Reducing energy consumption for heating water due to a lower flow rate from the showerhead can result from installing flow limiters or replacing showerheads with water-saving models. In extreme cases, it can also result from a reduction in the available pressure in the water supply network. Research conducted in Austria [41] indicated that actual energy consumption for water heating can be significantly lower than the design values. Therefore, during the planning stage of investing in a shower heat exchanger, it is crucial to take into account the investment risk related to reducing water consumption in the shower installation, especially since minimizing the use of potable water in buildings is a component of the circular economy model [42].

The SHAP analysis also highlighted the significant importance of initial investment outlay, which is one of the main weaknesses of shower heat exchangers [43]. The average SHAP values classify this parameter as third in the ranking. However, the analysis of individual observations revealed that depending on the magnitude of the projected financial savings, INV_0 can be either the most or the least significant parameter in the analysis. This proves that although global analysis is useful, only local analysis can provide the detailed information necessary to make optimal investment decisions. If the initial investment outlay is significant, it may be necessary to conduct a benefit–cost analysis [44].

It is worth noting that other parameters also influence the analysis, but their impact is generally significantly smaller. However, in some cases, their significance can prove to be more substantial. This confirms the need to develop tools that will enable future users of greywater heat recovery systems to precisely determine the *NPV* value in relation to the given specific conditions. Thanks to the developed artificial neural network model, it will be possible to create an application that allows potential users of shower heat exchangers to compare *NPV* values under different operating conditions. As a result, the financial forecasts will be tailored to the individual needs of the users. The application, developed as part of future research, should also provide the opportunity to assess how changes in input conditions will impact the financial efficiency of the investment. The development of such a tool will significantly increase the efficiency of investment decisions, contributing to better resource management and cost optimization.

It should also be taken into account that each study may encounter certain limitations that need to be taken into account when analyzing the results. In this analysis, this limitation is undoubtedly the analysis of only the horizontal heat exchanger with relatively low effectiveness. Therefore, the next step in the research should include an investigation of a vertical heat exchanger, which is characterized by higher heat recovery effectiveness [45]. Such a comprehensive assessment will provide a more complete picture of the influence of the considered parameters on the financial efficiency of investments in shower heat exchangers.

4.2. Impact on the Environment

An important aspect of analyzing the feasibility of installing a shower heat exchanger is undoubtedly the financial efficiency of such a project. Nevertheless, in times of growing ecological awareness of society, environmental issues are becoming more and more important, especially since, as noted by Szalay [46], the operation of buildings is responsible for over a quarter of energy-related CO_2 emissions. Even when the use of a shower heat exchanger turns out to be financially unprofitable, its use may be justified for environmental reasons. This is due to the fact that the direct effect of reducing energy consumption for heating domestic hot water is reducing emissions resulting from the combustion of fossil fuels. The amount of this reduction depends primarily on the type of primary energy source used in the DHW preparation installation. Under the established operating conditions of the shower installation, the expected reduction in CO_2 emission is almost three and a half times higher when the DHW is heated with an electric water heater than in the case of a gas water heater. Figures A2 and A3 (Appendix B) show the CO_2 emission reduction for selected shower use conditions in the building. The values of annual CO_2 emission reduction (E_r) presented in the plane diagrams were determined based on emission factors for the considered energy carriers, i.e., gas fuel [47] and electricity for end users [48]. It is clearly visible that the increase in water consumption related to the increase in water flow rate (q) and shower length (l_s) results in an increase in the annual reduction of CO_2 emission. With water consumption for showering at the level of 30 L per day and a minimum temperature difference between the media flowing through the heat exchanger, the annual reduction in CO_2 emission will not exceed 7 kg and 22 kg, respectively, for the gas and electric DHW heater. However, if more people use the shower and the shower head has a higher flow rate (q), the E_r value will be significantly higher. For example, for l_s = 50 min, the annual CO_2 emission reduction is from 32 to 398 kg for a gas water heater and from 106 to 1312 kg for an electric water heater. These differences result primarily from different water flow rates (q) but also from the wide range of water and greywater temperatures at the entrance to the shower heat exchanger. The greater the difference between these temperatures, the greater the reduction in CO_2 emission, which results from the greater energy demand for water heating and the higher effectiveness of the heat exchanger. The E_r values determined for T_{cw} = 8 °C and T_{dw} = 40 °C are more than three times higher than those determined for the lowest considered temperature difference between cold and greywater (T_{cw} = 20 °C and T_{dw} = 30 °C). It follows that the ecological benefits of using shower heat exchangers will be particularly visible in regions with a temperate or cold climate, where the temperature of cold tap water will be relatively low. If l_s = 90 min and the DHW is heated by an electric device, under the most favorable conditions, an annual reduction in CO_2 emission of almost 2400 kg can be expected. Analyzing Figures A2 and A3, one can also notice the impact of the unit bottom slope on the E_r value, which results from increasing effectiveness (ε) with increasing slope (i). This impact is not as pronounced as in the case of water flow rate (q) and shower length (l_s), but increasing the slope (i) from 0% to 4% can contribute to increasing the annual emission reduction by more than 50%.

Apart from the stage that covers the period of operation of the shower heat exchanger, the environmental analysis of the investment related to its installation should also consider its production and end of operation. In the production phase, emissions related to the extraction of raw materials, transport, material production, and assembly are analyzed. In the case of emissions related to the production of the analyzed shower heat exchanger, an important element of the environmental analysis is to take into account emissions related to the extraction and preparation of copper for its production, which is used due to its heat conduction properties and durability. The environmental impact analysis should also take into account emissions resulting from the shower exchanger production process, which involves forming the exchanger elements from copper, which are brazed using high-temperature techniques. In the end-of-life phase of the shower heat exchanger, emissions related to its dismantling and those related to the disposal, recycling, or storage processes should be considered. In the case of end-of-life emissions, the environmental impact of transporting the exchanger to its end-of-life location must also be taken into account. The collected data on the consumption of energy and raw materials necessary for production and end-of-life are converted into CO_2 equivalent emissions using appropriate factors. Analysis of the investment's impact on the environment allows for the identification and minimization of the largest emission sources already at the design stage. As Pei et al. [49] emphasize, considering the analysis of the life cycle impact of a given solution on the environment in the building sector is intended to guide designers to more sustainable solutions. It also contributes to the achievement of global goals of reducing greenhouse gas emissions [50].

5. Conclusions

This paper evaluated the financial efficiency of investing in the horizontal shower heat exchanger and examined the possibility of using artificial neural networks to assess the profitability of such a solution. Additionally, the magnitude of influence exerted by individual input parameters on predicting the net present value of this project was examined. Based on the analyses, the following conclusions were formulated:

- ANNs are effective for assessing the profitability of investing in a shower heat exchanger, providing accurate forecasts by considering key parameters like energy prices and initial investment, as well as variables with less significant impact;
- The profitability analysis showed that most of the analyzed cases were unprofitable. This was due to the wide range of input data, mainly energy prices and total daily shower length. Nevertheless, there was a certain group of scenarios that were highly profitable;
- Machine learning methods identify complex patterns, extract scenarios, and provide detailed analyses of parameter influences on *NPV* predictions. This advanced approach is necessary due to *NPV*'s sensitivity to key input changes, confirming the importance of detailed case analysis and the consideration of a wide range of input values for effective planning and risk management.

The manuscript also identifies directions for future research, which, according to the authors, should focus on the following:

- Assessment of the impact of input parameters on the profitability of investments for various models of shower heat exchangers with different characteristics affecting financial efficiency;
- Analyzing the feasibility of using shower heat exchangers in high water consumption buildings, such as sports facilities, campsites, or dormitories, and assessing the impact of input parameters on their effectiveness in these facilities;
- Conducting profitability analysis for devices with varying effectiveness and price, as well as buildings with specific shower usage conditions, to determine the most beneficial investment scenarios and adapt the technology to users' needs;
- Developing an application to compare *NPV* values under various conditions and for different devices, which is crucial for potential users;
- Conducting a detailed Life Cycle Assessment (*LCA*) analysis of the shower heat exchanger, considering the production, use, and disposal phases, and comparing this solution with other domestic hot water heating techniques.

Author Contributions: Conceptualization, S.K.-O., M.S. and B.P.; methodology, S.K.-O., M.S. and B.P.; software, S.K.-O., M.S. and B.P.; validation, S.K.-O., M.S. and B.P.; formal analysis, S.K.-O., M.S. and B.P.; investigation, S.K.-O., M.S. and B.P.; resources, S.K.-O., M.S. and B.P.; data curation, S.K.-O.,

M.S. and B.P.; writing—original draft preparation, S.K.-O., M.S. and B.P.; writing—review and editing, S.K.-O., M.S. and B.P.; visualization, S.K.-O., M.S. and B.P. All authors have read and agreed to the published version of the manuscript.

Funding: This research was financed by the Minister of Science and Higher Education of the Republic of Poland within the "Regional Excellence Initiative" program for the years 2024–2027 (RID/SP/0032/2024/01).

Institutional Review Board Statement: Not applicable.

Informed Consent Statement: Not applicable.

Data Availability Statement: Data are contained within this paper.

Acknowledgments: The authors would like to thank the reviewers for their feedback, which has helped improve the quality of the manuscript, and *Energies'* staff and editors for handling this paper.

Conflicts of Interest: The authors declare no conflicts of interest.

Appendix A

Figure A1 shows the temperatures of preheated water (T_{pw}) obtained during laboratory tests.

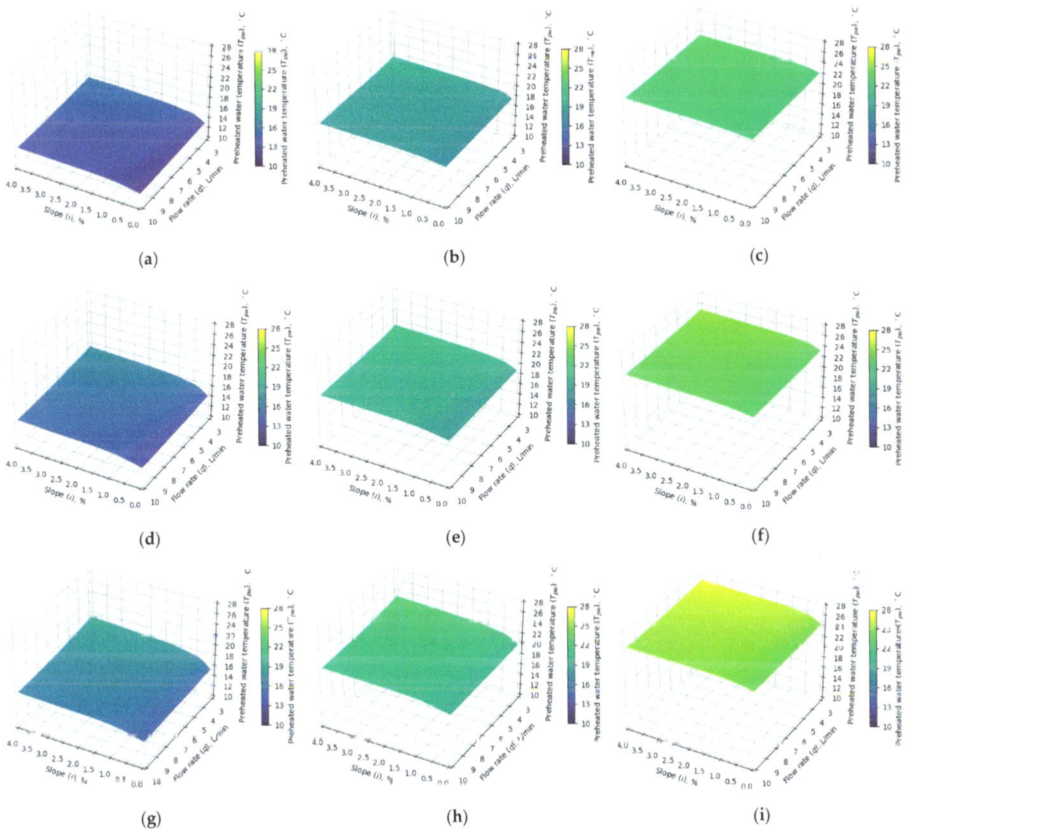

Figure A1. Temperatures of preheated water depending on the flow rate of water and greywater through the shower heat exchanger and the slope of its bottom: (**a**) T_{cw} = 8 °C, T_{dw} = 30 °C; (**b**) T_{cw} = 8 °C, T_{dw} = 35 °C; (**c**) T_{cw} = 8 °C, T_{dw} = 40 °C; (**d**) T_{cw} = 14 °C, T_{dw} = 30 °C; (**e**) T_{cw} = 14 °C, T_{dw} = 35 °C; (**f**) T_{cw} = 14 °C, T_{dw} = 40 °C; (**g**) T_{cw} = 20 °C, T_{dw} = 30 °C; (**h**) T_{cw} = 20 °C, T_{dw} = 35 °C; (**i**) T_{cw} = 20 °C, T_{dw} = 40 °C (designations as in the text).

Appendix B

Figure A2 shows the annual CO_2 emission reduction (E_r) for electric DHW heaters, and Figure A3 shows values of E_r for gas DHW heaters.

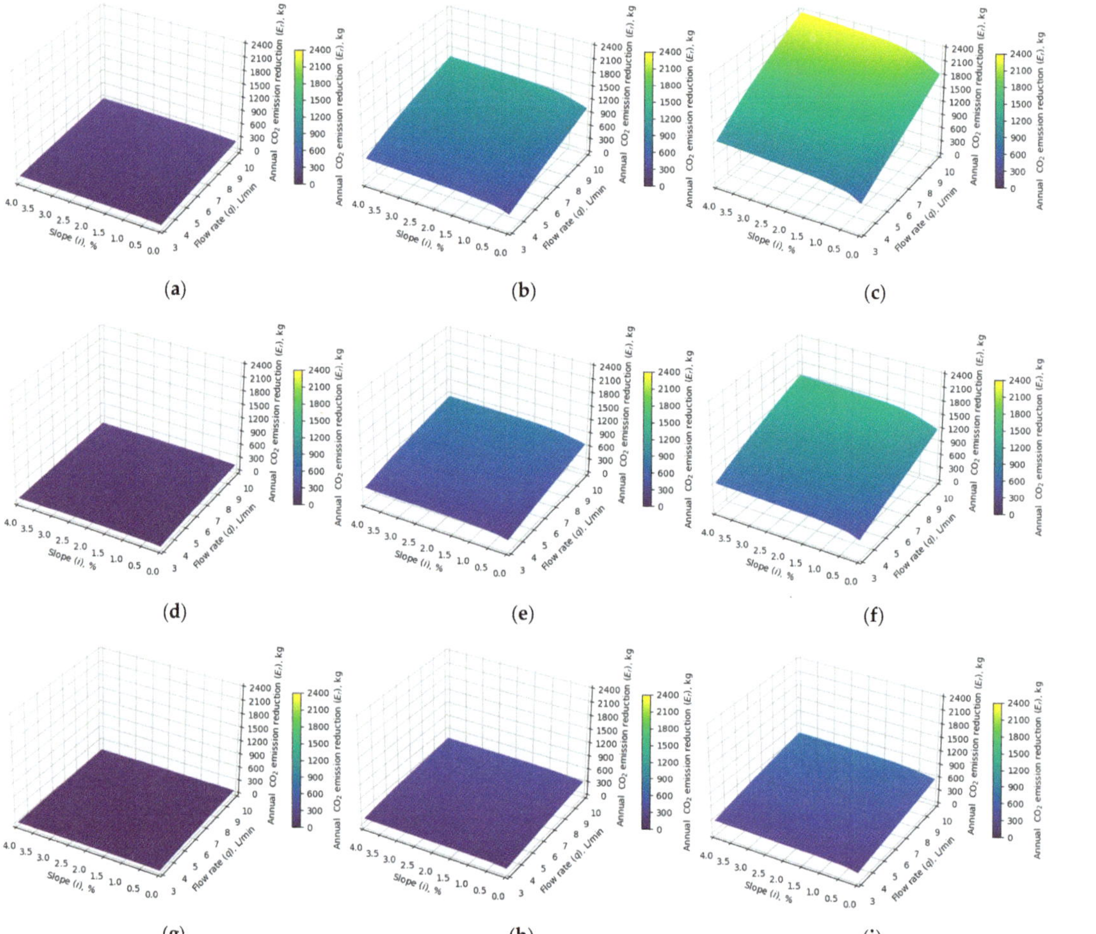

Figure A2. Annual CO_2 emission reduction for electric DHW heater depending on the flow rate of water and greywater through the shower heat exchanger and the slope of its bottom: (**a**) T_{cw} = 8 °C, T_{dw} = 40 °C, l_s = 10 min; (**b**) T_{cw} = 8 °C, T_{dw} = 40 °C, l_s = 50 min; (**c**) T_{cw} = 8 °C, T_{dw} = 40 °C, l_s = 90 min; (**d**) T_{cw} = 14 °C, T_{dw} = 35 °C, l_s = 10 min; (**e**) T_{cw} = 14 °C, T_{dw} = 35 °C, l_s = 50 min; (**f**) T_{cw} = 14 °C, T_{dw} = 35 °C, l_s = 90 min; (**g**) T_{cw} = 20 °C, T_{dw} = 30 °C, l_s = 10 min; (**h**) 20 °C, T_{dw} = 30 °C, l_s = 50 min; (**i**) 20 °C, T_{dw} = 30 °C, l_s = 90 min (designations as in the text).

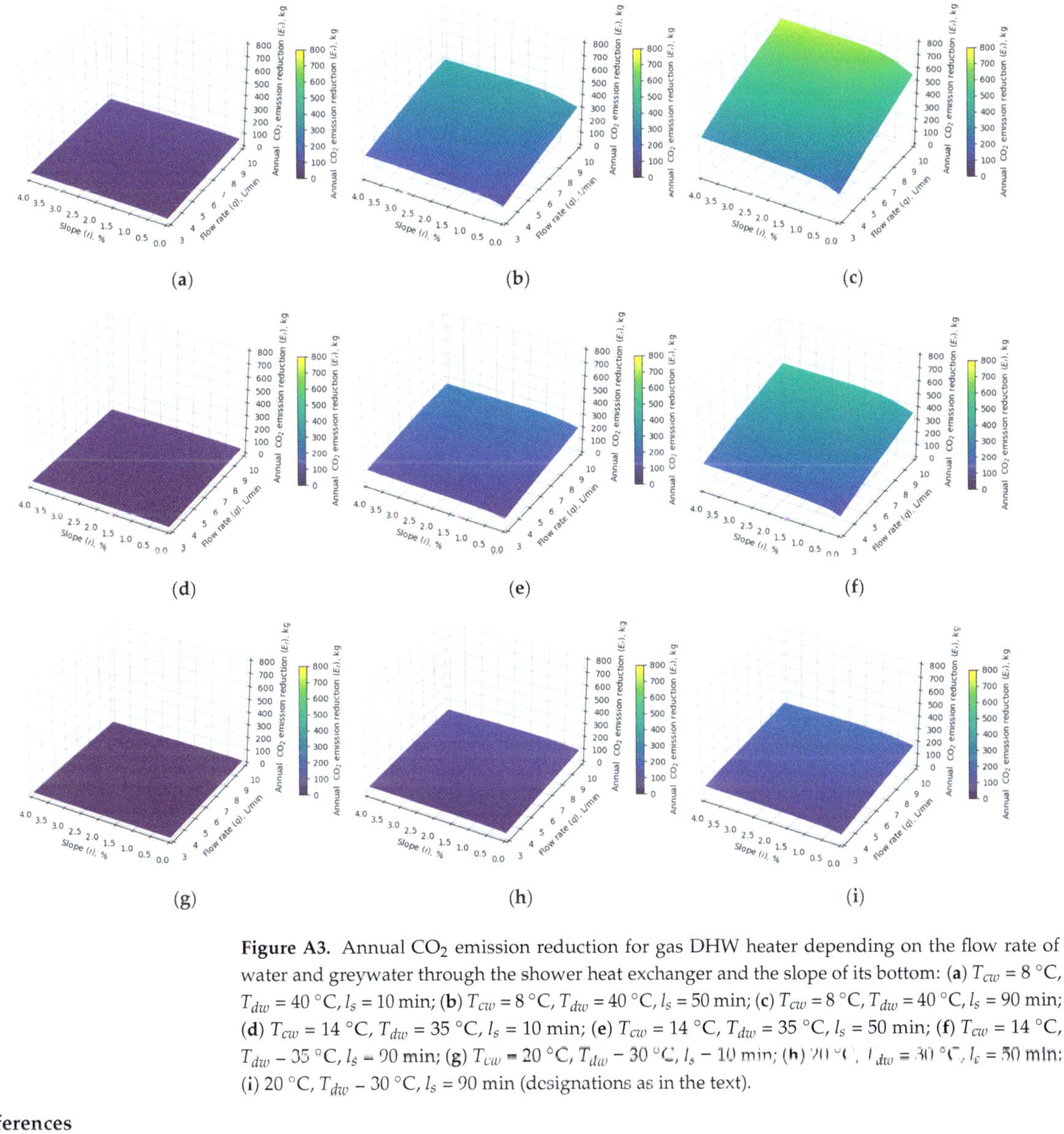

Figure A3. Annual CO_2 emission reduction for gas DHW heater depending on the flow rate of water and greywater through the shower heat exchanger and the slope of its bottom: (**a**) T_{cw} = 8 °C, T_{dw} = 40 °C, l_s = 10 min; (**b**) T_{cw} = 8 °C, T_{dw} = 40 °C, l_s = 50 min; (**c**) T_{cw} = 8 °C, T_{dw} = 40 °C, l_s = 90 min; (**d**) T_{cw} = 14 °C, T_{dw} = 35 °C, l_s = 10 min; (**e**) T_{cw} = 14 °C, T_{dw} = 35 °C, l_s = 50 min; (**f**) T_{cw} = 14 °C, T_{dw} = 35 °C, l_s = 90 min; (**g**) T_{cw} = 20 °C, T_{dw} = 30 °C, l_s = 10 min; (**h**) 20 °C, T_{dw} = 30 °C, l_s = 50 min; (**i**) 20 °C, T_{dw} = 30 °C, l_s = 90 min (designations as in the text).

References

1. Ruszel, M. The significance of the Baltic Sea Region for natural gas supplies to the V4 countries. *Energy Policy* **2020**, *146*, 111786. [CrossRef]
2. Mrozowski, M.; Bednarz, K. The Geopolitical Dependence of Hydrogen. *Energy Policy Stud.* **2023**, *2*, 39–59. [CrossRef]
3. Żywiec, J.; Szpak, D.; Wartalska, K.; Grzegorzek, M. The Impact of Climate Change on the Failure of Water Supply Infrastructure: A Bibliometric Analysis of the Current State of Knowledge. *Water* **2024**, *16*, 1043. [CrossRef]
4. Czajor, D.; Amanowicz, Ł. Methodology for Modernizing Local Gas-Fired District Heating Systems into a Central District Heating System Using Gas-Fired Cogeneration Engines—A Case Study. *Sustainability* **2024**, *16*, 1401. [CrossRef]
5. Ogarek, P.; Wojtoń, M.; Słyś, D. Hydrogen as a Renewable Energy Carrier in a Hybrid Configuration of Distributed Energy Systems: Bibliometric Mapping of Current Knowledge and Strategies. *Energies* **2023**, *16*, 5495. [CrossRef]
6. Seminario-Córdova, R.; Rojas-Ortega, R. Renewable Energy Sources and Energy Production: A Bibliometric Analysis of the Last Five Years. *Sustainability* **2023**, *15*, 10499. [CrossRef]
7. Neugebauer, G.; Lichtenwoehrer, P.; Huber, F.; Stoeglehner, G.; Kretschmer, F. Potentials and Integrated Suitability Pre-assessment of Wastewater Treatment Plants as Local Energy Cells. *Front. Environ. Sci.* **2022**, *9*, 785557. [CrossRef]

8. Voznyak, O.; Spodyniuk, N.; Antypov, I.; Dudkiewicz, E.; Kasynets, M.; Savchenko, O.; Tarasenko, S. Efficiency Improvement of Eco-Friendly Solar Heat Supply System as a Building Coating. *Sustainability* **2023**, *15*, 2831. [CrossRef]
9. IEA—International Energy Agency. Buildings. Available online: https://www.iea.org/energy-system/buildings (accessed on 6 May 2024).
10. Ratajczak, K.; Amanowicz, Ł.; Pałaszyńska, K.; Pawlak, F.; Sinacka, J. Recent Achievements in Research on Thermal Comfort and Ventilation in the Aspect of Providing People with Appropriate Conditions in Different Types of Buildings—Semi-Systematic Review. *Energies* **2023**, *16*, 6254. [CrossRef]
11. Statistics Poland. Energy Efficiency in Poland 2011–2021. Warsaw 2023. Available online: https://stat.gov.pl/en/topics/environment-energy/energy/energy-efficiency-in-poland-2011-2021,5,20.html (accessed on 7 May 2024).
12. Hamburg, A.; Kalamees, T. How well are energy performance objectives being achieved in renovated apartment buildings in Estonia? *Energy Build.* **2019**, *199*, 332–341. [CrossRef]
13. Stec, A.; Mazur, A. An Analysis of Eco-Technology Allowing Water and Energy Saving in an Environmentally Friendly House—A Case Study from Poland. *Buildings* **2019**, *9*, 180. [CrossRef]
14. Fuentes, E.; Arce, L.; Salom, J. A review of domestic hot water consumption profiles for application in systems and buildings energy performance analysis. *Renew. Sustain. Energy Rev.* **2018**, *81*, 1530–1547. [CrossRef]
15. Aleksić, N.; Šušteršič, V.; Jurišević, N.; Kowalik, R.; Ludynia, A. Reduction of wastewater pollution using the technologies for heat recovery from wastewater in buildings—A review of available cases. *Desalin. Water Treat.* **2023**, *301*, 242–255. [CrossRef]
16. Jadwiszczak, P.; Niemierka, E. Thermal effectiveness and NTU of horizontal plate drain water heat recovery unit—Experimental study. *Int. Commun. Heat Mass Transf.* **2023**, *147*, 106938. [CrossRef]
17. Piotrowska, B.; Słyś, D. Analysis of the Life Cycle Cost of a Heat Recovery System from Greywater Using a Vertical "Tube-in-Tube" Heat Exchanger: Case Study of Poland. *Resources* **2023**, *12*, 100. [CrossRef]
18. Zhang, D.; Mui, K.-W.; Wong, L.-T. Wise Choice of Showerhead Patterns: How to Save Energy during Showering While Maintaining Thermal Comfort. *Water* **2024**, *16*, 928. [CrossRef]
19. Tripathy, P.; Prakash, O.; Sharma, A.; Juneja, C.; Hiwrale, I.; Shukla, V.; Pal, S. Facets of cost-benefit analysis of greywater recycling system in the framework of sustainable water security. *J. Clean. Prod.* **2024**, *451*, 142048. [CrossRef]
20. Mazur, A. An assessment of the financial efficiency of a heat recovery system from graywater in a hotel. *E3S Web Conf.* **2018**, *45*, 00051. [CrossRef]
21. Starzec, M.; Kordana-Obuch, S.; Piotrowska, B. Evaluation of the Suitability of Using Artificial Neural Networks in Assessing the Effectiveness of Greywater Heat Exchangers. *Sustainability* **2024**, *16*, 2790. [CrossRef]
22. Selimli, S.; Eljetlawi, I.A.M. The experimental study of thermal energy recovery from shower greywater. *Energy Sources Part A* **2021**, *43*, 3032–3044. [CrossRef]
23. Słyś, D.; Stec, A. Centralized or Decentralized Rainwater Harvesting Systems: A Case Study. *Resources* **2020**, *9*, 5. [CrossRef]
24. Merino, C.; Castro, R. Optimization of a Hybrid Solar–Wind Microgrid for Sustainable Development: A Case Study in Antofagasta, Chile. *Sustainability* **2024**, *16*, 3668. [CrossRef]
25. Dudkiewicz, E.; Ludwińska, A. Family Dwelling House Localization in Poland as a Factor Influencing the Economic Effect of Rainwater Harvesting System with Underground Tank. *Sustainability* **2023**, *15*, 10687. [CrossRef]
26. Minister of Infrastructure and Development. Regulation of the Minister of Infrastructure of 12 April 2002 on the technical conditions to be met by buildings and their location. Journal of Laws of 2002, No. 75, item 690 (consolidated text: Journal of Laws of 2022, item 1225).
27. Chen, L.; Li, S.; Bai, Q.; Yang, J.; Jiang, S.; Miao, Y. Review of Image Classification Algorithms Based on Convolutional Neural Networks. *Remote Sens.* **2021**, *13*, 4712. [CrossRef]
28. Bhambu, A.; Gao, R.; Suganthan, P.N. Recurrent ensemble random vector functional link neural network for financial time series forecasting. *Appl. Soft Comput.* **2024**, *161*, 111759. [CrossRef]
29. Lagaros, N.D. Artificial Neural Networks Applied in Civil Engineering. *Appl. Sci.* **2023**, *13*, 1131. [CrossRef]
30. Huang, S.C.; Le, T.H. *Principles and Labs for Deep Learning*, 1st ed.; Academic Press: London, UK, 2021. [CrossRef]
31. Starzec, M.; Kordana-Obuch, S. Evaluating the Utility of Selected Machine Learning Models for Predicting Stormwater Levels in Small Streams. *Sustainability* **2024**, *16*, 783. [CrossRef]
32. Yau, H.T.; Kuo, P.H.; Hong, S.W. Milling wear prediction using an artificial neural network model. *Eng. Appl. Artif. Intell.* **2024**, *135*, 108686. [CrossRef]
33. Yue, Y.; Yan, Z.; Ni, P.; Lei, F.; Yao, S. Machine learning-based multi-performance prediction and analysis of Earth-Air Heat Exchanger. *Renew. Energy* **2024**, *227*, 120550. [CrossRef]
34. Cakiroglu, C.; Demir, S.; Hakar Ozdemir, M.; Latif Aylak, B.; Sariisik, G.; Abualigah, L. Data-driven interpretable ensemble learning methods for the prediction of wind turbine power incorporating SHAP analysis. *Expert Syst. Appl.* **2024**, *237*, 121464. [CrossRef]
35. Neubauer, A.; Brandt, S.; Kriegel, M. Relationship between feature importance and building characteristics for heating load predictions. *Appl. Energy* **2024**, *359*, 122668. [CrossRef]
36. Zhang, H.; Loaiciga, H.A.; Sauter, T. A Novel Fusion-Based Methodology for Drought Forecasting. *Remote Sens.* **2024**, *16*, 828. [CrossRef]

37. Etaati, B.; Jahangiri, A.; Fernandez, G.; Tsou, M.-H.; Ghanipoor Machiani, S. Understanding Active Transportation to School Behavior in Socioeconomically Disadvantaged Communities: A Machine Learning and SHAP Analysis Approach. *Sustainability* **2024**, *16*, 48. [CrossRef]
38. Su, Y.; Jin, Q.; Zhang, S.; He, S. A review on the energy in buildings: Current research focus and future development direction. *Heliyon* **2024**, *10*, e32869. [CrossRef]
39. Rajski, K.; Englart, S.; Sohani, A. Analysis of Greywater Recovery Systems in European Single-Family Buildings: Economic and Environmental Impacts. *Sustainability* **2024**, *16*, 4912. [CrossRef]
40. Ober, J.; Karwot, J.; Sitinjak, C. Modelling Possible Household Uses of Grey Water in Poland using Property Fitting Analysis. *Resources* **2024**, *13*, 25. [CrossRef]
41. Venturi, E.; Ochs, F.; Dermentzis, G. Identifying the influence of user behaviour on building energy consumption based on model-based analysis of in-situ monitoring data. *J. Build. Eng.* **2023**, *64*, 105717. [CrossRef]
42. Bąk, J. The Use of Precipitation in the Cities of the Future—Problems, Barriers and Challenges. *Sustainability* **2023**, *15*, 14381. [CrossRef]
43. Kordana-Obuch, S.; Wojtoń, M.; Starzec, M.; Piotrowska, B. Opportunities and Challenges for Research on Heat Recovery from Wastewater: Bibliometric and Strategic Analyses. *Energies* **2023**, *16*, 6370. [CrossRef]
44. Wittmanova, R.; Hrudka, J.; Meliska, M.; Stanko, S.; Skultetyova, I. Benefit-Cost Analysis of Small Domestic Wastewater Treatment Plants: A Review. In Proceedings of the 23rd International Multidisciplinary Scientific Geoconference: Water Resources. Forest, Marine and Ocean Ecosystems, SGEM 2023, Vienna, Austria, 28–30 November 2023; Volume 23, pp. 43–50. [CrossRef]
45. Dudkiewicz, E.; Ludwińska, A.; Rajski, K. Implementation of greywater heat recovery system in hospitals. *E3S Web Conf.* **2019**, *116*, 00018. [CrossRef]
46. Szalay, Z. A parametric approach for developing embodied environmental benchmark values for buildings. *Int. J. Life Cycle Assess.* **2024**. [CrossRef]
47. The National Centre for Emissions Management (KOBiZE). *Wskaźniki Emisji Zanieczyszczeń ze Spalania Paliw dla Źródeł o Nominalnej Mocy Cieplnej do 5 MW, Zastosowane do Automatycznego Wyliczenia Emisji w Raporcie do Krajowej Bazy za Rok 2022*; IOŚ-PIB: Warszawa, Poland, 2023. Available online: https://krajowabaza.kobize.pl/docs/Wska%C5%BAniki_ma%C5%82e_%C5%BAr%C3%B3dla_spalania_paliw_2022.pdf (accessed on 26 June 2024).
48. The National Centre for Emissions Management (KOBiZE). *Wskaźniki Emisyjności CO_2, SO_2, NOx, CO i Pyłu Całkowitego dla Energii Elektrycznej na Podstawie Informacji Zawartych w Krajowej Bazie o Emisjach Gazów Cieplarnianych i Innych Substancji za 2022 rok*; IOŚ-PIB: Warszawa, Poland, 2022. Available online: https://www.kobize.pl/en/file/wskazniki-emisyjnosci/id/198/wskazniki-emisyjnosci-dla-energii-elektrycznej-za-rok-2022-opublikowane-w-grudniu-2023-r (accessed on 26 June 2024).
49. Pei, L.; Schalbart, P.; Peuportier, B. Life cycle assessment of a residential building in China accounting for spatial and temporal variations of electricity production. *J. Build. Eng.* **2022**, *52*, 104461. [CrossRef]
50. Bahramian, M.; Yetilmezsoy, K. Life cycle assessment of the building industry: An overview of two decades of research (1995–2018). *Energy Build.* **2020**, *219*, 109917. [CrossRef]

energies

Article

Study of the Thermal Performance of Solar Air Collectors with and without Perforated Baffles

Ghizlene Boussouar [1,2], Brahim Rostane [2], Khaled Aliane [2], Dineshkumar Ravi [1], Michał Jan Gęca [1] and Arkadiusz Gola [1,*]

[1] Faculty of Mechanical Engineering, Lublin University of Technology, ul. Nadbystrzycka 36, 20-618 Lublin, Poland; d.ravi@pollub.pl (D.R.); m.geca@pollub.pl (M.J.G.)
[2] MECACOMP Laboratory, Department of Mechanic, Faculty of Technology, University Abou Bekr Belkaid, Tlemcen 13000, Algeria; khaled.aliane@univ-tlemcen.dz (K.A.)
* Correspondence: a.gola@pollub.pl; Tel.: +48-81-538-45-35

Abstract: Air plate solar collectors provide a sustainable and efficient solution for building heating. The absorber plate collects solar radiation and converts it into heat. Atmospheric air is then circulated through the collector plate with perforated baffles by forced convection. The heated air is then directed through ducts into the building's heating system. By significantly reducing reliance on fossil fuels for building heating, these collectors contribute to a lower life-cycle carbon footprint for buildings compared to conventional heating systems. While flat-plate solar collectors are widely used for renewable energy generation, their efficiency is frequently limited by the airflow path and the heat transfer efficiency within the collector. This study aims to quantify the impact of longitudinal, transverse, and perforated baffles with different hole diameters on the heat transfer characteristics and to identify the optimal design for maximizing thermal efficiency. This study also aims to integrate solar air collector in a conventional building and help reduce the overall energy demand of buildings and their associated carbon emissions. A three-dimensional numerical investigation was carried out on a flat-plate solar collector equipped with perforated transverse baffles with varying hole diameter and thickness. The results from the study predicted that perforated baffles with two holes with a diameter of 15 mm provided a maximum Nu of 79.56 and a pressure drop of 459 Pa for a Re of 8500.

Keywords: solar collector; perforated baffles; conjugative heat transfer; numerical simulation; carbon footprint reduction

Citation: Boussouar, G.; Rostane, B.; Aliane, K.; Ravi, D.; Gęca, M.J.; Gola, A. Study of the Thermal Performance of Solar Air Collectors with and without Perforated Baffles. *Energies* **2024**, *17*, 3812. https://doi.org/10.3390/en17153812

Academic Editor: Carlo Renno

Received: 26 June 2024
Revised: 16 July 2024
Accepted: 24 July 2024
Published: 2 August 2024

1. Introduction

Carbon emission is one of the main factors that influence climate change. The world is currently moving toward net-zero carbon emission by 2050 or at least a reduction in carbon emission to 45% by 2030. Keeping the world agenda in mind, the current research focuses on harnessing renewable energy resources, namely, solar energy as a source of room heating and ventilation in a conventional building. Air solar collectors, which use air for heating applications, are a promising renewable energy source. However, improving their efficiency is crucial to maximize their potential. This article investigates the performance of air solar collectors, a type of solar collector utilizing air as the heat transfer fluid. Typically used for space heating and hot water production, air solar collectors are the focus of this study.

In response to these challenges and the pursuit of enhanced solar collector efficiency, a substantial body of research has been undertaken to summarize and synthesize the findings from these efforts.

Solar collector design and construction have seen significant advancements through decades of research and development, drawing upon diverse scientific and engineering disciplines. A persistent challenge, however, lies in stagnant air pockets within the collector. These elements impede the transfer of heat from the absorber to the air, reducing the efficiency of the collector [1], which carried a theoretical and experimental study of a flat-plate air collector equipped with artificial roughness elements placed in the airflow between the absorber and the bottom plate. In addition, a flat-plate air collector without baffles was analyzed for reference purposes. The results of the study showed that the temperature of the absorber decreases as the airflow rate increases. It was also observed that the absorber temperature, outlet temperature, and collector efficiency were higher for the collector fitted with baffles.

Other studies have shown that staggered baffles improve the thermal performance of air collectors [2]. Two cases were considered: one with a 10 cm spacing between two rows and the other with a 20 cm spacing. The convective heat transfer coefficient, almost doubled in the first case when Re > 2100, is between 4 and 45 W/m^2 K for laminar flow and between 40 and 70 W/m^2 K for turbulent flow, with a similar observation for the Nusselt number (Nu).

Fins were introduced into the solar air collector to improve its performance. This approach was also investigated by Fakoor Pakdaman et al. [3]. The results indicated a 20% increase in heat transfer, attributed to a 66% increase in heat exchange surface area due to the fins. Finally, the study found that the angle of inclination of the unit had a minimal impact on heat transfer. Bensaci et al. [4] evaluated the effect of different baffle placements (smooth, 50% Up, 50% Down, 50% Middle, and 100%) on the local convective heat transfer coefficient, Nusselt number, and friction factor. The study identified a 50% bottom baffle (case 2) as the most effective configuration for achieving a good balance between thermal and hydraulic performance.

The efficiency of six types of solar air heaters fitted with baffles of different shapes was compared by Sharma et al. [5] with regard to their Reynolds numbers, ranging from 3000 to 18,000. Sinusoidal baffles were found to offer the best performance, with an increase in thermal–hydraulic efficiency of up to 2.05 times at Re = 15,000.

Previous research has examined the performance of an air collector incorporating semi-circular fins and baffles [6]. It has also been shown that the finned collector offers a significant increase in air temperature, heat transfer coefficient, and thermal efficiency over the conventional flat plate solar collector. These improvements make the hollow-fin baffled solar collector a promising solution for domestic and industrial applications. In this study, the performance of an air collector incorporating fins and baffles in the form of semi-circular loops was evaluated. The aim of this evaluation was to improve the heat transfer and thermal efficiency of the collector.

Two studies on the influence of changing solar intensity on flat-plate solar collectors were analyzed by Mohammadi et al. and Rani et al. [7,8]. The results showed that solar intensity has a significant impact on the collector's thermal parameters. These parameters include inlet and outlet temperatures, absorption temperature, and glass cover temperature. In addition, it was observed that increasing solar intensity leads to a minimal decrease in energy and effective efficiency.

Three fin-and-baffle configurations were investigated by Biswas et al. [9] to optimize heat transfer and pressure loss in a solar collector. The channel configuration was found to exhibit the best heat transfer but also the highest pressure loss. Pressure loss was reduced by the counter-flow and spiral configurations but at the expense of lower heat transfer. The 30° counter-flow configuration was determined to be the best compromise between thermohydraulic performance, pressure loss, and flow homogeneity within the collector.

The concept of an angle factor was also investigated by Promvonge and Skullog [10] and it was found that an angle of 30° resulted in the most favorable combination of heat transfer and friction factor.

An investigation into two active solar dryer designs, flat plate and evacuated tube, was conducted by Sai Kandukuri et al. [11]. The implementation of perforated baffle trays resulted in enhanced drying performance through improved air circulation and heat transfer. The evacuated tube dryer was found to achieve superior efficiency when compared to the flat plate dryer. Also, Hu et al. [12] found that perforating the deflectors in an SAC led to significant improvements. Efficiency increased by 3.6% and pressure drop reduction was observed to be 8.8%.

The crucial role of porous fins in enhancing the performance of a PCM solar collector was investigated by Khanlari et al. [13]. It was observed that an increase in the number of porous fins led to a corresponding increase in the average energy efficiency.

The study conducted by Abbas et al. [14] investigated the thermal performance of bare tube- and serpentine tube-concentrated solar collectors (CSCs). The results demonstrated that serpentine tube CSCs with a conductive material plate (C-HMA) exhibited superior thermal efficiency compared to bare tube CSCs with a non-conductive material plate (RHMA). Also, Zhang et al. [15] examined flat-plate solar collectors (FPSACs) with circular-rib absorbers that attained a thermal efficiency of 85%, surpassing the performance of other technologies evaluated. Evacuated-tube solar collectors (EVTSACs) equipped with micro-fins can achieve a thermal efficiency of 73%.

A novel triangular solar collector (TSAC) was proposed [16] to enhance the performance of air solar collectors. Three configurations of TSAC were investigated. The results of this study show that the new TSAC has improved performance in different climatic regions, especially in low-radiation regions. The optical performance of the three configurations was compared and the TSAC with a transparent plate on the side had the highest optical efficiency.

A study by Rajaseenivasan et al. [17] investigated the performance of a single-pass solar air heater (SAH) with circular and V-shaped turbulators. Two identical SAHs were tested: a conventional one and a modified one with varying Reynolds numbers (6000 to 12,000). The results indicated a direct correlation between system efficiency and both the Reynolds number and the number of turbulators employed. Notably, the thermal enhancement factor exhibited an inverse relationship with the Reynolds number across all configurations. In a similar vein, ref. [18] conducted an experimental study with V-shaped and offset rib artificial turbulators. They found that the Nusselt number and friction factor are maximized for relative offset rib lengths doubles of 3.5 and 2, respectively. Additionally, both V-shaped and offset rib artificial turbulators enhance heat transfer and the friction factor.

In another location study [19], the performances of two double-pass air solar collectors were compared. The results of the study showed that double-pass air solar collectors (DSACs) with a new type of absorbing surface, consisting of a mixture of matte black paint and graphene nanomaterial (NFAD), exhibited superior performance to conventional DSACs. DSACs with NFAD coating were a promising technology for the production of solar thermal energy. These collectors exhibited superior performance in terms of outlet temperature, thermal power, exergy efficiency, and durability index.

Another double-pass solar air heater with a porous medium was tested by [20] in Famagusta, Cyprus. The collector's average efficiency was 53.7% at an airflow rate of 0.037 kg/s. While increasing the airflow rate improves efficiency, it reduces the temperature difference between the inlet and outlet air.

Motivated by the critical role of efficient air circulation in optimizing thermal performance, research efforts are increasingly directed toward understanding the flow dynamics within solar air collectors. This pursuit aims to refine collector designs and thereby improve overall energy conversion efficiency. Early research on air solar collectors aimed to identify the optimal airflow path to maximize collector efficiency and temperature. Different airflow configurations were examined in studies conducted by [21]. It was found that the performance of airflow exclusively under the absorber plate collector was comparable to that of airflow over the absorber plate and then returning underneath it, in the absence of baffles.

Another study by [22] showed that the collector field configuration, flow rate, inlet temperature, and working fluid have a significant impact on the resistance and efficiency characteristics of direct-return flat-plate solar collector fields. Also, an experimental study by [23] was conducted on the impact of jet shapes and absorber spacing on the heat transfer and pressure drop performance of solar air collectors. The results show that circular jet collectors offer the best overall performance. They exhibit a higher outlet temperature, a higher useful heat gain, a higher Nusselt number, and a lower friction factor than square or triangular jet collectors. Of a similar nature, another experimental and transient study by [24] was conducted on two parallel-flow solar air collectors, one without modification and one equipped with aluminum cans filled with PCM. The results showed that the integration of aluminum cans filled with PCM in a parallel-flow solar air collector can significantly improve its energy and exergy efficiency.

The Machi et al. [25] study was conducted to examine the influence of inlet duct design on the efficiency of a single-pass air solar collector (SAC). Two identical collectors were evaluated, one with a side inlet duct (SESAC) and the other with a front inlet duct (FESAC). The results of the study suggest that collectors with a front inlet duct tend to be more efficient than collectors with a side inlet duct.

Other research on the interplay between sensor manufacturing equipment and sensor performance is gaining traction across diverse scientific disciplines.

Yousif et al. [26] conducted a study to compare the performance of two types of compound parabolic concentrator (CPC) solar air flat plate collectors with paraffin wax phase change material (PCM). Both models were tested under standard conditions in Mosul, Iraq. The results showed that the normal-shape model had a higher thermal efficiency than the involute-shape model for all air mass flow rates tested but that the outlet temperature of the normal-shape model was lower than that of the involute-shape model for different air mass flow rates. The results of this study show that the phase change material and the position of the receiver have a significant impact on the performance of both CPC collector models.

Yang et al. [27] conducted an experimental study of a new model of an air solar collector that utilizes a phase change material. The findings reveal that nighttime air heating is influenced by several factors, such as the quantity of heat stored within the phase change material, the velocity of airflow, and the ambient air temperature. Additionally, the collector's efficiency is contingent upon the start time of daytime air heating and exhibits variations between 34.51% and 44.49%. Zayed et al. [28] was implemented to assess the energy and exergy efficiencies of a wavy corrugated solar air collector (WCSAC) integrating phase change materials (PCMs), the outcomes unmistakably confirmed that the incorporation of PCMs resulted in a remarkable boost in the WCSAC's performance, preserving the outlet air temperature above ambient temperature even after sunset.

Recently Rawat et al. [29] analyzed a solar air heater with phase change material (SAH-PCM). The study explores how the thickness-to-length ratio (t/L) of the PCM container affects performance. Findings show that increasing the t/L ratio improves energy storage, delays PCM discharge, and enhances the overall system efficiency. The research demonstrates the importance of optimizing the t/L ratio for greater energy and cost efficiency in both single pass and double-pass flow configurations.

A thorough study was conducted to assess the thermal performance of an air solar collector (SAC) supplemented by the integration of obstacles embedded with a phase-change material (PCM) by [30]; the findings indicate that an increased solar irradiance induces a rise in the temperature gradient across the SAC simultaneously with an improved collector efficiency. However, this enhancement in efficiency is associated with a decline in the temperature difference and, recently, [31] investigated the impact of integrating a fan with an air-based solar collector (SAC) with a phase change material (PCM). The results showed that the fan significantly improved the thermal performance of the SAC, including increasing the outlet air velocity, outlet air temperature, and average indoor air temperature. The study conducted by Kumar Ahirwar and Kumar Arvind [32] investigated techniques for improving the efficiency of SAHs by disrupting the laminar sub-layer on the absorbing plate. It examines four methods: artificial roughness, jet impingement, piezo-electric fans, and phase change materials (PCMs). Research demonstrates that these techniques can significantly enhance the thermal–hydraulic performance factor of SAHs.

Also, the [33] study investigated the impact of metal foam on the thermal performance and entropy generation rate of a plate heat exchanger (IB-ETC-SAH) compared to a fin-and-tube heat exchanger (SAH). The results showed that the use of metal foam can improve the thermal performance of an IB-ETC-SAH but that it also leads to an increase in pressure drop (a maximum pressure drop of 26,338 Pa (0.26 atm)). A metal foam with a lower PPI and lower porosity is preferable to minimize pressure drop while maintaining high thermal performance.

Motivation and Objective of the Study

As per the author's knowledge, only limited studies have been carried out on optimizing the performance of perforated solar air collectors and improving their efficiency; it is clear that baffles increase solar collector efficiency but are also a key factor in increasing pressure drop. Current research is striving to improve collector efficiency while reducing the pressure drop and also reducing carbon consumption to preserve the environment.

The present study focuses on the impact of incorporating holes in the baffles to induce turbulence in the airflow; holes were introduced into the baffles to generate turbulence and reduce the pressure drop. A three-dimensional numerical simulation served to assess the performance of the cases with different configurations. The CFD simulations were carried out using ANSYS 2023 R2 at Reynolds numbers ranging from 2300 to 8500. The local convective heat transfer coefficient, thermal efficiency, Nusselt number (Nu), and pressure drop were discussed and compared.

2. Materials and Methods

2.1. Geometry

In the present work, a numerical study of a solar collector model is carried out, focusing on the airflow between the absorber plate and the insulation layer. The dimensions of the collector are fixed at 2000 mm × 1000 mm, with two types of baffles: longitudinal baffles with a length of L_LB = 800 mm and spacing of H = 333.33 mm and transverse baffles with a length of L_TB = 111.11 mm. The height of the baffles remains constant at t = 20 mm and, in part, by placing restricting baffles on the insulator to force the air to circulate through a small space left between the absorber and the baffles without touching the absorber [4], as shown in Figure 1, but with different thicknesses t = (1.0 mm, 1.5 mm, and 2.0 mm). The geometric parameters are summarized in Table 1.

Table 1. The geometric parameters of conventional and perforated baffles.

Models:	Figures	Length of Longitudinal Baffles L_{LB} (mm)	Length of Transverse Baffles L_{TB} (mm)	Height of the Baffle t_{LB} (mm)	Distance between the Longitudinal Baffles H (mm)	Distance between the Transverse Baffles D_B (mm)	Diameter of the Holes D (mm)	Thicknesses of the Baffles t_{TB} (mm)
Model 1 Smooth								

Table 1. *Cont.*

Models:	Figures	Length of Longitudinal Baffles L_{LB} (mm)	Length of Transverse Baffles L_{TB} (mm)	Height of the Baffle t_{LB} (mm)	Distance between the Longitudinal Baffles H (mm)	Distance between the Transverse Baffles D_B (mm)	Diameter of the Holes D (mm)	Thicknesses of the Baffles t_{TB} (mm)
Model 2 LT	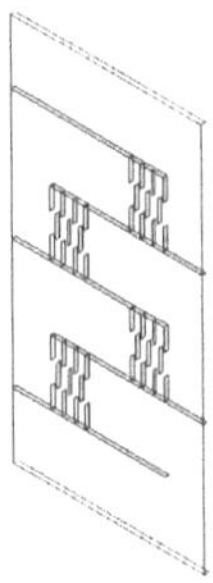	800	111.11	20	333.33	60		1, 1.5, 2
Model 3 LTH 1	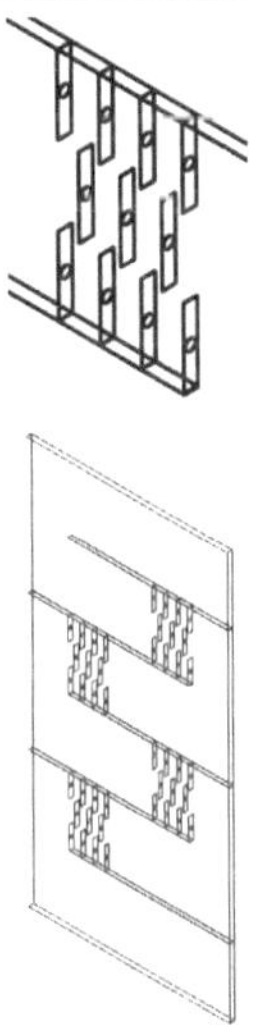	800	111.11	20	333.33	60	15, 10	1, 1.5, 2
Model 4 LTH 2	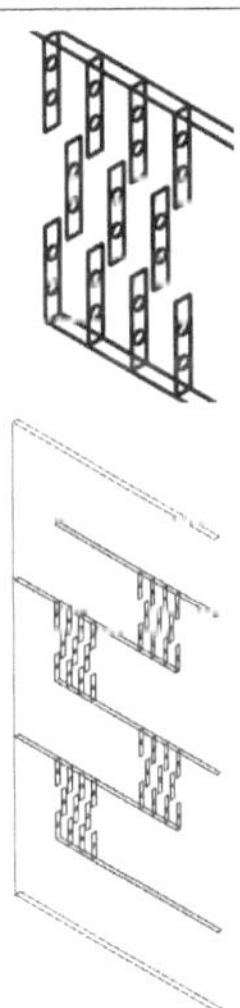	800	111.11	20	333.33	60	15, 10	1, 1.5, 2

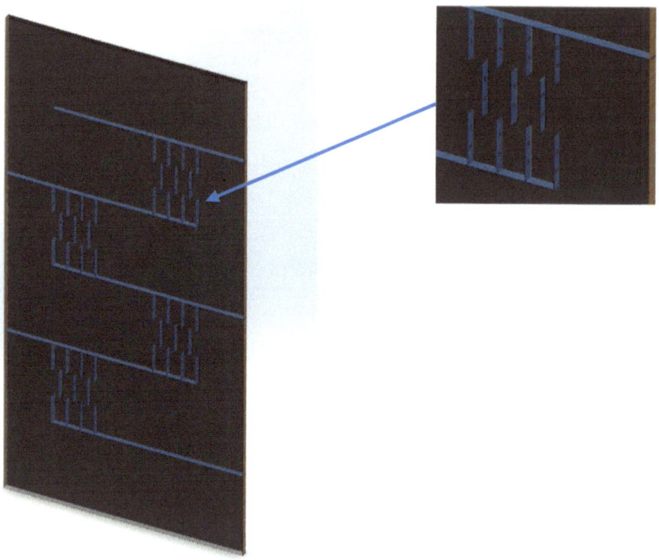

Figure 1. Visualization of a solar air collector with internal baffles (3D).

The present numerical study consists of six cases:

a. The first case: solar collector without baffles "Smooth".
b. The second case: solar collector with transverse and longitudinal baffles "LT".
c. The third case: a solar collector with a single hole in the middle of the transverse baffles, 10 mm of diameter "LTH 1".
d. The fourth case: a solar collector with a single hole in the middle of the transverse baffles 15 mm of diameter "LTH 1".
e. The fifth case: a solar collector with two holes in the transverse baffles, 10 mm in diameter "LTH 2".
f. The sixth case: a solar collector with two holes in the transverse baffles 15 mm in diameter "LTH 2".

2.2. Grid Mesh

The three-dimensional numerical model of the solar air heater was discretized into a finite number of small tetrahedral elements using the finite volume approach with the Ansys workbench meshing solver. A face sizing value of 5 mm was maintained on the absorber and insulator. The inlet, outlet, and side walls were meshed with element sizes of 1 mm and 2 mm for the smooth and baffled models, respectively. An inflation technique was employed to capture the variations within the velocity and thermal boundary layers. Ten inflation layers with a total thickness of 1 mm were used on the absorber, insulator, and baffles. The total number of elements ranged from 6 million to 8.3 million, with additional Mesh sizing for longitudinal and transverse baffles as well as baffles with holes and two holes, which depends more on the baffle configuration itself rather than any other factors. The discretized model of LTH baffles with 2 holes having a diameter of 15 mm, as shown in Figure 2.

2.3. Grid Independence Study

The grid independence study or mesh sensitivity test was carried out to verify that numerically simulated results are independent of the mesh size. In order to carry out the grid test, a smooth surface solar air collector was used for the study and the area-weighted outlet temperature was plotted for four different body/face sizing values, namely 10 mm, 7 mm, 5 mm, and 4 mm, which is equivalent to nearly 1.3 million, 24 million, 48 million, and 69 million elements, respectively, as shown in Figure 3. The Reynolds number of 5500 was maintained as constant for this study. The outcome of the grid test predicted that when the face sizing value is nearly 7 mm and below, the outlet temperature deviation is less than 0.01% and, hence, for the upcoming numerical simulations, a face sizing value of 5 mm and below was maintained for the study.

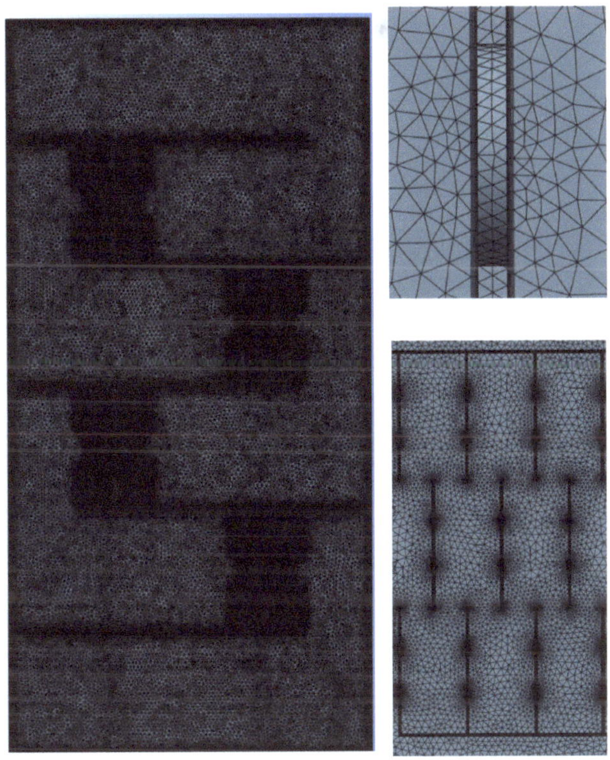

Figure 2. Discretized model of LTH 2 holes with a diameter of 15 mm.

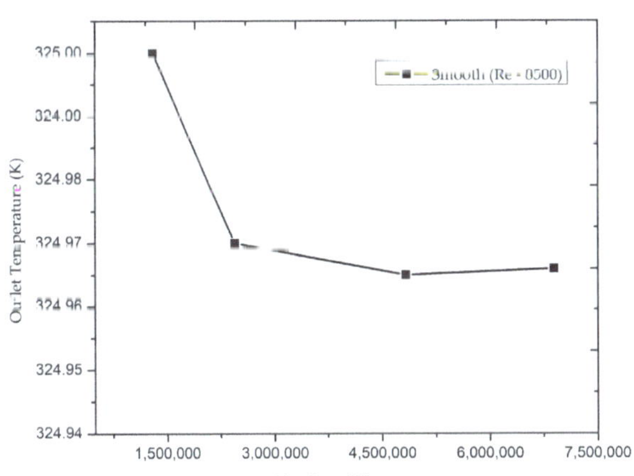

Figure 3. Outlet temperature plot vs. number of elements for a Re 5500.

2.4. Mathematical Modeling and Solution Technique

In this study, a computational fluid dynamics (CFD) approach was used to solve the flow equations and capture the relevant geometric parameters. The CFD tool ANSYS CFD 23-R2 was used to discretize and solve the conservation of mass, momentum, and energy equations.

The flow through and under the absorber was numerically simulated by solving the mass conservation equations as well as the velocity equations for all x, y, and z directions. Turbulence, an important phenomenon in this type of flow, was modeled using the evolving wall functions of the RGN k-epsilon model, known for their accuracy in the near-wall boundary layer region.

This numerical approach accurately captured the hydrodynamic characteristics of the flow, including the distribution of velocities, pressures, and turbulence fields, in the complex geometry of the absorber. The results provided valuable information for understanding the flow behavior and optimizing the absorber design.

Continuity Equation:

$$\frac{\partial u}{\partial x}+\frac{\partial v}{\partial y}+\frac{\partial w}{\partial z}=0 \tag{1}$$

Momentum Equation:

The velocity component's scalar momentum equation in the x direction is provided by

$$\rho\left(u\frac{\partial u}{\partial x}+v\frac{\partial u}{\partial y}+w\frac{\partial u}{\partial z}\right)=-\frac{\partial P}{\partial x}+\frac{\partial}{\partial x_j}\left[(\mu+\mu_T)\left(\frac{\partial u}{\partial x_j}+\frac{\partial u_j}{\partial x}\right)\right] \tag{2}$$

The velocity component's scalar momentum equation in the y direction is provided by

$$\rho\left(u\frac{\partial v}{\partial x}+v\frac{\partial v}{\partial y}+w\frac{\partial v}{\partial z}\right)=-\frac{\partial P}{\partial y}+\frac{\partial}{\partial x_j}\left[(\mu+\mu_T)\left(\frac{\partial v}{\partial x_j}+\frac{\partial u_j}{\partial y}\right)\right] \tag{3}$$

The velocity component's scalar momentum equation in the z direction is provided by

$$\rho\left(u\frac{\partial w}{\partial x}+v\frac{\partial w}{\partial y}+w\frac{\partial w}{\partial z}\right)=-\frac{\partial P}{\partial z}+\frac{\partial}{\partial x_j}[(\mu+\mu_T)] \tag{4}$$

Energy equation:

$$\frac{\partial T}{\partial t}+u\frac{\partial T}{\partial x}+v\frac{\partial T}{\partial y}+w\frac{\partial T}{\partial z}=\frac{k_{air}}{\rho c_p}\left(\frac{\partial^2 T}{\partial x^2}+\frac{\partial^2 T}{\partial y^2}+\frac{\partial^2 T}{\partial z}\right) \tag{5}$$

$$\frac{\partial \rho}{\partial t}+\frac{\partial}{\partial x}(\rho u)+\frac{\partial}{\partial y}(\rho v)+\frac{\partial}{\partial z}(\rho w)=0 \tag{6}$$

Transport equations for the standard k-ε model: The turbulent kinetic energy, k, and its turbulent eddy dissipation, ε, are obtained from the following transport equations:

$$\frac{\partial}{\partial x_i}(\rho k u_i)=\frac{\partial}{\partial x_j}\left[\left(\mu+\frac{\mu_t}{\sigma_k}\right)\frac{\partial k}{\partial x_j}\right]+G_k-\rho_\varepsilon \tag{7}$$

and

$$\frac{\partial}{\partial x_i}(\rho \varepsilon u_i)=\frac{\partial}{\partial x_j}\left[\left(\mu+\frac{\mu_t}{\sigma_\varepsilon}\right)\frac{\partial \varepsilon}{\partial x_j}\right]+C_{1\varepsilon}\frac{\varepsilon}{k}G_k-C_{2\varepsilon}\rho\frac{\varepsilon^2}{k} \tag{8}$$

The expression for the turbulent viscosity (μ_t) is expressed as

$$\mu_t=\rho C_\mu\frac{k^2}{\varepsilon} \tag{9}$$

The Reynolds number (Re) is based on the hydraulic diameter of the air duct, as follows:

$$Re = \frac{UD_h}{\upsilon} \quad (10)$$

where υ is kinematic.

U: The viscosity.

D_h: The hydraulic diameter (A is the cross-section area and P is the cross-section perimeter).

The local heat transfer coefficient from the CFD results is expressed as follows:

$$h_x = \frac{q}{T_{w,x} - T_{b,x}} \quad (11)$$

where q is the local heat flux.

$T_{w,x}$ is the local wall temperatures.

$T_{b,x}$ is the bulk fluid temperatures.

The local Nusselt number can be computed based on a local heat transfer coefficient, which is obtained directly from the Fluent panel, as follows:

$$Nu_x = \frac{h_x D_h}{\lambda} \quad (12)$$

A user-defined function (UDF) hooked in the Fluent panel was used to calculate the average Nusselt number as [4]

$$Nu = \frac{1}{S}\int Nu_x ds \quad (13)$$

The thermal efficiency of the collector is calculated by the equation

$$\eta_{th} = \frac{\dot{m}C_p(T_{out} - T_{in})}{I_T A_{abs}} \quad (14)$$

2.5. Validation

The simulation's validation was carried out by comparing its results with experimental data [4] for different flow Re. This was performed, in particular, through a three-dimensional analysis of the airflow in a flat-plate solar collector, as shown in Figure 4. This comparison revealed a remarkable agreement between the simulated results and the experimental observations, which confirms the validity of the simulation. Also, the maximum deviation is found to be less than 10% with that of the experimental data. Hence, for the present study, the Reynolds number varies from 2300 to 8500 and a solar heat flux value of 1000 W/m^2 was maintained for the entire study.

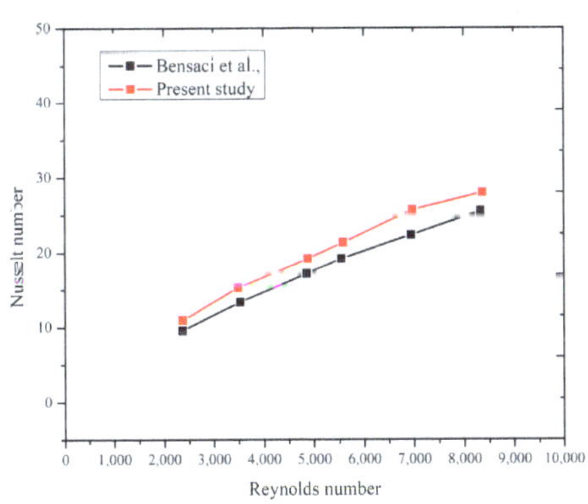

Figure 4. Validation of numerically predicted Nu with experimental data of [4].

3. The Boundary Conditions

The present numerical investigation was carried out using the following boundary conditions. The solar air collector inlet was maintained as a mass flow inlet ranging between 0.021807 kg/s and 0.080591 kg/s, which corresponds to an Re of 2300 to 8500, respectively. The collector outlet was maintained as atmospheric, which is equivalent to 101,325 Pa. The side walls, isolator, and baffles were maintained as adiabatic walls with no-slip conditions. A solar heat flux value of nearly 1000 W/m^2 was fixed to the absorber wall with a no-slip condition, as shown in Table 2. The above heat flux value was obtained from the literature [4].

Table 2. Boundary conditions opted for in the present study.

S.M	Boundary Name	Boundary Type	Value
1	Inlet	Mass flow	0.021807, to 0.080591 kg/s
2	Outlet	Pressure	0 Pa
3	Absorber	Wall heat flux	1000 W/m^2
4	Isolator	Adiabatic wall	Wall no slip
5	Baffles	Adiabatic wall	Wall no slip
6	Side wall	Adiabatic wall	Wall no slip

The thermophysical properties of the air and the absorber plate and baffles are presented in Table 3, assuming that they remain constant at the mean air temperature [4].

Table 3. Thermophysical properties of the air and the absorber plate and baffles: input data for CFD analysis.

Properties	Fluid (Air)	Absorber Plate and Baffles (Aluminum)
Density (kg/m^3)	1.167	2719
Viscosity (kg/ms)	1.85×10^{-5}	
Thermal conductivity (W/mk)	0.0262	202.4
Specific heat (J/kgk)	1006	871

4. Results and Discuss

4.1. Smooth Case

In this section, solar air collectors with smooth surfaces have been investigated numerically for varying Reynolds numbers ranging between 2300 and 8500. The results from the study indicated that by increasing Re from 2300 to 8500, the pressure drop increased from 0.45 Pa to 5.5 Pa, respectively. Also, the Nu was increased from 13.92 to 46.17 with an increase in Re. It is clear that an increase in Re promotes better heat transfer, as shown in Figure 5, and it also increases the pressure drop, caused by the increase in frictional resistance. The same phenomenon was observed in Figure 6, which clearly shows the pressure drop as a function of the Re. The results clearly depicted that the maximum value of the pressure drop, namely 5.5 Pa, which corresponds to the maximum value of Re (8500).

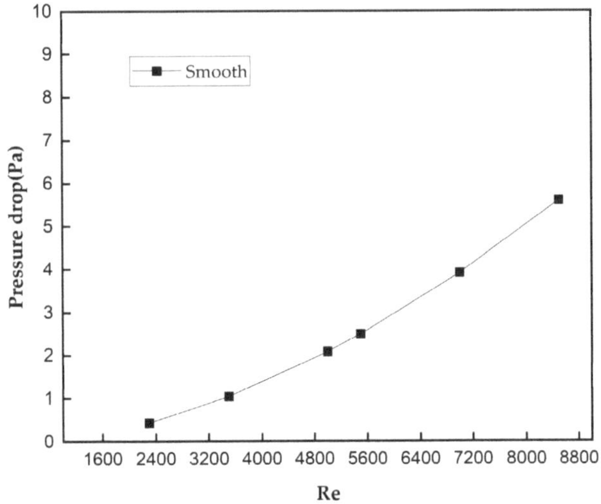

Figure 5. Variation in Re vs. pressure drop for a smooth surface solar collector.

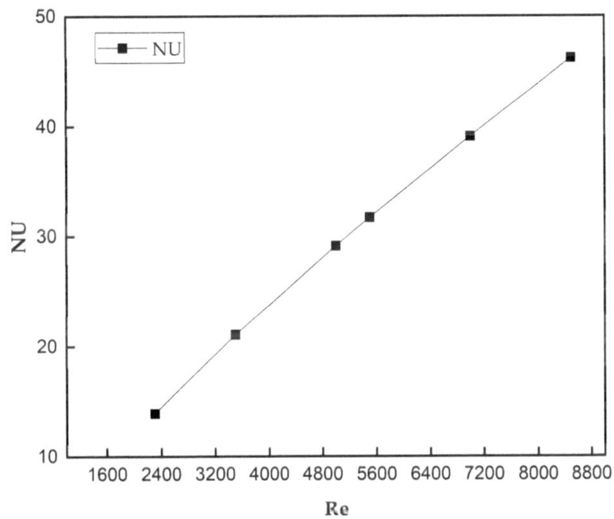

Figure 6. Variation in Re vs. Nu for a smooth surface solar collector.

Figures 7 and 8 indicate the temperature contour and pressure contour of a smooth solar air collector at the mid plane for a Re 8500. When the temperature of the air at the inlet is maintained at 300.15 K (ambient temperature), as the air enters the solar air collector, it absorbs heat from the absorber plate where the heat flux is maintained as 1000 W/m^2. The temperature rises gradually along the flow path and reaches a maximum temperature near the collector outlet. In this smooth surface, the air collector reaches about a maximum temperature of 325 K at the collector outlet for a Re 8500, as shown in Figure 7. Also, the pressure drop has increased significantly to a maximum value of 5.5 Pa from 0.45 Pa when the Re increased from 2300 to 8500, respectively, as shown in Figure 8.

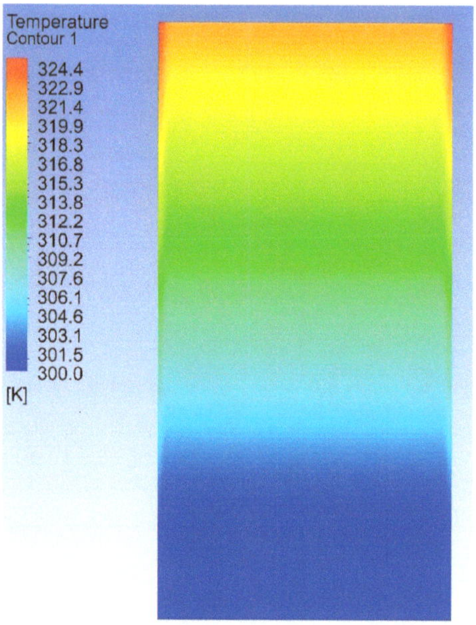

Figure 7. Temperature contour at the mid-plane for Re 8500.

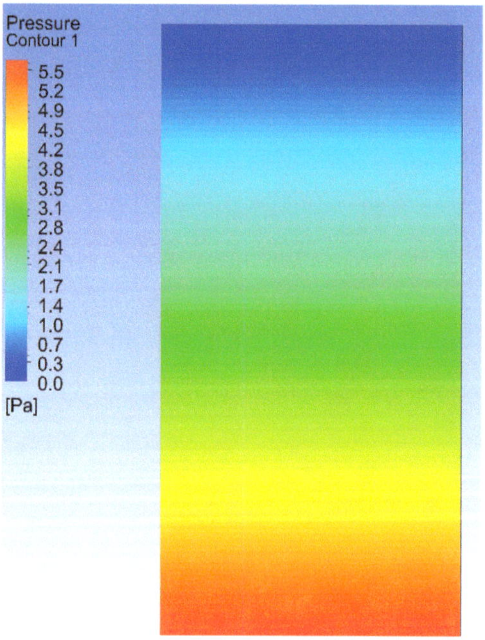

Figure 8. Pressure contour at the mid-plane for Re 8500.

4.2. Comparison of Longitudinal and Transverse Baffles with and without Perforation

In the second phase of the study, longitudinal and transverse baffles with and without perforation were investigated numerically. The effect of the thickness of baffles and hole diameter has also been studied extensively. The thickness variations considered for the study are 1.0 mm, 1.5 mm, and 2.0 mm and the maximum number of holes in the baffle is 2, with varying diameters of 10 mm and 12.5 mm and 15 mm. The longitudinal and transverse baffles combined together are called LT, with a single hole added to it, it is called LTH 1, and with two holes added to the LT, it is called LTH 2. In total, there are three different models, namely LT, LTH1, and LTH2, in combination with three different thicknesses of the baffles, namely, 1.0 mm, 1.5 mm, and 2.0 mm, and perforation with diameters of 10 mm and 15 mm. All the above-mentioned cases were simulated for various inlet Reynolds numbers of 2300, 3500, 5000, 5500, 7000, and 8500. Thus, a total number of 105 numerical simulations were carried out in this section.

Assessment of the Nusselt Number and Pressure Drop

In the case of a solar air collector, the introduction of baffles generally (a) increases turbulence, (b) results in more contact time with the absorber plate, and (c) increases the flow path; these factors would significantly improve the thermal performance of a solar air collector. On the contrary, the same baffles would increase the resistance to fluid flow, which would result in an increase in pressure drop to a greater extent. A tradeoff must be made between the baffle configuration and pressure drop in order to obtain the maximum performance from the solar air collector. The introduction of perforated baffles provided a promising solution for the pressure drop and heat transfer characteristics. In this study, initially, longitudinal and transverse baffles with varying thicknesses were investigated and the results from the study clearly indicated that compared to the smooth alternatives, LT baffles provided a maximum Nu of 63 from 46.17 whereas the pressure drop increased from 5.5 Pa to 496 Pa for a thickness of 2 mm and Inlet Re of 8500. Though the Nu has increased moderately, the pressure drop has swept over 496 Pa, as shown in Figure 9. This larger pressure drop would greatly affect the overall performance of a solar air collector. Furthermore, the study has been extended to reduce the thickness of the baffles from 2 mm to 1.5 mm and 1.0 mm, as shown in Figures 9–11.

The results predicted that Nu was increased to 64.34 from 63.25 when the thickness of baffles was reduced from 2 mm to 1 mm, as shown in Figures 12–14. Similarly, the pressure drop also decreased from 496 Pa to 486 Pa when baffle thickness was reduced from 2 mm to 1 mm, respectively, as shown in Figures 9 and 11.

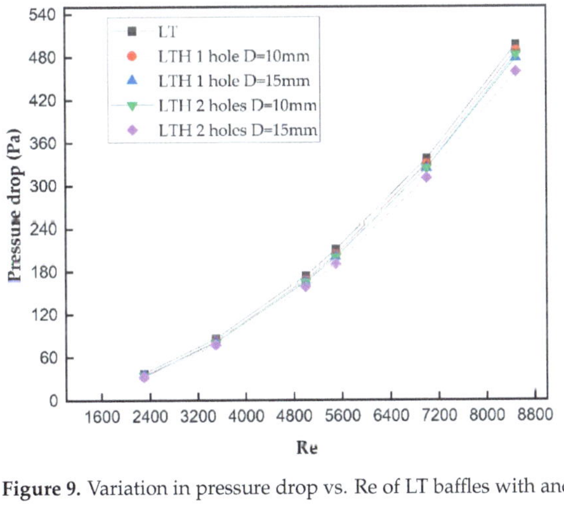

Figure 9. Variation in pressure drop vs. Re of LT baffles with and without perforation for t = 2.0 mm.

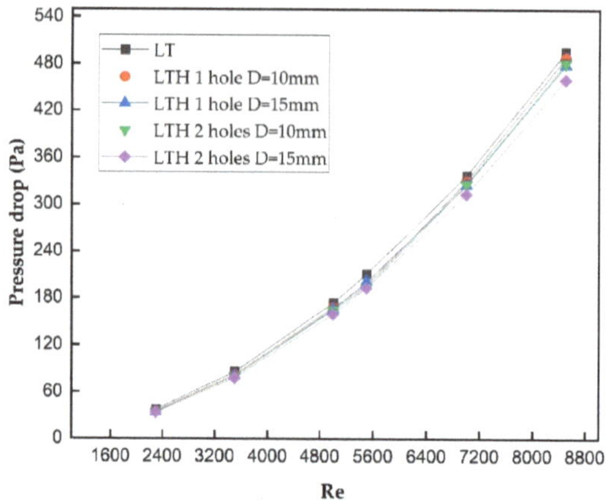

Figure 10. Variation in pressure drop vs. Re of LT baffles with and without perforation for t = 1.5 mm.

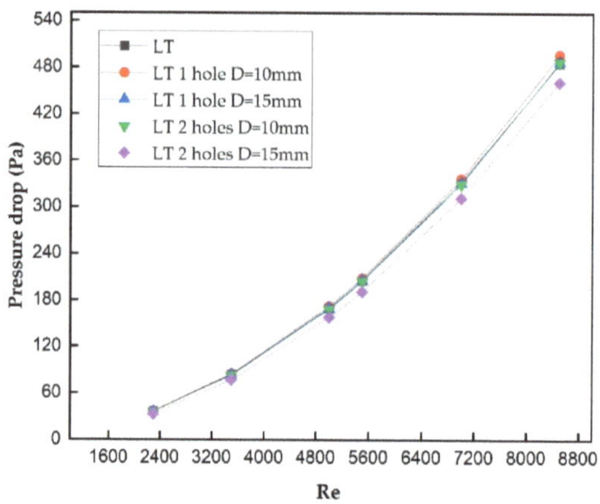

Figure 11. Variation in pressure drop vs. Re of LT baffles with and without perforation for t = 1.0 mm.

The introduction of perforated baffles with 1 hole and 2 holes with diameters of 10 mm and 15 mm provided a significant improvement in heat transfer characteristics and optimal pressure drop. When the transverse baffle was perforated with 1 hole of diameter 10 mm, this provided a small increase in Nu of 74 from 63, whereas the pressure drop was reduced to 489 Pa from 496 Pa. The comparison of LT and LTH 1 (t = 2 mm) with a diameter of 10 mm, as shown in Figure 14, has not provided a significant impact on the overall performance; a notable change was observed by just perforating the baffles. Furthermore, the diameter of the hole has been increased to 15 mm and the results estimated no significant improvement in Nu and the pressure drop. Also, the thickness of the baffles was reduced from 2 mm to 1 mm and the same trend was observed for all inlet Re, as shown in Figures 12–14.

Similarly, when the number of holes was increased to LTH 2 (t = 2 mm), the Nu increased to 76.71, as shown in Figure 14, and the pressure drop was reduced to 482 Pa, as shown in Figure 10. The thickness of the baffles was reduced from 2 mm to 1.5 mm and 1.0 mm, the Nu values were observed to be 77.29, 79.28, and 79.54, and the pressure drop values were observed to be 459.2, 460, and 460.1 Pa, respectively, for an Inlet Re of 8500. The results substantiated that the thickness of baffles does not provided any significant improvement to the pressure drop or the heat transfer characteristics. However, reducing the baffle thickness results in lower material costs, which makes it a cost-effective option without compromising the overall performance.

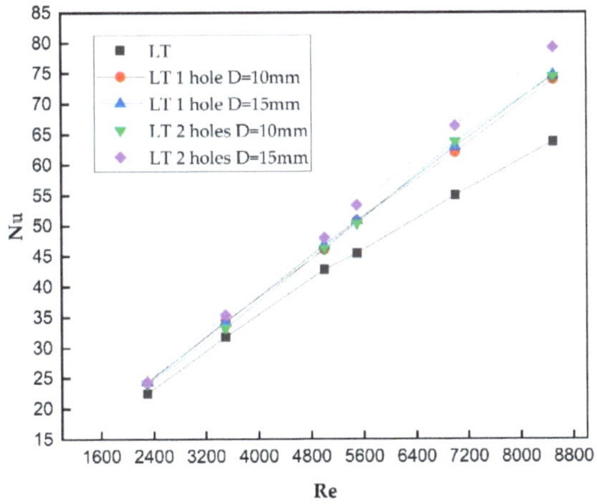

Figure 12. Variation in Nu vs. Re of LT baffles with and without perforation for t = 1.0 mm.

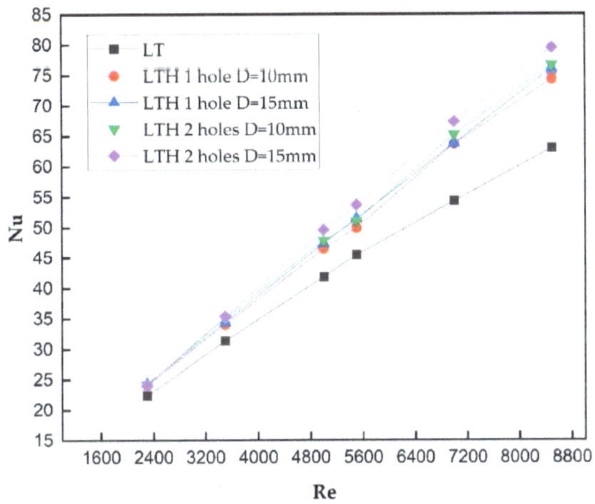

Figure 13. Variation in Nu vs. Re of LT baffles with and without perforation for t = 1.5 mm.

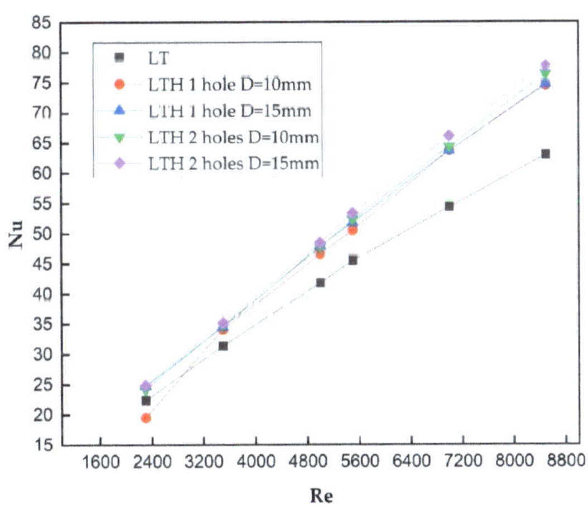

Figure 14. Variation in Nu vs. Re of LT baffles with and without perforation for t = 2.0 mm.

These perforated baffles allow the air to flow through the holes, as well as around the baffles, as shown in Figure 15 which reduces the airflow resistance and decreases the pressure drop. The velocity contour in the main flow direction x,y is shown in Figure 16; the velocity is highest in front of the baffles because of less resistance. Figure 17 indicates the temperature contour. As the flow moves, the temperature gradually rises, peaking at a maximum near the collector outlet.

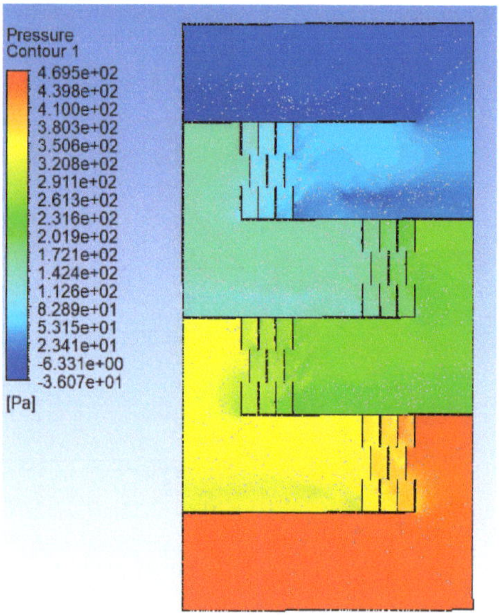

Figure 15. Pressure contour of LTH2 with a diameter of 15 mm for a Re 8500.

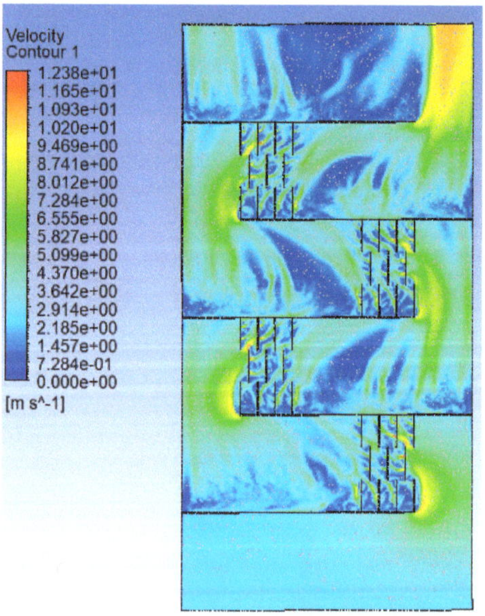

Figure 16. Velocity contour of LTH2 with a diameter of 15 mm for a Re 8500.

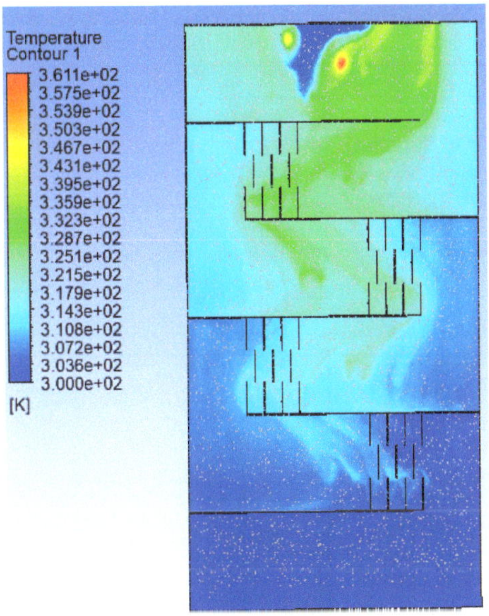

Figure 17. Temperature contour of LTH2 with a diameter of 15 mm for a Re 8500.

5. Conclusions

As part of this study, a detailed numerical simulation of a flat solar air collector with smooth, longitudinal, and transverse baffles was carried out. Baffles, both perforated and unperforated, were tested for different thicknesses and for all inlet Reynolds numbers using a detailed numerical simulation. The outcomes of the study have been presented as follows:

- The study revealed that utilizing longitudinal and transverse baffles led to improved heat transfer characteristics compared to a smooth surface. This improvement was manifested by an increase in the Nusselt number. However, it is important to note that a significant increase in pressure drop accompanied this enhanced performance. The influence of baffle thickness on the improvement in Nusselt number or the increase in pressure drop was found to be negligible;
- On comparison of baffles with and without perforation, results clearly predicted that perforated baffles with LTH 2 holes with a diameter of 15 mm provided better heat transfer characteristics and a better estimated maximum Nu and optimal pressure drop than LTH 1 holes of diameters 10 mm and 15 mm and LT without perforation for all Re, respectively. Also, variation in the thickness of baffles has no significant improvement in terms of the Nu and pressure drop of all the models considered. Even though thickness has no impact on the thermal performance of solar air collectors, it will be a deciding factor in terms of the selection of thickness of baffles.

Among the models studied, the perforated deflectors with a thickness of 1 mm and 2 holes with a diameter of 15 mm were awarded the highest Nusselt number (Nu) in terms of the optimum pressure drop.

Author Contributions: Conceptualization, G.B., B.R. and K.A.; methodology, M.J.G. and D.R.; software, G.B. and D.R.; formal analysis, G.B. and D.R.; investigation, G.B., D.R. and M.J.G.; data curation, G.B., D.R. and A.G.; writing—original draft preparation, G.B.; writing—review and editing, D.R., G.B. and M.J.G.; visualization, G.B., B.R. and K.A.; supervision, A.G.; project administration, A.G. funding acquisition, A.G. All authors have read and agreed to the published version of the manuscript.

Funding: This research received no external funding.

Data Availability Statement: The original contributions presented in the study are included in the article, further inquiries can be directed to the corresponding author.

Conflicts of Interest: The authors declare no conflicts of interest.

Nomenclature

C_p	Specific heat capacity ($J \cdot kg^{-1} \cdot K^{-1}$)
H	Heat transfer coefficient ($W \cdot m^{-2} \cdot K^{-1}$)
K	Turbulence kinetic energy ($m^2 \cdot s^{-2}$)
Nu	Nusselt number
Q	Absorbed heat flux ($W \cdot m^{-2}$)
Re	Reynolds number
T	Temperature (K)
V	Velocity (m/s)
P	Pressure (Pa)
$\dot{m}$	Air mass flow rate per meter (kg/m·s)
μ_t	The turbulent viscosity.
A_{abs}	Absorber plate surface area (m^2)
ρ	Fluid density ($kg \cdot m^{-3}$)

Abbreviations

SAH	Solar air heater.
SAC	Solar air collector.
RHMA	Reference hot-mix asphalt.
C-HMA	Conductive hot-mix asphalt.
CSCs	Concentrated solar collectors.
FPSACs	Flat-plate solar air collectors.
TSAC	Triangular solar air collector.
PCM	Phase change material.
DSAC	Double flow solar air collector.
NFAD	Nano-enhanced flexible aluminum air duct.
FESAC	Front entrance solar air collector.
SESAC	Side entrance solar air collector.
CPC	Compound parabolic concentrator.
WCSAC	Wavy corrugated solar air collector.
IB-ETC-SAH	Inserting a baffle inside the ETC-SAH.
PPI	Pores per inch.

References

1. Labed, A.; Moummi, N.; Aouès, K.; Zellouf, M.; Moummi, A. Etude Théorique et Expérimentale Des Performances d'un Capteur Solaire Plan à Air Muni d'une Nouvelle Forme de Rugosité Artificielle. *J. Renew. Energ.* **2009**, *12*, 551. [CrossRef]
2. Menasria, F.; Moummi, E.; Moummi, N.; Zedayria, M.; Guestal, M. Modélisation Des Échanges Convectifs Dans Le Conduit Utile d'un Capteur Solaire Plan à Air Muni de Rugosités Artificielles de Formes Rectangulaires. *J. Renew. Energ.* **2023**, *14*, 369–379. [CrossRef]
3. Fakoor Pakdaman, M.; Lashkari, A.; Basirat Tabrizi, H.; Hosseini, R. Performance Evaluation of a Natural-Convection Solar Air-Heater with a Rectangular-Finned Absorber Plate. *Energy Convers. Manag.* **2011**, *52*, 1215–1225. [CrossRef]
4. Bensaci, C.E.; Moummi, A.; Sanchez de la Flor, F.J.; Rodriguez Jara, E.A.; Rincon-Casado, A.; Ruiz-Pardo, A. Numerical and Experimental Study of the Heat Transfer and Hydraulic Performance of Solar Air Heaters with Different Baffle Positions. *Renew. Energy* **2020**, *155*, 1231–1244. [CrossRef]
5. Sharma, S.; Kumar Das, R.; Kulkarni, K. Sustainability Analysis of Solar Air Heater Roughned with Baffles Based on Exergy Efficiency. *Mater. Today Proc.* **2022**, *69*, 75–81. [CrossRef]
6. Rani, P.; Tripathy, P.P. Experimental Investigation on Heat Transfer Performance of Solar Collector with Baffles and Semicircular Loops Fins under Varied Air Mass Flow Rates. *Int. J. Therm. Sci.* **2022**, *178*, 107597. [CrossRef]
7. Mohammadi, K.; Sabzpooshani, M. Appraising the Performance of a Baffled Solar Air Heater with External Recycle. *Energy Convers. Manag.* **2014**, *88*, 239–250. [CrossRef]
8. Rani, P.; Tripathy, P.P. Thermal Characteristics of a Flat Plate Solar Collector: Influence of Air Mass Flow Rate and Correlation Analysis among Process Parameters. *Sol. Energy* **2020**, *211*, 464–477. [CrossRef]

9. Biswas, R.; Tripathy, P.P. Finite Element Based Computational Analysis to Study the Effects of Baffle and Fin on the Performance Assessment of Solar Collector. *Therm. Sci. Eng. Prog.* **2024**, *49*, 102431. [CrossRef]
10. Promvonge, P.; Skullong, S. Thermal Characteristics in Solar Air Duct with V-Shaped Flapped-Baffles and Chamfered-Grooves. *Int. J. Heat Mass Transf.* **2021**, *172*, 121220. [CrossRef]
11. Sai Kandukuri, K.; Kishor Sharma, P.; Kumar Arun, R. A Comparative Assessment of Distributive Mode Active Solar Dryers: Flat Plate Collector vs Evacuated Tube Collector with Thermal Energy Storage and Perforated Baffled Trays. *Sol. Energy* **2024**, *271*, 112421. [CrossRef]
12. Hu, J.; Liu, K.; Ma, L.; Sun, X. Parameter Optimization of Solar Air Collectors with Holes on Baffle and Analysis of Flow and Heat Transfer Characteristics. *Sol. Energy* **2018**, *174*, 878–887. [CrossRef]
13. Khanlari, A.; Aytaç, İ.; Tuncer, A.D.; Variyenli, H.İ.; Şahin, H.N. Improving the Performance of a PCM Integrated Solar Air Collector by Adding Porous Fins over the Bottom Side of the Absorber: A Transient CFD Study. *J. Energy Storage* **2024**, *90*, 111847. [CrossRef]
14. Abbas, F.A.; Alhamdo, M.H. Experimental and Numerical Analysis of an Asphalt Solar Collector with a Conductive Asphalt Mixture. *Energy Rep.* **2024**, *11*, 327–341. [CrossRef]
15. Zhang, J.; Zhu, T. Systematic Review of Solar Air Collector Technologies: Performance Evaluation, Structure Design and Application Analysis. *Sustain. Energy Technol. Assess.* **2022**, *54*, 102885. [CrossRef]
16. Jiang, Y.; Zhang, H.; Zhao, R.; Liu, Z.; Wang, Y.; You, S.; Wu, Z.; Wei, S. Thermal and Optical Performance Analysis of Triangular Solar Air Collectors and Regional Applicability in China. *Sol. Energy* **2023**, *249*, 288–300. [CrossRef]
17. Rajaseenivasan, T.; Srinivasan, S.; Srithar, K. Comprehensive Study on Solar Air Heater with Circular and V-Type Turbulators Attached on Absorber Plate. *Energy* **2015**, *88*, 863–873. [CrossRef]
18. Patel, S.S.; Lanjewar, A. Experimental Analysis for Augmentation of Heat Transfer in Multiple Discrete V-Patterns Combined with Staggered Ribs Solar Air Heater. *Renew. Energy Focus* **2018**, *25*, 31–39. [CrossRef]
19. Öztürk, M.; Çiftçi, E. Upgrading the Performance of a Solar Air Collector with Flexible Aluminum Air Ducts and Graphene Nanoplatelet-Enhanced Absorber Coating. *Therm. Sci. Eng. Prog.* **2023**, *40*, 101760. [CrossRef]
20. Nowzaria, R.; Aldabbagh, L.B.Y.; Mirzaei, N. Experimental Study on Double Pass Solar Air Heater with Mesh Layers as Absorber Plate. *Int. J. Electron. Mech. Mechatron. Eng.* **2011**, *3*, 673–682.
21. Bhargava, A.K.; Garg, H.P.; Sharma, V.K. Evaluation of the Performance of Air Heaters of Conventional Designs. *Sol. Energy* **1982**, *29*, 523–533. [CrossRef]
22. Wang, D.; Zhang, R.; Liu, Y.; Zhang, X.; Fan, J. Optimization of the Flow Resistance Characteristics of the Direct Return Flat Plate Solar Collector Field. *Sol. Energy* **2021**, *215*, 388–402. [CrossRef]
23. Elwekeel, F.N.; Nasr, A.A.E.; Radwan, M.I.; Aly, W.I. Influence of Impingement Jet Designs on Solar Air Collector Performance. *Renew. Energy* **2024**, *221*, 119757. [CrossRef]
24. Tuncer, A.D.; Amini, A.; Khanlari, A. Experimental and Transient CFD Analysis of Parallel-Flow Solar Air Collectors with Paraffin-Filled Recyclable Aluminum Cans as Latent Heat Energy Storage Unit. *J. Energy Storage* **2023**, *70*, 108009. [CrossRef]
25. Machi, M.H.; Farkas, I.; Buzas, J. Enhancing Solar Air Collector Performance through Optimized Entrance Flue Design: A Comparative Study. *Int. J. Thermofluids* **2024**, *21*, 100561. [CrossRef]
26. Yousif, M.N.; Alomar, O.R.; Saleem, A.M. Performance of Compound Parabolic Concentrator Solar Air Flat Plate Collector Using Phase Change Material. *Appl. Therm. Eng.* **2024**, *240*, 122224. [CrossRef]
27. Yang, L.; Zhang, N.; Yuan, Y.; Cao, X.; Qian, B.; Zeng, C. A New Operating Model for Improving Thermal Efficiency of Double-Glazed Solar Air-Phase Change Material Collector: An Experimental Study. *J. Energy Storage* **2023**, *58*, 106448. [CrossRef]
28. Zayed, M.E.; Kabeel, A.E.; Shboul, B.; Ashraf, W.M.; Ghazy, M.; Irshad, K.; Rehman, S.; Zayed, A.A.A. Performance Augmentation and Machine Learning-Based Modeling of Wavy Corrugated Solar Air Collector Embedded with Thermal Energy Storage: Support Vector Machine Combined with Monte Carlo Simulation. *J. Energy Storage* **2023**, *74*, 109533. [CrossRef]
29. Rawat, P.; Ashwni; Sherwani, A.F. Optimization of Single and Double Pass Solar Air Heater Phase Change Material (SAH-PCM) System Based on Thickness to Length Ratio. *Int. J. Heat Mass Transf.* **2024**, *224*, 125356. [CrossRef]
30. Fadhil, A.M.; Jalil, J.M.; Bilal, G.A. Experimental and Numerical Investigation of Solar Air Collector with Phase Change Material in Column Obstruction. *J. Energy Storage* **2024**, *79*, 110066. [CrossRef]
31. Bo, R.; Fu, W.; Meng, X.; Zhang, G. Employing a Fan to Improve the Wintery Heating Performance of Solar Air Collectors with Phase-Change Material: An Experimental Comparison. *Case Stud. Therm. Eng.* **2023**, *50*, 103425. [CrossRef]
32. Kumar Ahirwar, B.; Kumar, A. Review on Different Techniques Used to Enhance the Thermal Performance of Solar Air Heater. *Int. J. Heat Mass Transf.* **2024**, *220*, 124979. [CrossRef]
33. Abrofarakh, M.; Moghadam, H. Investigation of Thermal Performance and Entropy Generation Rate of Evacuated Tube Collector Solar Air Heater with Inserted Baffles and Metal Foam: A CFD Approach. *Renew. Energy* **2024**, *223*, 120022. [CrossRef]

energies

MDPI

Review

New Zero-Carbon Wooden Building Concepts: A Review of Selected Criteria

Agnieszka Starzyk [1], Kinga Rybak-Niedziółka [1], Aleksandra Nowysz [1], Janusz Marchwiński [2], Alicja Kozarzewska [1], Joanna Koszewska [1], Anna Piętocha [1], Polina Vietrova [1], Przemysław Łacek [1], Mikołaj Donderewicz [1], Karol Langie [1], Katarzyna Walasek [1], Karol Zawada [1], Ivanna Voronkova [1], Barbara Francke [1] and Anna Podlasek [1,*]

[1] Institute of Civil Engineering, Warsaw University Life Sciences, Nowoursynowska 166, 02-776 Warsaw, Poland; agnieszka_starzyk@sggw.edu.pl (A.S.); kinga_rybak@sggw.edu.pl (K.R.-N.); aleksandra_nowysz@sggw.edu.pl (A.N.); alicja_kozarzewska@sggw.edu.pl (A.K.); joanna_koszewska@sggw.edu.pl (J.K.); anna_pietocha@sggw.edu.pl (A.P.); przemyslaw_lacek@sggw.edu.pl (P.Ł.); mikolaj_donderewicz@sggw.edu.pl (M.D.); katarzyna_walasek@sggw.edu.pl (K.W.); karol_zawada@sggw.edu.pl (K.Z.); ivanna_voronkova@sggw.edu.pl (I.V.); barbara_francke@sggw.edu.pl (B.F.)

[2] Faculty of Architecture, University of Ecology and Management in Warsaw, Olszewska 12, 00-792 Warsaw, Poland; j.marchwinski@wseiz.pl

* Correspondence: anna_podlasek@sggw.edu.pl

Abstract: A Carbon Footprint (CF) is defined as the total emissions of greenhouse gases, primarily carbon dioxide, methane, and nitrous oxide, and is a specific type of Environmental Footprint that measures human impact on the environment. Carbon dioxide emissions are a major contributor to anthropogenic greenhouse gases driving climate change. Wood, as a renewable and ecological material, has relatively low carbon emissions. The study aimed to review and analyze the criteria influencing the feasibility of constructing modern zero-carbon wooden buildings. The review was conducted in two phases: (i) a literature review and (ii) an assessment of existing buildings. The preliminary research led to (i) narrowing the focus to the years 2020–2024 and (ii) identifying key criteria for analysis: sustainable material sourcing, carbon sequestration, energy efficiency, life cycle assessment (LCA), and innovative construction practices. The study's findings indicate that all these criteria play a vital role in the design and construction of new zero-carbon wooden buildings. They highlight the significant potential of wood as a renewable material in achieving zero-carbon buildings (ZCBs), positioning it as a compelling alternative to traditional construction materials. However, the research also underscores that despite wood's numerous potential benefits, its implementation in ZCBs faces several challenges, including social, regulatory, and financial barriers.

Keywords: carbon footprint; wooden construction; zero-carbon buildings (ZCBs); sustainable material sourcing; carbon sequestration; energy efficiency; life cycle assessment; innovative construction practices

Citation: Starzyk, A.; Rybak-Niedziółka, K.; Nowysz, A.; Marchwiński, J.; Kozarzewska, A.; Koszewska, J.; Piętocha, A.; Vietrova, P.; Łacek, P.; Donderewicz, M.; et al. New Zero-Carbon Wooden Building Concepts: A Review of Selected Criteria. *Energies* **2024**, *17*, 4502. https://doi.org/10.3390/en17174502

Academic Editors: Zsuzsa Szalay and Bruno Peuportier

Received: 21 August 2024
Revised: 5 September 2024
Accepted: 6 September 2024
Published: 8 September 2024

1. Introduction

The current climate crisis is the result of human activities over the past two centuries. The Environmental Footprint serves as a tool to measure the pressure humanity places on the environment, comparing the demand for renewable resources with nature's capacity to regenerate them. This concept encompasses all the negative impacts humans have on the natural world. Key quantitative indicators within this framework include the Ecological Footprint (EF), Carbon Footprint (CF), Water Footprint (WF), and Material Footprint (MF), which are central to understanding the extent of our environmental impact [1–3]. The Environmental Footprint also includes issues related to air, water, and soil pollution, as well as losses in biodiversity.

The CF originates from the concept of the EF, which dates back to the 1990s [4]. The CF measures the environmental impact of greenhouse gas emissions resulting from direct or indirect human activities, expressed in carbon dioxide equivalent (CO_2e). It includes emissions from various sources, such as production processes, transportation, energy consumption, food production, and waste generation [5–7]. Carbon dioxide emissions are the main component of anthropogenic greenhouse gases, responsible for over two-thirds of these emissions. They, along with methane and nitrous oxide, significantly contribute to climate change [8]. This is confirmed by the report of the Intergovernmental Panel on Climate Change (IPCC, United Nations body), which emphasizes the need to reduce global net anthropogenic CO_2 emissions by about 45% by 2030 compared to 1990 in order to limit global warming to 1.5 °C [9]. Regulation (EU) 2023/857 of the European Parliament and of the Council amending Regulation (EU) 2018/842 on binding annual greenhouse gas emission reductions by Member States from 2021 to 2030 resulting from the Paris Agreement and amending Regulation (EU) 2018/1999 sets out the obligations of Member States to achieve the EU's target of a 40% reduction in greenhouse gas emissions by 2030 compared to 2005. The guidelines in this Regulation are linked to the long-term objective of achieving climate neutrality in the EU by 2050 at the latest and a negative emissions balance thereafter.

In response to the climate crisis, the design and construction processes need to transition from conventional practices to lower-carbon approaches. The objective should be to maximize reductions in carbon dioxide emissions and work towards achieving zero-carbon buildings (ZCBs), including wooden structures. This approach effectively reduces the Environmental Footprint and contributes positively to climate change mitigation [10]. Zero-carbon buildings, including wooden structures, incorporate a life cycle assessment (LCA) approach to tackle CO_2 emissions. Wooden building materials exhibit lower emissions throughout their life cycle and offer substantial potential for carbon sequestration compared to other materials [11,12]. The LCA studies highlight the potential of wooden buildings to reduce anthropogenic carbon dioxide emissions, particularly by decreasing emissions during the construction phase [12–14]. The life cycle is defined as a series of interconnected stages of a product, beginning with the extraction or production of raw materials from natural resources and extending through to its final disposal [15]. When analyzing a product's life cycle, different scopes can be employed based on the assessment's goals: Cradle to Gate, Cradle to Grave, and Cradle to Cradle [16].

The construction sector, including both materials and operations, is responsible for over one-third of global CO_2 emissions. Life cycle assessment (LCA) analyses can significantly change this statistic. Reducing the CF of buildings involves developing materials that can sequester CO_2, such as engineered wood, which is a naturally occurring material with significant potential. Using wood in construction can act as a carbon sink, offering a way to offset CO_2 emissions. Choosing the right materials greatly impacts a building's CF, and wood provides a viable option for carbon compensation. Zero-carbon wooden buildings present numerous environmental benefits in the construction industry. Advances in technology are broadening the possibilities for designing sustainable, carbon negative wooden structures, which could significantly contribute to climate change mitigation [17–22].

New zero-carbon wooden building concepts play a crucial role in the sustainable development of urban areas. Wood, as a renewable material with minimal CF, is a strong contender for constructing sustainable buildings in urban environments, including high-rise structures [20,23–26]. Innovative construction and computational methods focus on reducing timber usage, minimizing transportation needs, and streamlining multiple stages of the building process, thereby lowering carbon dioxide emissions compared to traditional construction methods. Wooden buildings incorporating these advanced technologies are viewed as a highly competitive and sustainable alternative, providing a natural and efficient approach to design and construction [27]. Additionally, integrating energy-efficient systems and renewable energy sources can further reduce energy consumption and costs for a zero-energy, zero-carbon wooden building [28,29]. Moreover, zero-carbon buildings are intricate technical systems that are also shaped by social factors and influenced by local conditions [17,30–32].

A crucial aspect of modern design is using sustainable building certifications (Green Building Rating Systems—GBRS). These systems establish market standards, including those for zero-emission buildings. They aim to reduce greenhouse gas emissions, decrease energy consumption, and enhance the quality of life for occupants and residents [33]. There are numerous certification systems for sustainable building worldwide; the most commonly used are recognized methods, including Building Research Establishment Environmental Assessment Method (BREEAM), Leadership in Energy and Environmental Design (LEED), and Deutsche Gesellschaft für Nachhaltiges Bauen (DGNB) [34]. BREEAM acknowledges a range of low- and zero-carbon technologies (LZC), including solar energy, wind, geothermal and hydrothermal sources, hydropower, biomass, waste heat, and energy from waste incineration. A key requirement for the energy performance of new buildings is their reference emissions indicator. Within BREEAM certification, one assessed category is Ene 04 Low Carbon Design, which aims to minimize the building's energy demand. This involves reducing carbon dioxide emissions and maximizing the use of renewable energy sources. The evaluation criteria are divided into two main areas: passive design strategies and low- or zero-carbon technologies [35]. In LEED certification, references to CO_2 emissions are addressed across several key assessment areas: (i) Integrative Process, Planning, and Assessments, specifically under Carbon Assessment; (ii) Energy and Atmosphere (EA), focusing on Operational Carbon Projection and Decarbonization Plan; (iii) Materials and Resources (MR), which includes Assessing and Reducing Embodied Carbon [36]. The DGNB System is designed to support the creation of high-quality, sustainable buildings that meet environmental, economic, and social goals [37].

New concepts and trends in zero-emission construction, including European timber building, are closely aligned with European Union directives. These directives set the framework for future sustainable development, while regulations specific to timber construction are adopted nationally in each member state. The implementation of timber construction varies across EU countries due to differences in climate, history, building traditions, policies, and legal frameworks. The Directorate of the European Commission states that climate change is a reality and a serious problem for Europeans. According to a Europe-wide survey published in 2017, more than 9 out of 10 EU citizens (92%) consider climate change to be a serious problem [38]. One of the earliest directives setting the European Union's framework for nearly zero-energy buildings is Directive 2010/31/EU of the European Parliament and Council, dated 19 May 2010, on the energy performance of buildings (consolidated version). This directive leads member states towards the goal of achieving "nearly zero-energy buildings", which are characterized by very high energy performance. These buildings require minimal energy to operate, with most of the energy being sourced from renewable resources generated locally [39]. This directive was amended in 2012 with Directive 2012/27/EU on energy efficiency, followed by Directive 2018/844/EU in 2018. In 2024, the Directive (EPBD) (EU/2024/1275) was published, which introduced stricter requirements for the energy performance of new buildings [40,41]. The European Union has progressively tightened regulations on building energy performance to drive towards climate neutrality. In Article 9, the 2010 Directive stated that the Member States shall ensure that by 31 December 2020, all new buildings shall be nearly zero-energy buildings and that after 31 December 2018, all new buildings occupied and owned by public authorities shall be nearly zero-energy buildings. The 2024 Directive (EU/2024/1275) has increased the energy performance requirements for new buildings, mandating that all new buildings owned by public authorities must be zero-emission from 1 January 2028 and all other new buildings from 1 January 2030. These regulations aim to align building energy performance with the European Union's target of achieving climate neutrality by 2050 [41].

The ZCB technologies are now widely accessible, yet their successful implementation is hindered by a lack of industry knowledge and awareness. Effective deployment of these technologies requires enhancing stakeholders' understanding of carbon reduction strategies throughout a building's life cycle [10]. The slow adoption of zero-carbon practices, especially for wooden buildings, is attributed to various barriers, including social, regulatory, financial, and business model issues.

This study focuses on reviewing and analyzing the criteria that influence the feasibility of modern zero-carbon wooden buildings, highlighting the need to address these challenges to advance zero-emission construction.

2. Materials and Methods

The primary objective of this study is to analyze the benchmarks that influence the feasibility of constructing modern wooden buildings with zero carbon dioxide emissions. To achieve this goal, the paper is divided into two main stages: a literature review and building studies, as outlined below.

2.1. Stage 1—Literature Review

This stage involves analyzing the standards for constructing wooden buildings with zero CO_2 emissions. Two main criteria guided the selection of scientific papers for this analysis: (i) thematic scope, focus on zero CO_2 emissions in wooden buildings; (ii) temporal scope, including papers published between 2020 and 2024. The selected time range limits the review to the most recent scientific articles, ensuring that the analysis is focused on the latest developments and current trends in the field. This approach prioritizes up-to-date information, providing insights that reflect the most recent research findings and advancements.

The review of the selected papers, listed in Table 1, allowed us to analyze benchmarks determining the feasibility of constructing wooden buildings with zero CO_2 emissions. The benchmarks that have been taken into consideration are as follows:

1. Sustainable Material Sourcing:
    - ensuring that wood is sourced from sustainably managed forests that practice responsible harvesting;
    - local sourcing—minimizing transportation emissions.
2. Carbon Sequestration:
    - leveraging the natural carbon sequestration properties of wood, which stores carbon dioxide absorbed during the tree's growth;
    - innovative wood products that offer structural strength and carbon storage benefits.
3. Energy Efficiency:
    - implementing passive design strategies to reduce energy consumption;
    - integrating renewable energy sources;
    - using energy-efficient lighting, heating, ventilation, and air conditioning systems.
4. Life Cycle Assessment:
    - a comprehensive LCA to evaluate the building's CF from material extraction through to demolition or recycling;
    - designing for disassembly and recycling to ensure materials can be reused, reducing future CO_2 emissions.
5. Construction Practices:
    - utilizing low-carbon construction methods and technologies to minimize emissions during the building phase;
    - the review employed the following academic search engines: Scopus, Google Scholar, and ResearchGate, using the following keywords:
    - sustainable practices and material sourcing;
    - energy efficiency and renewable energy;
    - building life cycle assessment and environmental impact;
    - innovative modular and prefabricated wood constructions;

- carbon sequestration in wood architecture.

Table 1. List of papers according to selection criteria.

Example Ref.	Countries of Study	Thematic Scope				
		1	2	3	4	5
Abed et al. (2022) [42]	World	x		x		x
Ahmed et al. (2024) [43]	Kirkuk, Iraq			x		
Alvarez et al. (2023) [14]	World			x	x	
Andersen et al. (2021) [12]	World	x	x	x	x	
Andersen et al. (2024) [20]	World			x		x
Arlet (2021) [44]	Europe, Japan, Canada, New Zealand	x	x	x	x	x
Arumägi et al. (2020) [28]	World	x		x		x
Bai et al. (2018) [45]	World			x		x
Barclay et al. (2024) [46]	World					x
Besana et al. (2022) [18]	World			x	x	
Blay-Armah et al. (2023) [47]	World			x	x	x
Bøe et al. (2023) [48]	World					x
Bougiatioti et al. (2023) [49]	Greece					
Branchi et al. (2023) [1]	World			x		
Broda (2020) [50]	World			x		
Brogi et al. (2019) [51]	Europe	x		x	x	x
BuHamdan et al. (2021) [52]	World					x
Cabeza et al. (2021) [53]	World			x	x	
Calquin et al. (2024) [54]	World					x
Carletti et al. (2024) [24]	World			x	x	x
Chen et al. (2020) [55]	World			x	x	
Chen (2023) [56]	World	x		x		x
Ching et al. (2024) [6]	World			x		
Churkina et al. (2020) [57]	World				x	
Dai et al. (2023) [3]	World			x	x	
Defloor et al. (2022) [58]	Belgium	x		x		x
Devarajan et al. (2024) [59]	World				x	
Ding et al. (2022) [19]	World			x		x
Dong et al. (2021) [60]	China			x		x
Duan et al. (2022) [61]	World	x				
Dzhurko et al. (2024) [62]	Germany		x			
Elaouzy et al.(2022) [63]	World			x		
Elginoz et al. (2024) [64]	World	x				
El-Shorbagy (2020) [22]	World		x	x		x

Table 1. *Cont.*

Example Ref.	Countries of Study	Thematic Scope				
		1	2	3	4	5
European Innovation Agenda, European Commission, (2022) [65]	EU			x	x	x
Evans et al. (2022) [66]	World		x			
Feder (2023) [7]	World			x		
Fereidani et al. (2021) [67]	World			x		
Ferreira et al. (2023) [34]	World				x	
Furhana Shereen et al. (2023) [68]	World			x		
Garzon et al. (2020) [69]	Bulgaria, Turkey, North America	x				
Ghobadi et al. (2023) [70]	Australia		x			
Giridhar et al. (2022) [71]	World			x		
Grinham et al. (2021) [72]	World			x		x
Groll (2023) [8]	World			x		
Hamida et al. (2023) [73]	World			x		x
Hanifa et al. (2023) [74]	India		x			
He et al. (2024) [75]	World			x		x
Himes et al. (2020) [76]	USA		x			
Hoxha et al. (2020) [77]	World				x	
Hu (2023) [21]	World			x	x	x
Huang et al. (2024) [78]	China		x			
Hurmekoski et al. (2022) [79]	Finland		x			
Ibrahim et al. (2023) [80]	World			x		
Kazemian et al. (2023) [81]	World		x			
Keržič et al. (2021) [82]	World			x		
Király et al. (2022) [83]	Hungary		x			
Koval et al. (2023) [84]	World					x
Leszczyszyn et al. (2022) [85]	Europe, Chile		x	x		x
Li et al. (2022) [86]	Europe			x		
Lin et al. (2023) [13]	World			x	x	
Linkevičius et al. (2023) [87]	World			x		x
Lo et al. (2021) [88]	USA	x		x		x
Lou et al. (2024) [17]	World		x	x	x	
Lu et al. (2024) [89]	World			x		
Meleti et al. (2021) [9]	World			x		
Michalak et al. (2024) [90]	World					x
Michálková et al. (2022) [23]	World			x		x
Mirashk-Daghiyan et al. (2022) [91]	Tehran, Iran			x		
Motamedi et al. (2023) [92]	World					x

Table 1. *Cont.*

Example Ref.	Countries of Study	Thematic Scope				
		1	2	3	4	5
Mulya et al. (2024) [33]	World			x	x	x
Mushtaha et al. (2021) [93]	World			x		
Nidhin et al. (2023) [10]	World					x
Ouellet-Plamondon et al. (2023) [94]	Canada, Switzerland, Germany, Belgium, Australia, Sweden, Spain, Austria, Denmark, France, New Zealand, USA, Brazil, Norway		x			
Pasternack et al. (2022) [95]	World	x				
Pecio (2024) [96]	World					x
Pedreño-Rojas et al. (2024) [97]	World	x				
Phillips et al. (2020) [98]	USA			x		
Pilli et al. (2022) [99]	World	x		x		
Pomponi et al. (2020) [100]	World				x	
Porter et al. (2001) [101]	World	x		x		x
Premrov et al. (2023) [102]	World			x		x
Prieur-Richard et al. (2018) [103]	World			x		x
Prins et al. (2023) [104]	World	x				
Ridhosari et al. (2020) [5]	World			x		
Röck et al. (2020) [105]	World				x	
Salata et al. (2024) [106]	World			x		
Sandoli et al. (2021) [107]	World					x
Sasaki (2021) [108]	Thajland		x			
Schmidt et al. (2023) [109]	World					x
Schneider-Marin et al. (2020) [110]	World			x	x	
Schwarzschachner et al. (2024) [111]	Germany, Austria		x			
Scouse et al. (2020) [112]	USA	x				x
Sher et al. (2021) [113]	UK, Malaysia		x			
de Oliveira et al. (2023) [114]	World			x		x
Stanciu et al. (2024) [115]	World			x		
Starzyk et al. (2023) [25]	World			x		x
Tupenaite et al. (2023) [116]	Lithuania		x			
Udele et al. (2021) [117]	USA					x
Ürge-Vorsatz et al. (2020) [118]	USA, Canada, China, EU (Germany, Belgium ...)	x	x	x	x	x
Veichtlbauer et al. (2022) [119]	World			x		
Veloso et al. (2023) [120]	Belo Horizonte, Brazil			x		
Veillette et al. (2021) [121]	Quebec, Canada			x		
Wang et al. (2021) [51]	World					x
Wang et al. (2024) [122]	World			x		x
Warmling et al. (2022) [123]	World					x

Table 1. *Cont.*

Example Ref.	Countries of Study	Thematic Scope				
		1	2	3	4	5
Wilberforce et al. (2023) [124]	(EU), USA, China	x		x	x	x
Yang (2021) [125]	China	x			x	
Younis et al. (2022) [126]	World					x
Zawada et al. (2024) [127]	World					x
Zhan et al. (2023) [128]	World	x		x		x
Zhao et al. (2022) [129]	World	x		x		x
Zhao et al. (2015) [130]	World			x		x
Duan et al. (2022) [61]	China		x			

2.2. *Stage 2—Building Review*

This stage involves studying modern wooden buildings to evaluate whether the benchmarks identified in Stage 1 are applicable in practice. The selection of buildings for this analysis was based on (i) thematic scope: wooden buildings designed to achieve zero CO_2 emissions; (ii) temporal scope: buildings constructed between 2020 and 2024.

Out of a review of 98 near-zero and zero-emission buildings, 10 exemplary cases were chosen (Table 2). These examples include both residential and public utility buildings with various functions located in Europe and North America. The comparison criteria include the operational energy index, measured in $kgCO_2e/m^2/year$, and the embodied energy associated with material acquisition and construction, also measured in $kgCO_2e/m^2$.

Table 2. List of wooden buildings according to selection criteria.

Project (Authors)	City, Country (Date)	Thematic Scope				
		1	2	3	4	5
Lea Bridge Library Pavilion (Studio Weave) [131]	London, UK (2021)	x	x	x	x	x
Orueta Etxea (Emiliano López Mónica Rivera Arquitectos) [132]	Bilbao, Spain (2021)	x	x	x	x	x
Spruce House and studio (ao-ft) [133]	London, UK (2021)	x	x	x	x	-
Cooperative housing (Urban Power) [134]	Stavnsholt, Denamrk (2021)	x	x	x	-	-
The black and white building (Waugh Thistleton Architects) [135]	London, UK (2022)	x	x	x	x	
Workstack (dRMM) [136]	London, UK (2023)	x	x	x	x	-
EÑE House (Estudio Albar) [137]	Madrid, Spain (2023)	x	x	x	x	x
New Temple Complex (James Gorst Architects) [138]	Hampshire, UK (2023)	x	x	x	x	x
Durley Chine Environmental Hub (Footprint Architects) [139]	Bournemouth, UK (2024)	x	x	x	x	-
Humber Cultural Hub (Diamond Schmitt) [140]	Ontario, Canada (2024)	x	x	x	-	-

The analysis of the selected buildings, listed in Table 2, enabled an assessment of whether the criteria identified in the literature review align with real-world construction practices.

The results from both stages are then synthesized to discuss the connections between the literature review and the building studies concerning zero carbon dioxide emissions in modern wooden buildings.

3. Results

3.1. Stage 1: Literature Review

3.1.1. Sustainable Material Sourcing

Sustainable material sourcing is essential for minimizing the environmental impact of construction projects, especially when using wood. By sourcing wood from sustainably managed forests, we ensure responsible harvesting practices that maintain forest ecosystems and promote long-term ecological health. Certification schemes such as the Forest Stewardship Council (FSC) and the Programme for the Endorsement of Forest Certification (PEFC) offer credible assurances of these sustainable practices. Furthermore, prioritizing local sourcing helps reduce transportation emissions by reducing the distance that materials must travel, thereby decreasing the overall CF of construction activities. Collectively, these strategies foster a more sustainable approach to building, supporting both environmental protection and resource efficiency.

The embodied energy of building materials can represent between 10% and 30% of the total energy demand, depending on several factors. These include the building's anticipated lifespan and the energy required for heating or cooling, which is influenced by the local climate [61].

In net-zero carbon wooden buildings, the focus extends beyond just using wooden construction materials to incorporating various bio-based and wood-derived materials. These include structural components (such as beams, glued laminated panels like cross-laminated timber (CLT), glued laminated timber (Glulam), laminated veneer lumber (LVL), dowel-laminated timber (DLT), bamboo, and more), insulation materials (such as wood wool, cork, hemp, compressed straw, animal wool, and others), and finishing materials (including boards, plywood, OSB, MDF, HDF, LDF, woodblock flooring, wooden paving, thatch, etc.). Additionally, composite materials containing wood chips, sawdust, ash from burned wood, charcoal, and cut straw are also available, though they are beyond the scope of this discussion [97]. These materials are increasingly valued for their reduced CF, enhanced circularity, and growing societal acceptance of natural materials.

It is crucial to recognize that the CF of a building material includes stages A1 through A4 of the LCA as defined by EN 15978 [141] and EN 15804 [142] standards. These stages cover the extraction of raw materials, their transportation to the manufacturing facility, the processing of finished building components, and the final transportation to the construction site. Local sourcing of materials and minimizing transportation are key strategies for reducing a building's CF. Additionally, the type of materials and the logistics of producing mass timber elements play a significant role in minimizing this footprint. This is particularly important because transporting building materials to the site is often energy-intensive and typically involves diesel-powered trucks [143].

The origin of wooden and wood-based materials is critical due to the significant trade-offs inherent in forest management [104]. While using wood in buildings can lower their CF, it may impact other aspects of the environmental footprint of the construction project (ibid). Key factors include the "seven thematic elements of sustainable forest management" [144], which assess forest health and resources, such as forest existence (area, timber, and carbon stocks), forest condition (health and vitality), and the services provided (biodiversity support, timber and resource production, soil and water protection, and socio-economic benefits like livelihoods, employment, energy, recreation, and cultural value). It is essential to recognize the evolving definitions and indicators of sustainable forest management at various levels—from global to local forest management units—ranging from the continuous supply of wood to preserving biodiversity, ecosystems, and the full spectrum of ecosystem services [104].

Despite numerous international agreements aimed at promoting sustainable forest management, no unified system exists to guarantee the sustainable origin of timber. Various certification systems address this issue, with two of the most prominent global systems being the FSC (Forest Stewardship Council) and PEFC (Programme for the Endorsement of Forest Certification). These certifications cover both specific areas of sustainably managed forests and the timber products derived from them. By the end of Q1 2024, the PEFC certified 295 million hectares of forest across 42 countries and six continents, with approximately 62 million hectares also holding FSC certification [145]. Conversely, FSC certifications cover 160 million hectares in 2024, down from over 210 million hectares in 2021 [146].

A key difference between these systems is that the PEFC is tailored to national regulations, reflecting local environmental and socio-cultural conditions, with implementation and monitoring typically overseen by government agencies. In contrast, the FSC is managed by an independent organization. While the PEFC's approach varies by country, the FSC is often seen as more effective due to its rigorous standards, emphasis on procedural rights for non-commercial entities, Indigenous rights, and natural ecosystem protection [69]. A review covering 31 studies and 6 million hectares of certified forests (about 1.5% of the global area) found that the FSC has a more positive impact on ecosystem protection than the PEFC, including preventing deforestation in tropical regions [147].

The review also identified trade-offs between indicators for flora, fauna, and ecosystem services, such as dense forests being beneficial for flora and carbon sequestration versus forest clearings that enhance plant diversity. It also highlighted some limitations of the FSC system, including its focus on ecological connectivity, mainly for species with minimal habitat requirements (one hectare) and inconsistencies in national biodiversity protection indicators. Despite these environmental advantages recognized by forest managers, other certification systems, including the PEFC, are often selected for market access and other economic benefits [148].

3.1.2. Carbon Sequestration

Carbon sequestration in wooden construction reduces greenhouse gas emissions [66]. As a natural building material, wood effectively stores carbon throughout its lifespan [111], making wooden buildings valuable long-term carbon sinks crucial for combating climate change [81]. During photosynthesis, trees absorb atmospheric carbon dioxide, which is then stored as organic carbon within their complex tissue [62]. Utilizing wood in construction reduces emissions associated with producing energy-intensive materials like concrete and steel [74]. Furthermore, wood's renewability and carbon sequestration capabilities make it an environmentally friendly choice for sustainable building practices [116]. This contributes to lowering the global CF of the construction industry. The wood used in construction often comes from sustainably managed forests, which ensures a continuous cycle of carbon absorption and storage [108]. Compared to other building materials, wood has a lower CO_2 emission balance [119]. Additionally, wooden buildings not only store carbon but also offer improved energy efficiency due to their insulating properties [79]. The carbon sequestration in wood remains effective as long as the material is protected from decay and combustion [83]. Therefore, proper maintenance and protection of wooden structures are essential.

Innovative technologies like cross-laminated timber (CLT) and Glued Laminated Timber (Glulam) enable the use of wood in high-rise construction. CLT is an advanced engineered wood product composed of multiple layers of lumber arranged in alternating directions and bonded with adhesive. This construction method enhances its structural stability and load-bearing capacity, making CLT an ideal material for a range of applications, from residential buildings to high-rise towers. Glulam is another wood product that involves layering timber and bonding these layers with high-strength adhesives to form large, versatile structural components. Glulam provides remarkable strength and flexibility, allowing it to be used in diverse structural elements like beams, columns, arches, and bridges. One of CLT and Glulam's key environmental benefits is its ability to sequester carbon. By trapping the carbon captured by trees during their growth within the timber, CLT and Glulam effectively store this greenhouse gas throughout the lifetime of the building, helping to offset the carbon emissions typically associated with traditional construction materials such as concrete and steel [61].

The LCA demonstrates that wooden buildings have a lower environmental impact than those constructed from traditional materials [94]. Incorporating wood into construction projects supports sustainability goals and can stimulate local forestry economies, increasing demand for timber and expanding forested areas. Wood also positively impacts biodiversity and ecosystem stability [113]. Additionally, wood allows for creating energy-efficient structures and offers greater adaptability compared to concrete or steel buildings, extending their lifespan and enhancing carbon storage [76]. By contributing to "negative emissions", wood helps meet the targets of the Paris Agreement aimed at addressing climate change and limiting global warming [78].

Furthermore, promoting and educating the public about wooden construction can raise awareness of its carbon sequestration benefits. Supporting policies for wooden construction can accelerate the shift towards sustainable building practices and foster the development of healthier, more sustainable living spaces. Effective carbon sequestration in wood requires comprehensive management throughout its entire lifecycle, from harvesting and processing to recycling [70]. Continued research and development in this field are essential to further enhance the efficiency of carbon sequestration in wooden construction.

3.1.3. Energy Efficiency

Wood is a highly effective natural insulator, offering superior thermal performance compared to traditional building materials like concrete and steel. This insulating capability allows wooden structures to maintain stable indoor temperatures with less dependence on artificial heating and cooling systems, thereby reducing energy consumption. The cellular structure of wood naturally traps air, forming a barrier that minimizes heat transfer between a building's interior and exterior. This property is particularly beneficial in regions with extreme temperatures, where efficient climate control is essential. Additionally, the energy efficiency of wooden buildings can be further optimized through passive design strategies, such as careful sun orientation, natural ventilation, and the incorporation of shading elements [93]. These approaches work together with wood's inherent properties to significantly enhance a building's overall energy efficiency.

While wood is highly resistant to heat, it is susceptible to damage from direct sunlight and high humidity [71,82]. To maximize the durability of wooden structures, natural shading through trees or closely spaced neighboring buildings can be employed to protect the wood [63]. Additionally, natural-origin protective materials such as clay, natural waxes, vegetable seed oils, or natural oil paints offer an alternative to synthetic wood preservatives [50,115]. The choice of paint color should be adapted to the local climate; light colors are recommended in warm, dry climates to prevent overheating, while dark colors are preferable in moderate and cold climates to absorb solar energy and reduce heating costs. Furthermore, highly reflective roofs can significantly enhance energy efficiency [67]. It is, however, essential to balance energy efficiency measures with thoughtful design solutions [89,91]. Research by Salata et al. demonstrates that using genetic algorithms to identify optimal building designs for various European climate zones can effectively reduce energy consumption for heating and cooling [106].

When other structures or fences are positioned close to the sun-facing side of a building, ensuring adequate ventilation and allowing sufficient direct sunlight during daylight hours is crucial. This prevents excessive soil moisture accumulation, which could otherwise lead to premature decay or fungal damage, a subject that warrants further detailed research [49].

Building design should also take the prevailing climate into account, adjusting the percentage of glazing and window orientation accordingly. For instance, in temperate and subpolar climates, north-facing windows and small windows that provide standardized air conditioning are not recommended. In contrast, in subtropical and tropical climates, north-facing windows can improve indoor climate and humidity levels. Optimizing window orientation, U-value, and proportions can lead to energy savings of up to 25% [43]. Occupant behavior is another critical factor, with studies indicating that careful consideration can reduce a building's energy consumption by up to 40% or even 80% [121].

Households can partially or fully meet their energy needs through renewable energy sources [119]. However, it should be noted that the production of these sources is not entirely zero-emission, with exceptions for lightweight systems made from recycled materials, though these practices are not yet widespread [68].

A Net-Zero-Energy Building (NZEB) is defined as a structure that produces more energy than it consumes on average over a year. This is achieved by relying exclusively on renewable energy sources, such as solar, wind, and geothermal energy while avoiding fossil fuels in construction [150]. The energy independence and local renewable energy generation of such buildings make them reliable in the event of natural disasters or heightened threats. For example, during military conflicts, where large energy complexes are often targeted, a house with these features would serve as a dependable refuge. This is exemplified by the situation in Ukraine in 2024, where significant installations of individual heating systems and innovative methods of utilizing solar and wind energy are being developed [151].

One of the most effective ways to reduce electricity consumption for lighting is to maximize the use of natural sunlight and store it for later use. This principle is reflected in the architectural design by incorporating windows in all rooms, utilizing sunlight amplifiers (light lenses), and integrating secondary light sources, such as strategically placed window openings in interior walls or glazed doors in corridors and bathrooms. Lenses and fiber optic transmission of sunlight can reduce the energy needed for daytime lighting by 57% [80].

Thermal batteries, typically installed on the south-facing side of the roof, harness solar energy to heat water or air, transferring this heat to the building's interior and meeting heating requirements. The implementation of passive cooling systems has been shown to significantly reduce energy consumption while enhancing occupant comfort [86,98,120].

3.1.4. Life Cycle Assessment

Earlier efforts to mitigate the environmental impact of construction have concentrated on reducing primary energy consumption by enhancing building energy efficiency and incorporating renewable energy sources. These initiatives have culminated in the successful development of technologies for passive and nearly zero-energy buildings, delivering significant results [59,152,153]. Nevertheless, there is a growing focus on previously neglected stages of the construction process. This includes the extraction of raw materials, their processing and conversion into final products, transportation, the construction phase, ongoing maintenance, and the potential for demolition and reuse or disposal conditions [105,154]. In recent years, there has been a dramatic surge in research addressing greenhouse gas emissions within the framework of the LCA. The focus of these studies has largely been on two areas: theoretical exploration and the analysis of specific buildings. This growing body of research and the accompanying academic discourse have elevated the importance of these issues in legislative contexts, leading to a more effective dissemination of this knowledge within the professional community

Case studies have demonstrated that wooden buildings can significantly reduce carbon dioxide emissions compared to equivalent structures made from mineral materials. Studies indicate that the estimated total environmental impact of wooden buildings may be as much as about 50% of the value for buildings constructed using traditional technology (cement blocks and concrete), and this result is repeatable in both new and older studies [55,110,155,156]. This value decreases when buildings are low-energy, then embodied energy becomes important. Depending on the construction systems, different construction materials determine the share of embodied energy from 35% to 57%—minimum values accepted for wood; this result is also repeatable in both new and older studies [157,158]. It should be noted that the results of LCA studies for different materials may vary significantly due to differences in the methodologies used, assumptions made, building types, and technologies used [53,154].

The researchers' results are even divergent if the analyses are conducted under significantly different impact conditions. Therefore, it is important to clearly define the conditions of the analysis conducted and, in particular, the expected durability of the building. Wooden buildings, in comparison to traditional technology buildings, may have the disadvantage of having to import wood, while concrete production can be located near the construction site. Much also depends on the assumed service life of a wooden building, considering that concrete allows for long durability of the building. The geographical location and resources of the region have a very significant influence on the right choice of building materials. Despite the significant increase in the number of publications, the indication of the comparison of LCA for buildings made of different materials is still narrowly presented in the scientific literature, while at the same time, it is a broad issue enabling numerous new publications in this field.

Life cycle assessment (LCA) is a crucial tool for evaluating carbon dioxide emissions in construction, playing a key role in promoting sustainable design. It stands out by analyzing the entire life cycle of a building, encompassing (i) preparation for construction, (ii) the construction phase, (iii) operational use, and (iv) the end-of-life phase. This comprehensive approach ensures a thorough assessment of the environmental impact from the initial planning through to the building's eventual disposal [159,160].

The environmental impact of a building is fundamentally shaped by its design. At the outset of a project, architects are tasked with choosing building materials. This decision is particularly complex due to the extended lifespan of buildings and the ongoing need for maintenance and replacements, which are influenced by evolving consumption patterns and changing user needs [152]. Architects can make environmentally conscious decisions by accessing information about the CF of various products. While comprehensive data are not yet available for all products, the range of products with such information is steadily expanding [105]. Current information on CF does not yet account for carbon dioxide emissions over the entire lifespan of a product. To effectively optimize material choices, a comprehensive LCA is essential. It is crucial to note the complexity of this process—each project is typically unique and bespoke, necessitating a separate LCA for every individual building [152]. Since researchers primarily develop computational models, they often lack the practical adaptability needed for rapid and user-friendly applications [53].

It is crucial to recognize that utilizing wood and other natural materials, such as straw or hemp, plays a significant role in reducing CO_2 emissions. These materials contribute to carbon sequestration by capturing and storing carbon dioxide during their growth and maintaining it throughout their lifecycle. This process effectively removes CO_2 from the atmosphere, mitigating the impact of greenhouse gases [100,158,161,162]. This concept was explored in detail in Section 3.1.2. The fundamental advantage of using these materials in construction is their ability to act as carbon sinks, thus storing carbon within the building fabric. This characteristic underscores the potential of these materials to offset the carbon emissions associated with traditional construction methods. As indicated earlier, the result depends on the adopted calculation method; depending on the method, the difference at the level of the entire building between the final results was as much as 29%. The reference research by Hoxha et al. found that for multi-family massive wooden residential buildings, the impact of the building is approximately 20.7 kg-$CO_2e/m^2/yr$, which is 75% lower than the values for buildings constructed using traditional technology and approximately 50% lower than for buildings constructed using traditional low-energy technology. In his study, the author also warns that depending on the method adopted, the environmental impacts of building components present significant discrepancies, even up to 200% [77].

However, accurately evaluating the CF of biological products through LCA presents substantial challenges. One major difficulty lies in accounting for the highly variable growth rates of different plant species. These growth periods can vary widely based on environmental conditions, which complicates the assessment of carbon sequestration over the lifecycle of the materials. Furthermore, the LCA of biological products is often inconsistent due to variations in research methodologies and the complexities of different assessment models. As a result, achieving a reliable and uniform evaluation of the environmental benefits of using natural materials in construction remains a challenging task. The ongoing need for more refined and standardized assessment methods is essential to enhance the accuracy of LCA and fully understand the environmental impact of these materials [57,100,163].

The production of wooden components involves considerable energy consumption, primarily due to the drying and processing stages [158,164]. For wooden buildings, the carbon dioxide emissions associated with the pre-construction phase generally account for less than 10% of the total emissions throughout the building's lifecycle. The most notable difference in carbon emissions between wooden structures and those built with other materials is observed in the emissions related to the materials themselves. In traditional construction methods, the carbon emissions from materials can constitute more than 50% of the total lifecycle emissions [125].

The CF associated with the construction of a building is predominantly influenced by the chosen structural system, architectural design, and construction technology. Research by Yang et al. [125] reveals that for various types of wooden buildings, carbon dioxide emissions during the operational phase typically account for an average of 87.7% of the total lifecycle emissions. If a wooden building achieves the same thermal performance as structures made from other materials, the carbon emissions during the operational phase would be comparable, negating significant differences in emissions at this stage.

In the final stage of the LCA, which follows the building's operational phase, wooden components can be either repurposed for energy production, such as generating heat or electricity, or reused in new construction projects [165]. The reuse of building components is especially important given that recovery rates across all material types are still relatively low, hovering around 50%. This rate is considered unsatisfactory because reclaimed materials are mostly repurposed for energy recovery or for use in lower-quality applications. To enhance the chances of reusing wooden components, designers can adopt strategies such as reversible connections and components specifically designed for reuse [165]. In practice, even high-quality wood is typically discarded at the end of a building's life. Once wood components are used, they are immediately classified as wood waste to avoid the time-consuming and costly testing of treatments applied to them. This happens despite the fact that, theoretically, these treatments should not harm the environment or living organisms. As a result, wood is often regarded less favorably compared to other materials and is primarily used only as a raw material for energy production [165–167]. In addition to evaluating the impact of preservatives, wood should be assessed for its structural suitability, potential insect and fungal infestations, and the presence of metal fasteners. While these factors are significant, they are generally less critical than the issue of preservatives. Additionally, careful consideration is needed for the process of deconstruction to maximize the recovery of high-quality wood.

3.1.5. Innovative Construction Practices

The future of construction innovation may significantly advance the field of wooden and zero energy building design. Educating not only younger generations but also seasoned engineers, architects, and builders who adhere to outdated standards is crucial for popularizing new technologies [118]. Availability of models for low or zero CF properties is essential [45]. Emerging projects may focus not only on energy efficiency and reduced energy demand but also on buildings that produce energy, store it in batteries, and return it to the local energy grid (energy-plus buildings exceed net-zero energy buildings by generating more energy than they consume) (report "Advances Toward a Net-Zero Global Building Sector", in "The Annual Review of Environment and Resources" [118]). These innovations might also support local ecosystems, for instance, by incorporating mushroom or algae cultivation within the buildings.

This article synthesizes current knowledge and practical achievements in wooden architecture, particularly focusing on the design, construction, and evaluation of new wooden buildings with zero CO_2 emissions (net-zero-energy buildings—nZEB, with the term nearly-zero also appearing in the literature) [28]. Discussions in the literature on low-energy wooden construction (such as Cost and Energy Reduction of a New nZEB Wooden Building) cover aspects of minimizing construction and operational costs [28]. Transforming the construction sector to achieve net-zero-energy consumption is vital for climate neutrality, given that construction accounts for about 36% of final energy demand and 39% of greenhouse gas emissions related to technological processes [57]. Meeting new international energy standards involves exploring new technologies [65]. Large-scale actions necessitate a systemic approach and integration of models within local urban policies [103]. Ongoing research and development (R&D) includes laboratory testing of material strength and connection details [44]. Wood's properties are harnessed for carbon sequestration through innovative products like CLT and Glulam, which enhance structural strength and carbon storage benefits. Zero-emission wooden buildings also involve complex engineering calculations of energy efficiency and production, beginning with material procurement. Further assessments include building usage, project modifications, potential dismantling, or relocation. Comprehensive LCA is used to evaluate the total CF of buildings. The end of a building's life cycle involves demolition, disposal, or recycling of materials and may include design modifications. These innovations are described in the previous chapters. Moreover, new processes may affect design software, assembly, construction logistics, and usage methods [124].

Innovation in construction is a systematic process within firms and organizations, transitioning from an innovative idea through development to implementation and potential commercialization [88]. Innovations can be categorized into technological–programming, material–production, process–organizational, service–economic, and social branches. The literature offers various interpretations of innovation, which extends beyond scientific discovery to include the transformation of knowledge into commercial products, processes, and services [101]. Innovations are essential for maintaining global market competitiveness and driving economic and technological progress.

Construction is a significant sector with a broad spatial impact and considerable influence on environmental pollution. Addressing the climate crisis and environmental changes is a major focus of new construction practices [58]. Measuring innovation outcomes involves evaluating diverse aspects: companies' strategies for generating new ideas and knowledge, customer needs, market conditions, learning and knowledge sharing, and employee engagement and creativity [88]. These factors significantly affect the performance of construction companies. The concept of social capital, referring to the pool of knowledge and the communication networks among employees engaged in innovation, plays a crucial role [168].

In the field of management, there are studies on performance indicators for organizational and national innovation. However, to apply these to the construction industry, the specific characteristics of the sector must be taken into account. Innovation indicators depend on growth and inhibition factors, and measuring progress remains challenging. Lo and Kam [88] explored the conditions required for precisely assessing innovation progress in design. It emphasizes that these indicators are essential for advancing design and construction firms. However, many organizations in this sector lack a comprehensive framework for measuring innovation outcomes. To better characterize a specific resource, such as an organization, it is crucial to define key performance indicators for innovation. These indicators help manage and foster the creation of new solutions. The article presents findings from a literature review and expert consultations to develop performance metrics for innovation within the Architecture, Engineering, and Construction (AEC) industry.

Organizations could gain a competitive edge through ongoing management while fostering future innovations [51]. However, conventional building preferences can delay innovation. Conventional building practices are often preferred because they adhere to rigid procedures and administrative regulations that rely heavily on paper documentation [88]. Consequently, end-users frequently choose conventional building designs over innovative ones. Due to time and cost constraints, investors, designers, and contractors typically follow standard building codes to ensure compliance and secure construction permits. As a result, performance-based approaches are not given priority, as highlighted in some papers [88].

3.2. Stage 2: Building Review

This section delves into contemporary examples of sustainable wooden architecture, highlighting innovative approaches to reducing carbon emissions at both the construction and operational stages. Exploring a range of projects across Europe and North America showcases how architects and designers are leveraging renewable materials, advanced construction techniques, and collaborative design processes to create structures that are both environmentally responsible and adaptable to future needs. These case studies illustrate the potential of wooden architecture to not only meet but exceed sustainability goals, setting new standards for the industry.

From a review of 98 near-zero and zero-emission buildings, 10 examples were chosen. The timeframe spans from 2020 to 2024. These examples encompass both residential and public utility buildings with various functions across Europe and North America. The comparison criteria include the operational energy indicator, measured in $kgCO_2e/m^2/year$, and the embodied energy involved in material procurement and construction, also measured in $kgCO_2e/m^2$. These parameters were chosen because of their versatility in estimating the carbon footprint of buildings in relation to 1 m^2 of building. Upfront carbon emissions, as well as operational energy, are components of the LCA which allows a proper comparison of the carbon footprint of buildings. Table 3 presents a summary of selected buildings constructed entirely from wooden structures, as detailed in the following paragraphs.

Table 3. List of example wooden buildings.

Project (Author)	Building Type	Location	Year	Area	Operational Energy [$kgCO_2e/m^2/y$]	Upfront Carbon Emissions (A1–A5) [$kgCO_2e/m^2$]	Criteria *
Lea Bridge Library Pavilion (Studio Weave) [131]	Public building (library)	London, UK	2021	250 m^2	9.69	147	1, 2, 3, 4, 5
Orueta Etxea (Emiliano López Mónica Rivera Arquitectos) [132]	Residential	Bilbao, Spain	2021	308 m^2	3.83	-	1, 2, 3, 4, 5
Spruce house and studio (ao-ft) [133]	Residential	London, UK	2021	132 m^2	-	336	1, 2, 3, 4
Cooperative housing (Urban Power) [134]	Residential	Stavnsholt, Denamrk	2021	3100 m^2	8.70	-	1, 2, 3

Table 3. *Cont.*

Project (Author)	Building Type	Location	Year	Area	Operational Energy [$kgCO_2e/m^2/y$]	Upfront Carbon Emissions (A1–A5) [$kgCO_2e/m^2$]	Criteria *
The black and white building (Waugh Thistleton Architects) [135]	Commercial, office	London, UK	2022	4480 m^2	9.11	329	1, 2, 3, 4
Workstack (dRMM) [136]	Commercial	London, UK	2023	1583 m^2	48.41	271	1, 2, 3, 4
EÑE House (Estudio Albar) [137]	Residential	Madrid, Spain	2023	250 m^2	14.50	-	1, 2, 3, 4, 5
New Temple Complex (James Gorst Architects) [138]	Public building	Hampshire, UK	2023	585 m^2	42.60	407	1, 2, 3, 4, 5
Durley Chine Environmental Hub (Footprint Architects) [139]	Public building	Bournemouth, UK	2024	887 m^2	15.73	210	1, 2, 3, 4
Humber Cultural Hub (Diamond Schmitt) [140]	Public building	Ontario, Canada	2024	23,244 m^2	6.63	-	1, 2, 3

* Criteria: 1—Sustainable sourcing of materials; 2—Carbon sequestration; 3—Energy efficiency; 4—Life cycle assessment; 5—Innovative construction practices.

Studio Weave's Lea Bridge Library Pavilion extends the existing library, connecting it with a garden to create a multifunctional space. The construction uses LVL (Laminated Veneer Lumber) beams, partially anchored into the existing structure, reducing the number of necessary support columns. The wood used was reclaimed from trees felled or damaged in London, resulting in a diverse range of species, including European spruce, chestnut, poplar, sycamore, and European oak [131].

Orueta Etxea is a single-family house project by Spanish studio Emiliano López Mónica Rivera Arquitectos. It is an important example of a comprehensive approach to design based on the principles of sustainability and a closed-loop economy. The designers used only local materials with low CO_2 emissions. The house was built to passive building standards and has been Passivhaus Classic certified. Larch was used as structural and finishing timber [132].

The Spruce House and Studio, designed by ao-ft, is a project that reflects a commitment to renewable materials and sustainable construction techniques. This addition to an existing neighborhood features a main structure made of prefabricated CLT panels, chosen for their environmental benefits and aesthetic appeal as an exposed interior finish. The design minimizes the use of steel and incorporates prefabrication to enhance precision and reduce waste. Additionally, the building and its connections are designed for easy disassembly, facilitating future reuse. The timber used includes Siberian larch, spruce, and birch [133].

Urban Power's Cooperative Housing project in Denmark is a residential development created in collaboration with future residents and local authorities to optimize the design of shared spaces. The project includes communal kitchens, dining areas, laundry rooms, guest rooms, and storage, allowing for smaller individual homes and reducing the overall CF. The construction uses local wood, with prefabricated CLT elements ensuring both affordability and energy efficiency [134].

The black and white building by Waugh Thistleton Architects is a workspace with an elastic approach to design. The layout of the building was designed to increase the proportion of shared spaces and to allow the space to be easily rearranged in the future. At the time of completion, the building was the tallest timber building in London, reaching 17.8 m. The entire structure, floors, partitions, and curtain walls were made from timber. Species such as pine and beech were used [135].

Workstack, a project by dRMM, is designed to provide affordable workspaces for small businesses in the Charlton Riverside industrial zone in Greenwich. The building offers a consolidated alternative to scattered workshops typically housed in metal sheds. By bringing multiple businesses under one roof, the design reduces space usage and improves energy efficiency, cutting heating and cooling costs. The structure is made from prefabricated CLT panels, using locally sourced spruce and birch wood [136].

EÑE House, designed by Estudio Albar, is a project that, from the outset, aimed for near-zero energy consumption and minimal environmental impact. The house was designed according to the Passivhaus certification standard. The building is largely made of prefabricated pine timber. An unusual design solution is the use of cork wood as the façade cladding [137].

The New Temple Complex, designed by James Gorst Architects, serves as a multi-functional space for a spiritual organization. The complex includes a temple, chapel, library, community hall, kitchen, and foyer, designed to be shared by multiple religions, thus reducing the need for separate buildings and minimizing the CF. The structure is built with European spruce, and the exterior cladding is made from Siberian larch. Interior woodwork, cladding, and furniture are crafted from locally sourced ash [138].

Footprint Architects' Durley Chine Environmental Hub is a building dedicated to promoting environmental stewardship through workshops and training on waste segregation and reuse. The construction uses reclaimed wood, with beams sourced from a decommissioned pier and breakwaters at a local naval base. Insulation is made from shredded newspapers, with all materials sourced locally, emphasizing reclaimed content. The timber frame construction uses a mix of reclaimed woods, including basralocus, ekki, opepe, and accoya, originally used in marine structures [139].

The Humber Cultural Hub, designed by Diamond Schmitt, is a multi-purpose building on the Lakeshore Humber College campus. The design adheres to industry standards such as the Toronto Green Standard, Zero Carbon Building—Design Standard, and LEED Platinum certification. During the planning stage, the project team considered the building's operations, maintenance requirements, and lifecycle costs. The structure is built with CLT panels, and the exterior cladding uses high-performance panels to ensure energy efficiency. The building was awarded the 2023 World Federation of Colleges and Polytechnics (WFCP) Construction Award for its innovative construction practices and commitment to sustainability [140].

These examples illustrate varied approaches to designing wooden buildings, focusing on reducing carbon emissions during construction and subsequent operation. Designers achieve this by reducing the demand for new spaces through shared usage, utilizing reclaimed and locally sourced materials, employing prefabrication to minimize construction waste, and planning for easy disassembly to facilitate future reuse. Residential buildings tend to have the highest embedded CF due to their dense, multi-functional spaces, while buildings with open floor plans achieve lower values.

4. Discussion

New concepts for wooden buildings with zero CO_2 emissions are increasingly popular in response to growing demands for sustainability and combating global warming [72,122,130,169]. This trend is driven by advancements in material manufacturing technologies [17,170]. Among the most prominent solutions in this field are Cross Laminated Timber (CLT) and Glued Laminated Timber (GLT) technologies [56,171].

CLT is a modern construction material made from layers of wood glued in a crosswise arrangement, which provides exceptional strength and stability. The cross-laminated configuration enhances the material's load-bearing capacity and rigidity, making it suitable for structural applications in high-rise buildings [46,126,172]. CLT naturally resists fire, as the outer wood layers carbonize, creating a protective barrier that slows further burning [48,173]. Additionally, its excellent insulating properties contribute to the energy efficiency of buildings. CLT is a renewable material that sequesters CO_2, thereby reducing the CF of buildings [60,174]. The production of CLT generates less waste compared to traditional building materials. Examples include the Dalston Works Compact Residential Complex in Hackney, London—referred to as "The World's Largest CLT Building" (Design: Waugh Thistleton Architects, 2017)—and the Puukuokka Complex in Jyväskylä [90] (Design: OOPEAA, 2011).

GLT is another innovative material consisting of wood layers glued together longitudinally. This allows for the creation of large, open spaces without numerous supports, offering greater design flexibility. GLT's flexibility enables the design of arches, curves, and other complex forms and provides high fire resistance. Notable examples include the Wood Innovation Design Centre in Prince George (Design: Michael Green Architecture, 2014) and the "Perspective" Office Building in Bordeaux [90] (Design: Laisné Roussel, 2018).

Both CLT and GLT are advanced, sustainable building materials that offer significant environmental, economic, and design benefits [175]. Their unique properties make them suitable for various types of construction. The choice between CLT and GLT depends on specific project requirements such as strength, flexibility, assembly speed, and aesthetics [92]. Integrating these materials with Building Information Modeling (BIM) provides numerous advantages over traditional methods, including enhancements in design, execution, performance, analysis, process monitoring, and facility management [127]. Combining BIM with intelligent systems and wooden technologies like CLT and GLT is considered a forward-thinking approach. Thanks to modern CNC machines, the high degree of prefabrication achievable with CLT and GLT allows for rapid and precise assembly on-site. This automation reduces labor costs and project timelines, making construction faster and more cost-effective. The monolithic nature of these materials minimizes additional construction work, thus lowering overall investment costs. Additionally, reduced material use and improved energy efficiency contribute to decreased operational costs [52,54,87,123,176].

Hybrid wooden constructions, which combine wood with materials such as steel and concrete, offer an intriguing solution by merging the benefits of various materials while minimizing their drawbacks [42]. This approach can enhance building strength and enable more complex designs. The growing popularity of hybrid wooden and steel structures offers advantages in terms of zero-emission and material efficiency. Combining wood with steel, which has high recycling potential, helps reduce CO_2 emissions in the construction process [47,75]. This hybrid approach provides an alternative when purely wooden structures are impractical due to factors such as legal or fire safety considerations that vary internationally [84,96].

The recycling efficiency of wooden materials used in CLT or GLT constructions remains a subject for further research and discussion. The adhesives used in the manufacturing process can limit the potential for complete recycling of wood after its use [64,177]. Efforts should focus on minimizing obstacles in the recycling process of glued wood materials. Overcoming this issue could enhance the sustainability of glued wood as a low-emission material. Modern wooden buildings worldwide demonstrate how innovative concepts can be practically applied [42]. One notable example is the HoHo Vienna Tower (Design: RLP Rüdiger Lainer + Partner, 2016), one of the tallest wooden buildings globally, combining wood and concrete in a hybrid construction. This 24-story building exemplifies sustainable design, focusing on energy efficiency and renewable energy use [102]. Sustainable design and execution are crucial components of zero-emission wooden buildings [178]. Passive design strategies, which use natural energy sources such as sunlight and ventilation to reduce energy consumption, are recommended for these structures. Buildings designed with passive principles are more energy-efficient and environmentally friendly [114]. The integration of renewable energy sources such as solar panels, wind turbines, and geothermal systems can help buildings achieve a zero-energy balance. Incorporating these technologies into wooden structures further reduces their environmental impact [95,107,172]. Modular and prefabricated construction techniques enable faster and more efficient execution, reducing material waste and transportation emissions. Prefabricated modules produced under controlled conditions ensure higher quality and precision [164,179].

Despite these benefits, zero-emission wooden construction faces several challenges that must be addressed to fully realize its potential. Regulatory barriers, discussed in the certification chapter, are a significant issue. Many countries need to update their building codes to support the broad use of modern wooden technologies. New standards and financial incentives could accelerate the adoption of sustainable building practices [99,180]. Technical challenges related to wood in contemporary construction also need to be addressed. While wood offers many advantages, it requires specific fire protection and measures against moisture and pests [109,117,181,182]. Investment in research and development is crucial to overcoming these barriers. Market acceptance is another critical factor in promoting wood as a modern and durable material [129,183]. Education and promotion of the benefits of wooden construction are key to shifting perceptions and increasing the acceptance of wood as a standard building material [73,128,158]. Modern zero-emission wooden buildings provide practical solutions for sustainable urban development. With innovative materials and technologies, sustainable design, and global examples, wood has the potential to become a key material in the future of construction [112].

5. Conclusions

The article provides a comprehensive review and analysis of the criteria essential for the feasibility of contemporary zero-carbon wooden buildings (ZCBs), which constituted its primary research aim. The study is augmented by an examination of selected wooden buildings, focusing on how these criteria are integrated to minimize the CF.

The findings reveal that all defined criteria—sustainable material sourcing, carbon sequestration, energy efficiency, LCA, and innovative construction practices—are crucial in designing and constructing new zero-carbon wooden buildings. These results underscore the significant potential of wood as a renewable material for ZCBs and suggest it could be a compelling alternative to traditional construction materials. However, the findings also highlight the complexity and multifaceted nature of effectively incorporating wood to achieve this objective.

The study shows that optimal outcomes are achieved by addressing all five criteria, though ranking their importance is challenging. The LCA stands out as the most measurable criterion for evaluating the ecological benefits of buildings comprehensively. Nonetheless, each criterion emphasizes different aspects that should not be considered in isolation. Based on these criteria, the study identifies the following key recommendations:

1. Sustainable Material Sourcing: Sustainable sourcing is foundational for zero-carbon wooden buildings. This involves using wood from responsibly managed forests, as certified by organizations like the FSC or PEFC. Local wood sourcing reduces transportation-related emissions, which is crucial for minimizing the building's CF. Additionally, the use of bio-based and composite materials, such as Cross Laminated Timber (CLT) or other advanced wooden products, enhances carbon sequestration;
2. Carbon Sequestration: Maximizing wood's role as both a structural and finishing material is vital due to its ecological benefits. Wood not only helps in reducing CO_2 emissions but also serves as a long-term carbon sink;
3. Energy Efficiency: Implementing passive design strategies, including natural ventilation, effective insulation, and optimal building orientation, significantly lower energy demand. Incorporating renewable energy sources, such as solar panels and geothermal systems, meets the building's energy needs and, combined with energy-efficient heating and ventilation systems, further reduces CO_2 emissions;

4. LCA: Conducting a thorough ecological assessment throughout all stages of a building's life (from construction to demolition) is essential to understanding its total carbon footprint. Designing for disassembly and material recycling further minimizes future CO_2 emissions;
5. Innovative Construction Practices: Employing low-carbon construction technologies, such as prefabrication and modular construction, reduces waste and construction time. These innovative methods help minimize emissions during the construction phase, which is critical for achieving zero CO_2 emissions.

The literature review revealed a significant gap and delay in bridging advanced scientific research with practical professional knowledge. This underscores the urgent need to enhance research efforts and increase the dissemination of publications. Such measures are crucial for the broader implementation of advanced solutions and application models and for the effective development of climate and energy policies.

The study also finds that despite the potential benefits of using wood, several barriers impede its implementation in ZCBs, including social, regulatory, and financial challenges. A significant issue is the low level of knowledge among construction industry stakeholders regarding ZCB design and construction. Overcoming this requires enhancing awareness and education on sustainable building practices. Technological challenges persist as well, such as the increased energy embedded and the CF associated with more processed wood products. Furthermore, the recycling rate of wood in construction remains lower compared to materials like steel.

The analysis highlights several research gaps that need addressing, including the following:

- Use of Bio-Based Materials: Further research on bio-based materials like bamboo and wooden composites, which could be used in wooden construction and offer carbon sequestration benefits;
- Long-Term Carbon Sequestration Effects: Investigation into the long-term effects of carbon sequestration in wooden buildings, including their impact on the LCA and end-of-life scenarios such as recycling or disposal;
- Energy Efficiency in Various Climates: Studies to understand how different climatic conditions influence energy demand and design strategies for wooden buildings;
- Social and Regulatory Barriers: Identifying and addressing social and regulatory barriers to adopting ZCBs, which will aid in developing effective strategies to promote sustainable construction;
- Stakeholder Knowledge and Awareness: Assessing the knowledge and awareness levels among construction industry stakeholders about zero-carbon buildings and determining effective educational and training methods;
- Cost–Benefit Analysis: Evaluating the costs and benefits associated with constructing and operating ZCBs to better understand their economic feasibility compared to traditional construction methods.

Addressing these research gaps can advance the development and implementation of zero-carbon wooden buildings, supporting sustainable development within the construction sector.

Author Contributions: Conceptualization, A.S.; methodology, A.N. and A.S.; software, A.S.; validation, A.S.; formal analysis, A.S., K.R.-N., J.M., A.N., A.K., J.K., A.P. (Anna Piętocha), P.V., P.Ł., M.D., K.L., K.W. and K.Z.; investigation, A.S., K.R.-N., J.M., A.N., A.K., J.K., A.P. (Anna Piętocha), P.V., P.Ł., M.D., K.L., K.W. and K.Z.; resources, A.S., K.R.-N., J.M., A.N., A.K., J.K., A.P. (Anna Piętocha), P.V., P.Ł., M.D., K.L., K.W., K.Z., I.V., B.F. and A.P. (Anna Podlasek); data curation, A.S, A.N. and P.Ł.; writing—original draft preparation, A.S., K.R.-N., J.M., A.N., A.K., J.K., A.P. (Anna Piętocha), P.V., P.Ł., M.D., K.L., K.W., K.Z., I.V., B.F. and A.P. (Anna Podlasek); writing—review and editing, A.S., K.R.-N., J.M. and A.N.; supervision, A.S.; project administration, A.S. and A.P. (Anna Podlasek). All authors have read and agreed to the published version of the manuscript.

Funding: This research received no external funding.

Conflicts of Interest: The authors declare no conflicts of interest.

References

1. Branchi, B.A.; Ferreira, D.H.L.; Barbosa, A.M.; Ferreira, A.L. Footprints' Effectiveness as Decision-Making Tools for Promoting Sustainability. In *Proceedings of the 8th Brazilian Technology Symposium (BTSym'22)*; Iano, Y., Saotome, O., Kemper Vásquez, G.L., De Moraes Gomes Rosa, M.T., Arthur, R., Gomes De Oliveira, G., Eds.; Smart Innovation, Systems and Technologies; Springer International Publishing: Cham, Switzerland, 2023; Volume 353, pp. 472–478, ISBN 978-3-031-31006-5.
2. Wackernagel, M.; Lin, D.; Hanscom, L.; Galli, A.; Iha, K. Ecological Footprint. In *Encyclopedia of Ecology*; Elsevier: Amsterdam, The Netherlands, 2019; pp. 270–282, ISBN 978-0-444-64130-4.
3. Dai, J.; Ouyang, Y.; Hou, J.; Cai, L. Long-Time Series Assessment of the Sustainable Development of Xiamen City in China Based on Ecological Footprint Calculations. *Ecol. Indic.* **2023**, *148*, 110130. [CrossRef]
4. Fang, K.; Heijungs, R.; De Snoo, G.R. Theoretical Exploration for the Combination of the Ecological, Energy, Carbon, and Water Footprints: Overview of a Footprint Family. *Ecol. Indic.* **2014**, *36*, 508–518. [CrossRef]
5. Ridhosari, B.; Rahman, A. Carbon Footprint Assessment at Universitas Pertamina from the Scope of Electricity, Transportation, and Waste Generation: Toward a Green Campus and Promotion of Environmental Sustainability. *J. Clean. Prod.* **2020**, *246*, 119172. [CrossRef]
6. Ching, S.L.; Sari, K.A.M.; Muslim, R. Analysis of Carbon Footprint of Transportation, Food, and Manufactured Product in Industrial Manufacture. *AIP Conf. Proc.* **2024**, *2991*, 020063.
7. Feder, T. Scientists Take Steps in the Lab toward Climate Sustainability. *Phys. Today* **2023**, *76*, 20–23. [CrossRef]
8. Groll, M. Can Climate Change Be Avoided? Vision of a Hydrogen-Electricity Energy Economy. *Energy* **2023**, *264*, 126029. [CrossRef]
9. Meleti, V.; Delitheou, V. Smart Cities and the Challenge of Cities' Energy Autonomy. In *Handbook of Smart Cities*; Augusto, J.C., Ed.; Springer International Publishing: Cham, Switzerland, 2021; pp. 563–592, ISBN 978-3-030-69697-9.
10. Nidhin, B.K.S.N.; Domingo, N.; Bui, T.T.P.; Wilkinson, S. Construction Stakeholders' Knowledge on Zero Carbon Initiatives in New Zealand. *Int. J. Build. Pathol. Adapt.* **2023**. [CrossRef]
11. Talvitie, I.; Amiri, A.; Junnila, S. Climate Benefits of Wooden Construction in Urban Context. *IOP Conf. Ser. Earth Environ. Sci.* **2022**, *1101*, 022048. [CrossRef]
12. Andersen, C.E.; Rasmussen, F.N.; Habert, G.; Birgisdóttir, H. Embodied GHG Emissions of Wooden Buildings—Challenges of Biogenic Carbon Accounting in Current LCA Methods. *Front. Built Environ.* **2021**, *7*, 729096. [CrossRef]
13. Lin, C.-L.; Chiang, W.-H.; Weng, Y.-S.; Wu, H.-P. Assessing the Anthropogenic Carbon Emission of Wooden Construction: An LCA Study. *Build. Res. Inf.* **2023**, *51*, 138–157. [CrossRef]
14. Alvarez, D.; Kouda, R.; Ho, A.D.; Kubota, T. Scenario Analysis of Embodied Energy and CO_2 Emissions for Multistory Apartments in Indonesia. *E3S Web Conf.* **2023**, *396*, 04015. [CrossRef]
15. *ISO 14040:2006*; Environmental Management—Life Cycle Assessment—Principles and Framework. International Organization for Standardization: Geneva, Switzerland, 2006.
16. McDonough, W.; Braungart, M. *Cradle to Cradle: Remaking the Way We Make Things*; North Point Press: New York, NY, USA, 2002.
17. Lou, H.-L.; Hsieh, S.-H. Towards Zero: A Review on Strategies in Achieving Net-Zero-Energy and Net-Zero-Carbon Buildings. *Sustainability* **2024**, *16*, 4735. [CrossRef]
18. Besana, D.; Tirelli, D. Reuse and Retrofitting Strategies for a Net Zero Carbon Building in Milan: An Analytic Evaluation. *Sustainability* **2022**, *14*, 16115. [CrossRef]
19. Ding, Y.; Pang, Z.; Lan, K.; Yao, Y.; Panzarasa, G.; Xu, L.; Lo Ricco, M.; Rammer, D.R.; Zhu, J.Y.; Hu, M.; et al. Emerging Engineered Wood for Building Applications. *Chem. Rev.* **2023**, *123*, 1843–1888. [CrossRef] [PubMed]
20. Andersen, C.E.; Hoxha, E.; Nygaard Rasmussen, F.; Grau Sørensen, C.; Birgisdóttir, H. Evaluating the Environmental Performance of 45 Real-Life Wooden Buildings: A Comprehensive Analysis of Low-Impact Construction Practices. *Build. Environ.* **2024**, *250*, 111201. [CrossRef]
21. Hu, M. Exploring Low-Carbon Design and Construction Techniques: Lessons from Vernacular Architecture. *Climate* **2023**, *11*, 165. [CrossRef]
22. El-Shorbagy, A.-M. Wood Shapes the Future of Sustainable Architecture. In Proceedings of the 2020 Advances in Science and Engineering Technology International Conferences (ASET), Dubai, United Arab Emirates, 4 February–9 April 2020; pp. 1–6.
23. Michálková, D.; Ďurica, P. Natural Materials in Building Construction—Annual Evaluation. *APP* **2022**, *38*, 222–227. [CrossRef]
24. Carletti, C.; Piselli, C.; Sciurpi, F. Are Design Strategies for High-Performance Buildings Really Effective? Results from One Year of Monitoring of Indoor Microclimate and Envelope Performance of a Newly Built nZEB House in Central Italy. *Energies* **2024**, *17*, 741. [CrossRef]
25. Starzyk, A.; Donderewicz, M.; Rybak-Niedziółka, K.; Marchwiński, J.; Grochulska-Salak, M.; Łacek, P.; Mazur, Ł.; Voronkova, I.; Vietrova, P. The Evolution of Multi-Family Housing Development Standards in the Climate Crisis: A Comparative Analysis of Selected Issues. *Buildings* **2023**, *13*, 1985. [CrossRef]
26. Amiri, A.; Ottelin, J.; Sorvari, J.; Junnila, S. Cities as Carbon Sinks—Classification of Wooden Buildings. *Environ. Res. Lett.* **2020**, *15*, 094076. [CrossRef]

27. Avellan, K.C.; Belopotocanova, E.; Ghobakhlou, M. Massive Wood Elements and Modular Housing Technology as Innovative Building Concept of Sustainable Urban Planning. In Proceedings of the IABSE Conference–Engineering the Developing World, Kuala Lumpur, Malaysia, 25–27 April 2018; pp. 1085–1090.
28. Arumägi, E.; Kalamees, T. Cost and Energy Reduction of a New nZEB Wooden Building. *Energies* **2020**, *13*, 3570. [CrossRef]
29. Moschetti, R.; Brattebø, H.; Sparrevik, M. Exploring the Pathway from Zero-Energy to Zero-Emission Building Solutions: A Case Study of a Norwegian Office Building. *Energy Build.* **2019**, *188–189*, 84–97. [CrossRef]
30. Parkin, A.; Herrera, M.; Coley, D.A. Energy or Carbon? Exploring the Relative Size of Universal Zero Carbon and Zero Energy Design Spaces. *Build. Serv. Eng. Res. Technol.* **2019**, *40*, 319–339. [CrossRef]
31. *Net Zero Energy Buildings (NZEB)*; Elsevier: Amsterdam, The Netherlands, 2018; ISBN 978-0-12-812461-1.
32. Sultanuzzaman, M.R.; Yahya, F.; Lee, C.-C. Exploring the Complex Interplay of Green Finance, Business Cycles, and Energy Development. *Energy* **2024**, *306*, 132479. [CrossRef]
33. Mulya, K.S.; Ng, W.L.; Biró, K.; Ho, W.S.; Wong, K.Y.; Woon, K.S. Decarbonizing the High-Rise Office Building: A Life Cycle Carbon Assessment to Green Building Rating Systems in a Tropical Country. *Build. Environ.* **2024**, *255*, 111437. [CrossRef]
34. Ferreira, A.; Pinheiro, M.D.; Brito, J.D.; Mateus, R. A Critical Analysis of LEED, BREEAM and DGNB as Sustainability Assessment Methods for Retail Buildings. *J. Build. Eng.* **2023**, *66*, 105825. [CrossRef]
35. *BREEAM International New Construction. Version 6.0 Technical Manual—SD250*; BRE Group: Watford, UK, 2021.
36. *Leed V5, Rating System Building Design and Construction: New Construction First Public Comment Draft*; U.S. Green Building Council: Washington, DC, USA, 2024.
37. *DGNB Criteria Set New Construction Buildings, Version 2023*; DGNB: Stuttgart, Germany, 2023.
38. European Commission. *Directorate General for Climate Action. Going Climate-Neutral by 2050: A Strategic Long Term Vision for a Prosperous, Modern, Competitive and Climate Neutral EU Economy*; Publications Office: Luxembourg, 2019.
39. DIRECTIVE 2010/31/EU of the European Parliament and of the Council on the Energy Performance of Buildings. Available online: https://eur-lex.europa.eu/eli/dir/2010/31/oj (accessed on 10 June 2024).
40. Directive (EU) 2018/844 of the European Parliament and of the Council of 30 May 2018 Amending Directive 2010/31/EU on the Energy Performance of Buildings and Directive 2012/27/EU on Energy Efficiency (Text with EEA Relevance); EUR-Lex. 2018. Available online: http://data.europa.eu/eli/dir/2018/844/oj (accessed on 10 June 2024).
41. Directive (EU) 2024/1275 of the European Parliament and of the Council of 24 April 2024 on the Energy Performance of Buildings (Recast) (Text with EEA Relevance); EUR-Lex. 2024. Available online: http://data.europa.eu/eli/dir/2024/1275/oj (accessed on 10 June 2024).
42. Abed, J.; Rayburg, S.; Rodwell, J.; Neave, M. A Review of the Performance and Benefits of Mass Timber as an Alternative to Concrete and Steel for Improving the Sustainability of Structures. *Sustainability* **2022**, *14*, 5570. [CrossRef]
43. Ahmed, A.E.; Suwaed, M.S.; Shakir, A.M.; Ghareeb, A. The Impact of Window Orientation, Glazing, and Window-to-Wall Ratio on the Heating and Cooling Energy of an Office Building: The Case of Hot and Semi-Arid Climate. *J. Eng. Res.* **2023**, S230718772300295X. [CrossRef]
44. Arlet, J.L. Innovative Carpentry and Hybrid Joints in Contemporary Wooden Architecture. *Arts* **2021**, *10*, 64. [CrossRef]
45. Bai, X.; Dawson, R.J.; Ürge-Vorsatz, D.; Delgado, G.C.; Salisu Barau, A.; Dhakal, S.; Dodman, D.; Leonardsen, L.; Masson-Delmotte, V.; Roberts, D.C.; et al. Six Research Priorities for Cities and Climate Change. *Nature* **2018**, *555*, 23–25. [CrossRef]
46. Barclay, S.; Salem, S. Behaviour of Cross-Laminated Timber Slabs Subjected to Fire—A State-Of-The-Art Review. In *Proceedings of the Canadian Society of Civil Engineering Annual Conference 2022*; Gupta, R., Sun, M., Brzev, S., Alam, M.S., Ng, K.T.W., Li, J., El Damatty, A., Lim, C., Eds.; Lecture Notes in Civil Engineering; Springer Nature: Cham, Switzerland, 2024; Volume 367, pp. 183–198, ISBN 978-3-031-35470-0.
47. Blay-Armah, A.; Mohebbi, G.; Bahadori-Jahromi, A.; Fu, C.; Amoako-Attah, J.; Barthorpe, M. Evaluation of Embodied Carbon Emissions in UK Supermarket Constructions: A Study on Steel, Brick, and Timber Frameworks with Consideration of End-of-Life Processes. *Sustainability* **2023**, *15*, 14978. [CrossRef]
48. Bøe, A.S.; Friquin, K.L.; Brandon, D.; Steen-Hansen, A.; Ertesvåg, I.S. Fire Spread in a Large Compartment with Exposed Cross-Laminated Timber and Open Ventilation Conditions: #FRIC-02—Exposed Wall and Ceiling. *Fire Saf. J.* **2023**, *141*, 103986. [CrossRef]
49. Bougiatioti, F.; Alexandrou, E.; Katsaros, M. Sustainable Refurbishment of Existing, Typical Single-Family Residential Buildings in Greece. *Int. J. Build. Pathol. Adapt.* **2023**. *ahead of print*. [CrossRef]
50. Broda, M. Natural Compounds for Wood Protection against Fungi—A Review. *Molecules* **2020**, *25*, 3538. [CrossRef]
51. Brogi, S.; Menichini, T. Do the ISO 14001 Environmental Management Systems Influence Eco-Innovation Performance? Evidences from the EU Context. *Eur. J. Sustain. Dev.* **2019**, *8*, 292. [CrossRef]
52. BuHamdan, S.; Duncheva, T.; Alwisy, A. Developing a BIM and Simulation-Based Hazard Assessment and Visualization Framework for CLT Construction Design. *J. Constr. Eng. Manag.* **2021**, *147*, 04021003. [CrossRef]
53. Cabeza, L.F.; Rincón, L.; Vilariño, V.; Pérez, G.; Castell, A. Life Cycle Assessment (LCA) and Life Cycle Energy Analysis (LCEA) of Buildings and the Building Sector: A Review. *Renew. Sustain. Energy Rev.* **2014**, *29*, 394–416. [CrossRef]
54. Lobos Calquin, D.; Mata, R.; Correa, C.; Núñez, E.; Bustamante, G.; Caicedo, N.; Blanco Fernandez, D.; Díaz, M.A.; Pulgar—Rubilar, P.; Roa, L. Implementation of Building Information Modeling Technologies in Wood Construction: A Review of the State of the Art from a Multidisciplinary Approach. *Buildings* **2024**, *14*, 584. [CrossRef]

55. Chen, Z.; Gu, H.; Bergman, R.; Liang, S. Comparative Life-Cycle Assessment of a High-Rise Mass Timber Building with an Equivalent Reinforced Concrete Alternative Using the Athena Impact Estimator for Buildings. *Sustainability* **2020**, *12*, 4708. [CrossRef]
56. Chen, Q. Sustainable Future: Development and Potential of Modern Timber Structures. *Highlights Sci. Eng. Technol.* **2023**, *75*, 86–93. [CrossRef]
57. Churkina, G.; Organschi, A.; Reyer, C.P.O.; Ruff, A.; Vinke, K.; Liu, Z.; Reck, B.K.; Graedel, T.E.; Schellnhuber, H.J. Buildings as a Global Carbon Sink. *Nat. Sustain.* **2020**, *3*, 269–276. [CrossRef]
58. Defloor, B.; Bleys, B.; Verhofstadt, E.; Van Ootegem, L. How to Reduce Individuals' Ecological Footprint without Harming Their Well-Being: An Application to Belgium. *Sustainability* **2022**, *14*, 5232. [CrossRef]
59. Parthiban Devarajan; Alicja Kozarzewska; Dhanasingh Sivalinga Vijayan; Sanjay Kumar; Sivasuriyan, A. Wiktor Sitek Transformational Green Sustainable Concepts in the Field of Infrastructure. *Acta. Sci. Pol. Archit.* **2024**, *23*, 56–78. [CrossRef]
60. Dong, Y.; Wang, R.; Xue, J.; Shao, J.; Guo, H. Assessment of Summer Overheating in Concrete Block and Cross Laminated Timber Office Buildings in the Severe Cold and Cold Regions of China. *Buildings* **2021**, *11*, 330. [CrossRef]
61. Duan, Z.; Huang, Q.; Zhang, Q. Life Cycle Assessment of Mass Timber Construction: A Review. *Build. Environ.* **2022**, *221*, 109320. [CrossRef]
62. Dzhurko, D.; Haacke, B.; Haberbosch, A.; Köhne, L.; König, N.; Lode, F.; Marx, A.; Mühlnickel, L.; Neunzig, N.; Niemann, A.; et al. Future Buildings as Carbon Sinks: Comparative Analysis of Timber-Based Building Typologies Regarding Their Carbon Emissions and Storage. *Front. Built Environ.* **2024**, *10*, 1330105. [CrossRef]
63. Elaouzy, Y.; El Fadar, A. Impact of Key Bioclimatic Design Strategies on Buildings' Performance in Dominant Climates Worldwide. *Energy Sustain. Dev.* **2022**, *68*, 532–549. [CrossRef]
64. Elginoz, N.; Van Blokland, J.; Safarian, S.; Movahedisaveji, Z.; Yadeta Wedajo, D.; Adamopoulos, S. Wood Waste Recycling in Sweden—Industrial, Environmental, Social, and Economic Challenges and Benefits. *Sustainability* **2024**, *16*, 5933. [CrossRef]
65. *Communication from the Commission to the European Parliament, the Council, the European Economic and Social Committee and the Committee of The Regions A New European Innovation Agenda*; European Commission: Brussels, Belgium, 2022.
66. Evans, P.D.; Matsunaga, H.; Preston, A.F.; Kewish, C.M. Wood Protection for Carbon Sequestration—A Review of Existing Approaches and Future Directions. *Curr For. Rep* **2022**, *8*, 181–198. [CrossRef]
67. Azimi Fereidani, N.; Rodrigues, E.; Gaspar, A.R. A Review of the Energy Implications of Passive Building Design and Active Measures under Climate Change in the Middle East. *J. Clean. Prod.* **2021**, *305*, 127152. [CrossRef]
68. Furhana Shereen, M.; Vishal Malolan, V.; Devanesan, M.G.; Sudalai, S.; Arumugam, A. A Critical Analysis of Renewable and Sustainable Energy Technologies: Energy Concept and Conversion Techniques. In *Recent Advances in Recycling Engineering*; Siddiqui, N.A., Baxtiyarovich, A.S., Nandan, A., Mondal, P., Eds.; Lecture Notes in Civil Engineering; Springer Nature: Singapore, 2023; Volume 275, pp. 117–137, ISBN 978-981-19393-0-3.
69. Gutierrez Garzon, A.R.; Bettinger, P.; Siry, J.; Abrams, J.; Cieszewski, C.; Boston, K.; Mei, B.; Zengin, H.; Yeşil, A. A Comparative Analysis of Five Forest Certification Programs. *Forests* **2020**, *11*, 863. [CrossRef]
70. Ghobadi, M.; Sepasgozar, S.M.E. Circular Economy Strategies in Modern Timber Construction as a Potential Response to Climate Change. *J. Build. Eng.* **2023**, *77*, 107229. [CrossRef]
71. Giridhar, B.N.; Pandey, K.K. Wood Modification for Wood Protection. In *Science of Wood Degradation and its Protection*; Sundararaj, R., Ed.; Springer: Singapore, 2022; pp. 647–663, ISBN 9789811687969.
72. Grinham, J.; Fjeldheim, H.; Yan, B.; Helge, T.D.; Edwards, K.; Hegli, T.; Malkawi, A. Zero-Carbon Balance: The Case of HouseZero. *Build. Environ.* **2022**, *207*, 108511. [CrossRef]
73. Hamida, A.; Zhang, D.; Ortiz, M.A.; Bluyssen, P.M. Indicators and Methods for Assessing Acoustical Preferences and Needs of Students in Educational Buildings: A Review. *Appl. Acoust.* **2023**, *202*, 109187. [CrossRef]
74. Hanifa, M.; Agarwal, R.; Sharma, U.; Thapliyal, P.C.; Singh, L.P. A Review on CO_2 Capture and Sequestration in the Construction Industry: Emerging Approaches and Commercialised Technologies. *J. CO_2 Util.* **2023**, *67*, 102292. [CrossRef]
75. He, J.; Fu, L.; Hu, J.; Lv, Y.; Chen, S.; He, Z.; Miao, W. Optimization Analysis of Ultra-high-rise Steel Structure Construction Based on Carbon Emission. *Eng. Rep.* **2024**, *6*, e12833. [CrossRef]
76. Himes, A.; Busby, G. Wood Buildings as a Climate Solution. *Dev. Built Environ.* **2020**, *4*, 100030. [CrossRef]
77. Hoxha, E.; Passer, A.; Saade, M.R.M.; Trigaux, D.; Shuttleworth, A.; Pittau, F.; Allacker, K.; Habert, G. Biogenic Carbon in Buildings: A Critical Overview of LCA Methods. *Build. Cities* **2020**, *1*, 504–524. [CrossRef]
78. Huang, Z.; Huang, Y.; Zhang, S. The Possibility and Improvement Directions of Achieving the Paris Agreement Goals from the Perspective of Climate Policy. *Sustainability* **2024**, *16*, 4212. [CrossRef]
79. Hurmekoski, E.; Seppälä, J.; Kilpeläinen, A.; Kunttu, J. Contribution of Wood Based Products to Climate Change Mitigation. In *Forest Bioeconomy and Climate Change*; Hetemäki, L., Kangas, J., Peltola, H., Eds.; Managing Forest Ecosystems; Springer International Publishing: Cham, Switzerland, 2022; Volume 42, pp. 129–149, ISBN 978-3-030-99205-7.
80. Ibrahim, B.S.; Soomro, D.M.; Sundarajoo, S.; Nordin, Z. Natural Lighting System Using Fiber Optics for Energy Efficiency. In Proceedings of the 2023 IEEE 8th International Conference on Engineering Technologies and Applied Sciences (ICETAS), Bahrain, Bahrain, 25–27 October 2023; pp. 1–6.
81. Kazemian, M.; Shafei, B. Carbon Sequestration and Storage in Concrete: A State-of-the-Art Review of Compositions, Methods, and Developments. *J. CO_2 Util.* **2023**, *70*, 102443. [CrossRef]

82. Keržič, E.; Humar, M. Studies on the Material Resistance and Moisture Dynamics of Wood after Artificial and Natural Weathering. *Wood Mater. Sci. Eng.* **2022**, *17*, 551–557. [CrossRef]
83. Király, É.; Börcsök, Z.; Kocsis, Z.; Németh, G.; Polgár, A.; Borovics, A. Carbon Sequestration in Harvested Wood Products in Hungary an Estimation Based on the IPCC 2019 Refinement. *Forests* **2022**, *13*, 1809. [CrossRef]
84. Koval, R.; Yemelyanenko, S.; Kuzyk, A.; Starodub, Y. Assessing the Risk of Material Damage of Building Construction of High-Rise Rooms Due to Fires and Emergencies. *Constr. Technol. Archit.* **2023**, *9*, 49–57.
85. Leszczyszyn, E.; Heräjärvi, H.; Verkasalo, E.; Garcia-Jaca, J.; Araya-Letelier, G.; Lanvin, J.-D.; Bidzińska, G.; Augustyniak-Wysocka, D.; Kies, U.; Calvillo, A.; et al. The Future of Wood Construction: Opportunities and Barriers Based on Surveys in Europe and Chile. *Sustainability* **2022**, *14*, 4358. [CrossRef]
86. Li, C.Z.; Zhang, L.; Liang, X.; Xiao, B.; Tam, V.W.Y.; Lai, X.; Chen, Z. Advances in the Research of Building Energy Saving. *Energy Build.* **2022**, *254*, 111556. [CrossRef]
87. Linkevičius, E.; Žemaitis, P.; Aleinikovas, M. Sustainability Impacts of Wood- and Concrete-Based Frame Buildings. *Sustainability* **2023**, *15*, 1560. [CrossRef]
88. Lo, J.T.Y.; Kam, C. Innovation Performance Indicators for Architecture, Engineering and Construction Organization. *Sustainability* **2021**, *13*, 9038. [CrossRef]
89. Lu, J.; Luo, X.; Cao, X. Research on Geometry Optimization of Park Office Buildings with the Goal of Zero Energy. *Energy* **2024**, *306*, 132179. [CrossRef]
90. Michalak, H.; Michalak, K. Selected Aspects of Sustainable Construction—Contemporary Opportunities for the Use of Timber in High and High-Rise Buildings. *Energies* **2024**, *17*, 1961. [CrossRef]
91. Mirashk-Daghiyan, M.; Dehghan-Touran-Poshti, A.; Shahcheragi, A.; Kaboli, M.H. The Effect of Surrounding Buildings' Height and the Width of the Street on a Building's Energy Consumption. *Int. J. Energy Environ. Eng.* **2022**, *13*, 207–217. [CrossRef]
92. Motamedi, S.; Rousse, D.R.; Promis, G. The Evolution of Crop-Based Materials in the Built Environment: A Review of the Applications, Performance, and Challenges. *Energies* **2023**, *16*, 5252. [CrossRef]
93. Mushtaha, E.; Salameh, T.; Kharrufa, S.; Mori, T.; Aldawoud, A.; Hamad, R.; Nemer, T. The Impact of Passive Design Strategies on Cooling Loads of Buildings in Temperate Climate. *Case Stud. Therm. Eng.* **2021**, *28*, 101588. [CrossRef]
94. Ouellet-Plamondon, C.M.; Ramseier, L.; Balouktsi, M.; Delem, L.; Foliente, G.; Francart, N.; Garcia-Martinez, A.; Hoxha, E.; Lützkendorf, T.; Nygaard Rasmussen, F.; et al. Carbon Footprint Assessment of a Wood Multi-Residential Building Considering Biogenic Carbon. *J. Clean. Prod.* **2023**, *404*, 136834. [CrossRef]
95. Pasternack, R.; Wishnie, M.; Clarke, C.; Wang, Y.; Belair, E.; Marshall, S.; Gu, H.; Nepal, P.; Dolezal, F.; Lomax, G.; et al. What Is the Impact of Mass Timber Utilization on Climate and Forests? *Sustainability* **2022**, *14*, 758. [CrossRef]
96. Pecio, M. Replacement Fire Protection Solutions for a Pick Tower Building—Case Study. *Inżynieria Bezpieczeństwa Obiektów Antropog.* **2024**, *1*, 23–34. [CrossRef]
97. Pedreño-Rojas, M.A.; Porras-Amores, C.; Villoria-Sáez, P.; Morales-Conde, M.J.; Flores-Colen, I. Characteruniteization and Performance of Building Composites Made from Gypsum and Woody-Biomass Ash Waste: A Product Development and Application Study. *Constr. Build. Mater.* **2024**, *419*, 135435. [CrossRef]
98. Phillips, R.; Troup, L.; Fannon, D.; Eckelman, M.J. Triple Bottom Line Sustainability Assessment of Window-to-Wall Ratio in US Office Buildings. *Build. Environ.* **2020**, *182*, 107057. [CrossRef]
99. Pilli, R.; Alkama, R.; Cescatti, A.; Kurz, W.A.; Grassi, G. The European Forest Carbon Budget under Future Climate Conditions and Current Management Practices. *Biogeosciences* **2022**, *19*, 3263–3284. [CrossRef]
100. Pomponi, F.; Hart, J.; Arehart, J.H.; D'Amico, B. Buildings as a Global Carbon Sink? A Reality Check on Feasibility Limits. *One Earth* **2020**, *3*, 157–161. [CrossRef]
101. Porter, M.E.; Stern, S. Innovation: Location Matters. *MITSloan Manag. Rev.* **2001**, *42*, 28–36.
102. Premrov, M.; Kozem Šilih, E. Numerical Analysis of the Racking Behaviour of Multi-Storey Timber-Framed Buildings Considering Load-Bearing Function of Double-Skin Façade Elements. *Sustainability* **2023**, *15*, 6379. [CrossRef]
103. Anne-Hélène Prieur-Richard; Walsh, B.; Craig, M.; Megan, L.; Melamed, M.; Colbert, L.; Pathak, M.; Connors, S.; Xuemei, B.; Aliyu, B.; et al. Extended Version: Global Research and Action Agenda on Cities and Climate Change Science. In Proceedings of the Cities & Climate Change Science Conference, Edmonton, AB, Canada, 5–7 March 2018. [CrossRef]
104. Prins, K.; Köhl, M.; Linser, S. Is the Concept of Sustainable Forest Management Still Fit for Purpose? *For. Policy Econ.* **2023**, *157*, 103072. [CrossRef]
105. Röck, M.; Saade, M.R.M.; Balouktsi, M.; Rasmussen, F.N.; Birgisdottir, H.; Frischknecht, R.; Habert, G.; Lützkendorf, T.; Passer, A. Embodied GHG Emissions of Buildings—The Hidden Challenge for Effective Climate Change Mitigation. *Appl. Energy* **2020**, *258*, 114107. [CrossRef]
106. Salata, F.; Ciardiello, A.; Dell'Olmo, J.; Ciancio, V.; Ferrero, M.; Rosso, F. Geometry Optimization in the Schematic Design Phase of Low-Energy Buildings for All European Climates through Genetic Algorithms. *Sustain. Cities Soc.* **2024**, *112*, 105639. [CrossRef]
107. Sandoli, A.; D'Ambra, C.; Ceraldi, C.; Calderoni, B.; Prota, A. Sustainable Cross-Laminated Timber Structures in a Seismic Area: Overview and Future Trends. *Appl. Sci.* **2021**, *11*, 2078. [CrossRef]
108. Sasaki, N. Timber Production and Carbon Emission Reductions through Improved Forest Management and Substitution of Fossil Fuels with Wood Biomass. *Resour. Conserv. Recycl.* **2021**, *173*, 105737. [CrossRef]

109. Schmidt, L.; Hilditch, R.; Ervine, A.; Madden, J. Explicit fire safety for modern mass timber structures—from theory to practice. In Proceedings of the World Conference on Timber Engineering (WCTE 2023), Oslo, Norway, 19–22 June 2023; pp. 1738–1747.
110. Schneider-Marin, P.; Harter, H.; Tkachuk, K.; Lang, W. Uncertainty Analysis of Embedded Energy and Greenhouse Gas Emissions Using BIM in Early Design Stages. *Sustainability* **2020**, *12*, 2633. [CrossRef]
111. Schwarzschachner, H.; Hernandez, S. Prolonged Carbon Storage and CO_2 Reduction by Circular Design with Wood. *J. Sustain. Archit. Civ. Eng.* **2024**, *35*, 23–33. [CrossRef]
112. Scouse, A.; Kelley, S.S.; Liang, S.; Bergman, R. Regional and Net Economic Impacts of High-Rise Mass Timber Construction in Oregon. *Sustain. Cities Soc.* **2020**, *61*, 102154. [CrossRef]
113. Sher, F.; Curnick, O.; Azizan, M.T. Sustainable Conversion of Renewable Energy Sources. *Sustainability* **2021**, *13*, 2940. [CrossRef]
114. De Oliveira, R.S.; De Oliveira, M.J.L.; Nascimento, E.G.S.; Sampaio, R.; Nascimento Filho, A.S.; Saba, H. Renewable Energy Generation Technologies for Decarbonizing Urban Vertical Buildings: A Path towards Net Zero. *Sustainability* **2023**, *15*, 13030. [CrossRef]
115. Stanciu, M.-C.; Teacă, C.-A. Changes of Wood Surfaces Treated with Natural-Based Products—Structural and Properties Investigation. *BioResources* **2024**, *19*, 5895–5915. [CrossRef]
116. Tupenaite, L.; Kanapeckiene, L.; Naimaviciene, J.; Kaklauskas, A.; Gecys, T. Timber Construction as a Solution to Climate Change: A Systematic Literature Review. *Buildings* **2023**, *13*, 976. [CrossRef]
117. Udele, K.E.; Morrell, J.J.; Sinha, A. Biological Durability of Cross-Laminated Timber—The State of Things. *For. Prod. J.* **2021**, *71*, 124–132. [CrossRef]
118. Ürge-Vorsatz, D.; Khosla, R.; Bernhardt, R.; Chan, Y.C.; Vérez, D.; Hu, S.; Cabeza, L.F. Advances Toward a Net-Zero Global Building Sector. *Annu. Rev. Environ. Resour.* **2020**, *45*, 227–269. [CrossRef]
119. Veichtlbauer, A.; Praschl, C.; Gaisberger, L.; Steinmaurer, G.; Strasser, T.I. Toward an Effective Community Energy Management by Using a Cluster Storage. *IEEE Access* **2022**, *10*, 112286–112306. [CrossRef]
120. Veloso, A.C.O.; Filho, C.R.A.; Souza, R.V.G. The Potential of Mixed-Mode Ventilation in Office Buildings in Mild Temperate Climates: An Energy Benchmarking Analysis. *Energy Build.* **2023**, *297*, 113445. [CrossRef]
121. Veillette, D.; Rouleau, J.; Gosselin, L. Impact of Window-to-Wall Ratio on Heating Demand and Thermal Comfort When Considering a Variety of Occupant Behavior Profiles. *Front. Sustain. Cities* **2021**, *3*, 700794. [CrossRef]
122. Wang, Q.; Zhu, K.; Guo, P.; Zhang, J.; Xiong, Z. Key Issues and Solutions in the Study of Quantitative Mechanisms for Tropical Islands Zero Carbon Buildings. *Appl. Sci.* **2024**, *14*, 1659. [CrossRef]
123. Warmling, J.G.; Espindola, L.D.R.; Abreu, A.L.P.D. *Elaboração de Projeto BIM de Uma Habitação Em CLT*; Encontro Nacional de Tecnologia do Ambiente Construído: Canela, Brasil, 2022; pp. 1–10.
124. Wilberforce, T.; Olabi, A.G.; Sayed, E.T.; Elsaid, K.; Maghrabie, H.M.; Abdelkareem, M.A. A Review on Zero Energy Buildings—Pros and Cons. *Energy Built Environ.* **2023**, *4*, 25–38. [CrossRef]
125. Yang, X.; Zhang, S.; Wang, K. Quantitative Study of Life Cycle Carbon Emissions from 7 Timber Buildings in China. *Int. J. Life Cycle Assess.* **2021**, *26*, 1721–1734. [CrossRef]
126. Younis, A.; Dodoo, A. Cross-Laminated Timber for Building Construction: A Life-Cycle-Assessment Overview. *J. Build. Eng.* **2022**, *52*, 104482. [CrossRef]
127. Zawada, K.; Rybak-Niedziółka, K.; Donderewicz, M.; Starzyk, A. Digitization of AEC Industries Based on BIM and 4.0 Technologies. *Buildings* **2024**, *14*, 1350. [CrossRef]
128. Zhan, T.; Li, R.; Liu, Z.; Peng, H.; Lyu, J. From Adaptive Plant Materials toward Hygro-Actuated Wooden Building Systems: A Review. *Constr. Build. Mater.* **2023**, *369*, 130479. [CrossRef]
129. Zhao, J.; Wei, X.; Li, L. The Potential for Storing Carbon by Harvested Wood Products. *Front. For. Glob. Change* **2022**, *5*, 1055410. [CrossRef]
130. Zhao, X.; Pan, W. Delivering Zero Carbon Buildings: The Role of Innovative Business Models. *Procedia Eng.* **2015**, *118*, 404–411. [CrossRef]
131. Lea Bridge Library Pavilion. Available online: https://timberdevelopment.uk/case-studies/lea-bridge-library-pavilion/ (accessed on 1 July 2024).
132. Orueta Etxea. Available online: https://egoin.com/projects/orueta-etxea/ (accessed on 1 July 2024).
133. Spruce House Studio. Available online: https://timberdevelopment.uk/case-studies/spruce-house-studio/ (accessed on 5 July 2024).
134. Cooperative Housing. Available online: https://worldgbc.org/case_study/cooperative-housing/ (accessed on 11 July 2024).
135. The Black and White Building. Available online: https://timberdevelopment.uk/case-studies/the-black-and-white-building/ (accessed on 5 July 2024).
136. Workstack. Available online: https://timberdevelopment.uk/case-studies/workstack/ (accessed on 21 July 2024).
137. EÑE House. Available online: https://plataforma-pep.org/ejemplos-ph/casa-ene/ (accessed on 23 July 2024).
138. The New Temple Complex. Available online: https://timberdevelopment.uk/case-studies/new-temple-complex/ (accessed on 23 July 2024).
139. Durley Chine Environmental Hub. Available online: https://timberdevelopment.uk/case-studies/durley-chine-environmental-hub/ (accessed on 27 July 2024).

140. The Humber Cultural Hub. Available online: https://www.cagbc.org/green-building-showcase/green-building-spotlight/case-studies/humber-cultural-hub/ (accessed on 27 July 2024).
141. *EN 15978:2011*; Sustainability of Construction Works—Assessment of Environmental Performance of Buildings—Calculation method. European Committee for Standardization: Brussels, Belgium, 2011.
142. *EN 15804:2012*; Sustainability of Construction Works—Environmental Product Declarations—Core Rules for the Product Category of Construction Products. European Committee for Standardization: Brussels, Belgium, 2012.
143. Padilla-Rivera, A.; Amor, B.; Blanchet, P. Evaluating the Link between Low Carbon Reductions Strategies and Its Performance in the Context of Climate Change: A Carbon Footprint of a Wood-Frame Residential Building in Quebec, Canada. *Sustainability* **2018**, *10*, 2715. [CrossRef]
144. United Nations. Forum on Forests Report of the Seventh Session (24 February 2006 and 16 to 27 April 2007). Available online: https://www.un.org/esa/forests/wp-content/uploads/2013/09/E-2007-42-UNFF7Report.pdf (accessed on 12 June 2024).
145. PEFC. Available online: www.pefc.org/discover-pefc/facts-and-figures (accessed on 10 July 2024).
146. Area of Certified Forest Stewardship Council (FSC) Worldwide from 2016 to 2024. Available online: www.statista.com/statistics/807548/global-forest-stewardship-council-land-area/ (accessed on 10 July 2024).
147. Di Girolami, E.; Arts, B. *Environmental Impacts of Forest Certifications*; Forest and Nature Conservation Policy Group—Wageningen University and Research: Wageningen, The Netherlands, 2018.
148. Cashore, B.; Van Kooten, G.C.; Vertinsky, I.; Auld, G.; Affolderbach, J. Private or Self-Regulation? A Comparative Study of Forest Certification Choices in Canada, the United States and Germany. *For. Policy Econ.* **2005**, *7*, 53–69. [CrossRef]
149. Ayikoe Tettey, U.Y.; Dodoo, A.; Gustavsson, L. Carbon Balances for a Low Energy Apartment Building with Different Structural Frame Materials. *Energy Procedia* **2019**, *158*, 4254–4261. [CrossRef]
150. Jaysawal, R.K.; Chakraborty, S.; Elangovan, D.; Padmanaban, S. Concept of Net Zero Energy Buildings (NZEB)—A Literature Review. *Clean. Eng. Technol.* **2022**, *11*, 100582. [CrossRef]
151. Чорна, Н.А. Перспективи Застосування Водневих Технологій Для Автономних Енергетичних Комплексів На Основі Відновлюваних Джерел Енергії. *Vidnovluvana Energ.* **2021**, *3*, 18–32. [CrossRef]
152. Meex, E.; Hollberg, A.; Knapen, E.; Hildebrand, L.; Verbeeck, G. Requirements for Applying LCA-Based Environmental Impact Assessment Tools in the Early Stages of Building Design. *Build. Environ.* **2018**, *133*, 228–236. [CrossRef]
153. Passer, A.; Kreiner, H.; Maydl, P. Assessment of the Environmental Performance of Buildings: A Critical Evaluation of the Influence of Technical Building Equipment on Residential Buildings. *Int. J. Life Cycle Assess.* **2012**, *17*, 1116–1130. [CrossRef]
154. Mirabella, N.; Röck, M.; Ruschi Mendes Saade, M.; Spirinckx, C.; Bosmans, M.; Allacker, K.; Passer, A. Strategies to Improve the Energy Performance of Buildings: A Review of Their Life Cycle Impact. *Buildings* **2018**, *8*, 105. [CrossRef]
155. Peuportier, B.L.P. Life Cycle Assessment Applied to the Comparative Evaluation of Single Family Houses in the French Context. *Energy Build.* **2001**, *33*, 443–450. [CrossRef]
156. Buchanan, A.H.; Honey, B.G. Energy and Carbon Dioxide Implications of Building Construction. *Energy Build.* **1994**, *20*, 205–217. [CrossRef]
157. Venkatarama Reddy, B.V.; Jagadish, K.S. Embodied Energy of Common and Alternative Building Materials and Technologies. *Energy Build.* **2003**, *35*, 129–137. [CrossRef]
158. Asdrubali, F.; Ferracuti, B.; Lombardi, L.; Guattari, C.; Evangelisti, L.; Grazieschi, G. A Review of Structural, Thermo-Physical, Acoustical, and Environmental Properties of Wooden Materials for Building Applications. *Build. Environ.* **2017**, *114*, 307–332. [CrossRef]
159. Akhimien, N.G.; Latif, E.; Hou, S.S. Application of Circular Economy Principles in Buildings: A Systematic Review. *J. Build. Eng.* **2021**, *38*, 102041. [CrossRef]
160. Larsen, V.G.; Tollin, N.; Sattrup, P.A.; Birkved, M.; Holmboe, T. What Are the Challenges in Assessing Circular Economy for the Built Environment? A Literature Review on Integrating LCA, LCC and S-LCA in Life Cycle Sustainability Assessment, LCSA. *J. Build. Eng.* **2022**, *50*, 104203. [CrossRef]
161. Pittau, F.; Krause, F.; Lumia, G.; Habert, G. Fast-Growing Bio-Based Materials as an Opportunity for Storing Carbon in Exterior Walls. *Build. Environ.* **2018**, *129*, 117–129. [CrossRef]
162. Peñaloza, D.; Erlandsson, M.; Falk, A. Exploring the Climate Impact Effects of Increased Use of Bio-Based Materials in Buildings. *Constr. Build. Mater.* **2016**, *125*, 219–226. [CrossRef]
163. Sandin, G.; Peters, G.M.; Svanström, M. *Life Cycle Assessment of Forest Products*; Springer Briefs in Molecular Science; Springer International Publishing: Cham, Switzerland, 2016; ISBN 978-3-319-44026-2.
164. Ramage, M.; Burridge, H.; Busse-Wicher, M.; Fereday, G.; Reynolds, T.; Shah, D.; Wu, G.; Yu, L.; Fleming, P.; Densley-Tingley, D.; et al. The Wood from the Trees: The Use of Timber in Construction. *Renew. Sustain. Energy Rev.* **2017**, *68*, 333–359. [CrossRef]
165. Klinge, A.; Roswag-Klinge, E.; Radeljic, L.; Lehmann, M. Strategies for Circular, Prefab Buildings from Waste Wood. *IOP Conf. Ser. Earth Environ. Sci.* **2019**, *225*, 012052. [CrossRef]
166. Ximenes, F.A.; Grant, T. Quantifying the Greenhouse Benefits of the Use of Wood Products in Two Popular House Designs in Sydney, Australia. *Int. J. Life Cycle Assess.* **2013**, *18*, 891–908. [CrossRef]
167. Sandberg, D.; Kutnar, A.; Mantanis, G. Wood Modification Technologies—A Review. *iForest* **2017**, *10*, 895–908. [CrossRef]

168. Wang, Q.; Zhao, L.; Chang-Richards, A.; Zhang, Y.; Li, H. Understanding the Impact of Social Capital on the Innovation Performance of Construction Enterprises: Based on the Mediating Effect of Knowledge Transfer. *Sustainability* **2021**, *13*, 5099. [CrossRef]
169. Pan, W. System Boundaries of Zero Carbon Buildings. *Renew. Sustain. Energy Rev.* **2014**, *37*, 424–434. [CrossRef]
170. Attia, S. Towards Regenerative and Positive Impact Architecture: A Comparison of Two Net Zero Energy Buildings. *Sustain. Cities Soc.* **2016**, *26*, 393–406. [CrossRef]
171. Harte, A.M. Mass Timber—The Emergence of a Modern Construction Material. *J. Struct. Integr. Maint.* **2017**, 2, 121–132. [CrossRef]
172. Izzi, M.; Casagrande, D.; Bezzi, S.; Pasca, D.; Follesa, M.; Tomasi, R. Seismic Behaviour of Cross-Laminated Timber Structures: A State-of-the-Art Review. *Eng. Struct.* **2018**, *170*, 42–52. [CrossRef]
173. Hadden, R.M.; Bartlett, A.I.; Hidalgo, J.P.; Santamaria, S.; Wiesner, F.; Bisby, L.A.; Deeny, S.; Lane, B. Effects of Exposed Cross Laminated Timber on Compartment Fire Dynamics. *Fire Saf. J.* **2017**, *91*, 480–489. [CrossRef]
174. Liu, Y.; Guo, H.; Sun, C.; Chang, W.-S. Assessing Cross Laminated Timber (CLT) as an Alternative Material for Mid-Rise Residential Buildings in Cold Regions in China—A Life-Cycle Assessment Approach. *Sustainability* **2016**, *8*, 1047. [CrossRef]
175. Cabeza, L.F.; Boquera, L.; Chàfer, M.; Vérez, D. Embodied Energy and Embodied Carbon of Structural Building Materials: Worldwide Progress and Barriers through Literature Map Analysis. *Energy Build.* **2021**, *231*, 110612. [CrossRef]
176. Marriage, G.; Sutherland, B. *New Digital Housing Typologies: CNC Fabrications of CLT Structure and BIM Cladding*; University of Genoa: Genoa, Italy, 2022.
177. Risse, M.; Weber-Blaschke, G.; Richter, K. Eco-Efficiency Analysis of Recycling Recovered Solid Wood from Construction into Laminated Timber Products. *Sci. Total Environ.* **2019**, *661*, 107–119. [CrossRef] [PubMed]
178. Zhao, C.; Zhou, J.; Liu, Y. Financial Inclusion and Low-Carbon Architectural Design Strategies: Solutions for Architectural Climate Conditions and Architectural Temperature on New Buildings. *Environ. Sci Pollut Res* **2023**, *30*, 79497–79511. [CrossRef] [PubMed]
179. Suresh Ramanan, S.; Arunachalam, A.; Handa, A.K. Timber Production Potential of Trees on Farmlands. *Small-Scale For.* **2023**, *22*, 371–380. [CrossRef]
180. Pilli, R.; Grassi, G.; Kurz, W.A.; Fiorese, G.; Cescatti, A. The European Forest Sector: Past and Future Carbon Budget and Fluxes under Different Management Scenarios. *Biogeosciences* **2017**, *14*, 2387–2405. [CrossRef]
181. Östman, B.; Brandon, D.; Frantzich, H. Fire Safety Engineering in Timber Buildings. *Fire Saf. J.* **2017**, *91*, 11–20. [CrossRef]
182. Wang, J.Y.; Stirling, R.; Morris, P.I.; Taylor, A.; Lloyd, J.; Kirker, G.; Lebow, S.; Mankowski, M.; Barnes, H.M.; Morrell, J.J. Durability of mass timber structures: A review of the biological risks. *Wood Fiber Sci.* **2018**, *50*, 110–127. [CrossRef]
183. Johnston, C.M.T.; Radeloff, V.C. Global Mitigation Potential of Carbon Stored in Harvested Wood Products. *Proc. Natl. Acad. Sci. USA* **2019**, *116*, 14526–14531. [CrossRef]

MDPI AG
Grosspeteranlage 5
4052 Basel
Switzerland
Tel.: +41 61 683 77 34

Energies Editorial Office
E-mail: energies@mdpi.com
www.mdpi.com/journal/energies

www.ingramcontent.com/pod-product-compliance
Lightning Source LLC
Chambersburg PA
CBHW042011260926
51850CB00020B/3171
9783725824830